Lesch/Dettmar/Mebold/Schlickeiser (eds.)
The Physics of Galactic Halos

Lesch/Dettmar/Mebold/Schlickeiser (eds.)
The Physics of Galactic Halos

The Physics
of Galactic Halos

Proceedings of the 156th WE-Heraeus Seminar,
Bad Honnef, Germany, February 11–14, 1996

edited by
Harald Lesch
Ralf-Jürgen Dettmar
Ulrich Mebold
Reinhard Schlickeiser

Akademie Verlag

Authors:

Prof. Dr. Harald Lesch, Ludwig-Maximilians-Universität München, Institut für Astronomie
und Astrophysik, Munich/Germany

Prof. Dr. Ralf-Jürgen Dettmar, Ruhr-Universität Bochum, Astronomisches Institut,
Bochum/Germany

Prof. Dr. Ulrich Mebold, Rheinische Friedrich-Wilhelms-Universität Bonn, Radioastronomisches
Institut, Bonn/Germany

Prof. Dr. Reinhard Schlickeiser, Max-Planck-Institut für Radioastronomie Bonn, Bonn/Germany

With 89 figures and 14 tables

1st edition

Die Deutsche Bibliothek – CIP-Einheitsaufnahme

The **physics of galactic halos** : proceedings of the 156th
WE-Heraeus Seminar Bad Honnef, Germany, February 11–14, 1996 ;
[with 14 tables] / ed by Harald Lesch ... – 1. ed. –
Berlin : Akad. Verl., 1997
  ISBN 3-05-501752-8
NE: Lesch, Harald [Hrsg.]; W.-E.-Heraeus-Seminar
  <156, 1996, Honnef>

Printing: GAM Media GmbH, Berlin
Bookbinding: Verlagsbuchbinderei Mikolai GmbH, Berlin

Printed in the Federal Republic of Germany

Akademie Verlag GmbH
Mühlenstraße 33–34 · D-13187 Berlin
Federal Republic of Germany

Preface

The physics of the Galactic Halo has remained of top scientific interest for more than 15 years. A first workshop entitled "Gaseous Halos of Galaxies" was held in 1985 at the National Radioastronomy Observatory in Green Bank, USA. A second major conference on that subject entitled "The Interstellar Disk-Halo Connection" was held in 1990 in Leiden, The Netherlands. Comparing the titles of the papers given at these meetings with those in the present book we find that quite a number of subjects discussed then were fairly similar to those discussed here. This has lead to the provocative statement of one of our conference participants that there has hardly been any progress in the meantime. We hope that we can convince even the critical reader that there rather has been significant new input from new data sets, notably from the Dwingeloo/Leiden HI survey of the northern sky and the orbiting X-ray observatory ROSAT.

The story of this book began in fact with the emergence of new observational facts derived from these data, which we did not know how to fit into standard wisdom. Some of these facts are:

- enhanced X-ray emission from the so called HI high velocity clouds (HVCs), neutral clouds in the halo of our Galaxy which appear too cold to emit this energetic radiation,

- soft X-ray emission from a layer in the halo of our Galaxy which we started to call the "Kerp" layer,

- emission from a layer of HI gas which has an extremely large velocity dispersion (LVD gas) of about 60 km/s and which we started to call the "Kalberla" layer, and finally

- rather energetic disturbances of HVCs which show up as strong velocity gradients at there boundaries. These hint at a fast collision with a heavy counterpart, a counterpart that we do not expect to meet in the galactic halo.

In relation to the first apparent paradox, the X-ray emission from HVCs, we wondered whether conversion of the kinetic energy of the HVCs into thermal energy might work. Checking the numbers, we require that all of the kinetic energy of HVCs is converted into heat. But this is difficult to achieve. The plasma in the halo is hardly dense enough to stop the HVCs at all. And if it were dense enough, the shocks induced by inelastic collisions are too inefficient to create the large observed temperatures. Radiation losses limit the temperatures in shocks to values much smaller than those seen in observations.

Therefore we sought refuge to unconventional ideas. Since 1994 we started speculating about the influence of magnetic fields upon the impact of HVCs into the halo. Is it conceivable that the dissipation of frozen-in magnetic fields – also known as "magnetic reconnection" – in HVCs on the one hand side and their halo-environment on the other provides the required heating energy? Carrying out a thorough investigation of the micro-physics of magnetic reconnection in the boundary layers between impacting HVCs and and magnetized plasma in the galactic halo it was found that in the parameter space regarded here magnetic compression and subsequent reconnection provides a more efficient mechanism to convert kinetic in to thermal energy than hydrodynamic shocks could possibly do. Whether or not that is the heating mechanism for the HVCs was not obvious at that stage. At least we had a clue in what direction to search for the solution of the energy problem of the X-ray radiation from HVCs.

We have used that clue and – of course – other evidence to try and understand the second and third apparent paradox, the co-existence of warm neutral and hot ionized plasmas along the same line of sight: if magnetic fields play a major role during the infall of HVCs in the halo, they might also help to insulate hot and ionized plasma cells from cold and neutral once. The spatial relation of ionized and neutral gas is fairly well established in and close to the galactic plane. But how does the mixture of neutral and ions look away from the plane? How does the fraction of ionized gas vary as a function of the distance from the galactic plane. What is the source of ionization? Where is that source located? And of course we always asked ourselves what is observable in extragalactic systems?

At one stage we arrived at the question which got us pretty close to a solution of all of our problems – a very unstable one, however: what is the interrelation of all those famous layers that we have been reading about in the literature for years and those infamous layers that we know about only since very recently: how relate the "Reynolds" layer of electrons,

- the "Lockman" layer of diffuse HI gas,

- the "Kalberla" layer of large-velocity-dispersion HI gas,

- the "IUE" layer of highly ionized atoms,

- and the "Kerp" layer of soft X-ray emitting plasma?

The answer of course was: why don't we ask the experts? "Ulrich, why don't you ring up some of the people you know". Then we started inviting colleagues who have been working on these problems for years.

This is how the idea of our workshop on galactic halos was born. The rest was fairly easy since we have the HERAEUS Stiftung in Germany. This foundation is the perfect sponsor for such workshop-like conferences – if the case can be presented in a convincing proposal. We got their support because the subject was timely and of high scientific importance. The great fun of working with HERAEUS is that they do not only provide funds but also take over most of the organization. It is a great pleasure to thank Frau Hartmann and Herrn Dr. Schaefer for the perfect support before, during, and in the aftermath of the workshop. It would certainly not have been as pleasant and smooth if one of us - or even worse - all of us in an concerted effort had tried to organize that meeting.

As with all conference proceedings, our book cannot present the essence of the meeting: it does not talk about meal conversations, coffee discussions, and the great late-night show-down which always took place after dry throats got well wetted with the excellent wines upon which the fame of our meeting place is founded. It was probably the most significant meeting place down there in the catacombs of the Physics Center in Bad Honnef on the Rhine. These cellars created the charming atmosphere that stopped the gun men from shooting and got them to listen to each other. Here the talks of our great heroes which were held during daylight became the starting points of discussions lasting far into the night. This created the atmosphere where new and innovative ideas did in fact emerge.

We like to thank all participants, in particular those visiting from overseas, for joining us in Bad Honnef and for devoting their precious time and their research activities to our conference.

Apart from the scientific harvest, the workshop provided an appropriate frame for celebrating the 60th birthday of a scientist who has significantly contributed to the exploration of magnetic fields in the universe: Richard Wielebinski, Director of the 'Max Planck Institut für Radioastronomie' in Bonn. His impressive scientific achievements were praised in an evening talk by Phil Kronberg from the University of Toronto. The organizers are happy that they had the opportunity to celebrate one of the pioneers of modern Radioastronomy during this workshop.

We are particularly grateful to Dr. Götz Golla, who volunteered to edit the submitted contributions. He was assisted by Andrea Stolte and Uwe-Willy Geiersbach at Ruhr-Universität Bochum.

We also like to thank the team of the Physics-Center in Bad Honnef. Their quiet but efficient work in the background provided the perfect basis for a smooth and undisturbed working environment on the one hand side and the very friendly and personal touch of the service to us guest on the other. Finally, we like to thank the Akademie Verlag in Berlin for the patience in waiting for our manuscript. Last not least we thank all the readers who will help to promote the results of our workshop by ultimately using them in their own work.

Bonn and München, Fall 1996

U. Mebold and H. Lesch for the Editors

Contents

Our conference was held at the 'Physik Zentrum' in Bad Honnef. The
building belongs to the Elly Hölterhoff-Böcking Foundation, an organisation
which ran a housekeeping school for daughters of wealthy families between
1908 and 1940.

HI and HVCs:
galactic and extragalactic

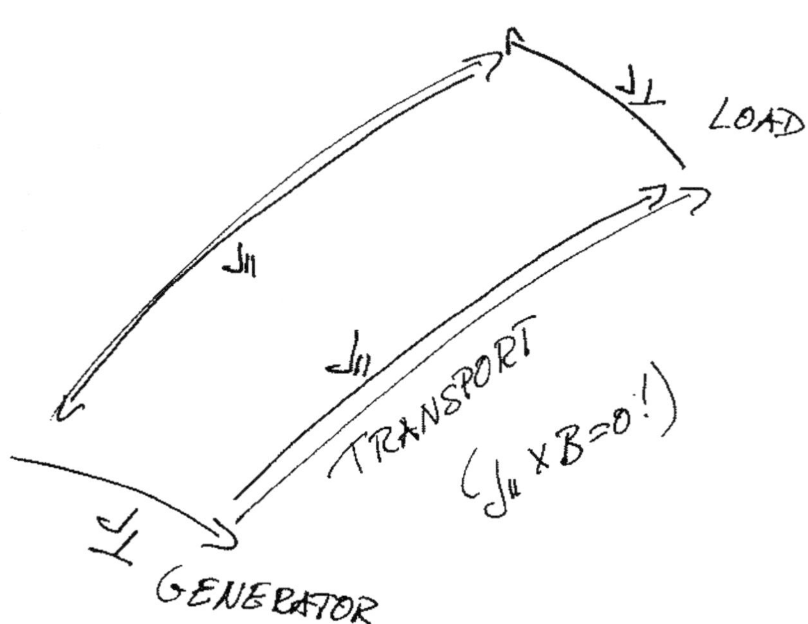

Haralds mystical vision of the Galactic Halo

HI Gas in the Galactic Halo

P.M.W. Kalberla, G. Westphalen, J. Pietz, U. Mebold,
Dap Hartmann & W.B. Burton

1 Abstract

The Leiden/Dwingeloo Survey (LDS), which becomes available soon, offers new possibilities for analyzing Galactic HI with an outstanding sensitivity. The survey data have been carefully corrected for sidelobe contamination of the antenna and for baseline effects in the velocity range -450 km s^{-1} $< V_{lsr} <$ $+400$ km s^{-1}. At present this survey is the most reliable database for analysis of faint, large-scale HI features which may be attributed to an HI halo of the Galaxy. This motivated investigations which are briefly described in this paper.

We find evidence for weak HI emission connecting high-velocity clouds (HVC) with gas at intermediate velocities (IV) or low velocities (LV). These weak emission features, which we call 'velocity bridges', are very common at least in HVC complex C. It is difficult to interpret such a wide-spread phenomenon in the context of the current theory placing the HVCs at distances of several kpc above the Galactic plane.

HI gas with large velocity dispersions is another phenomenon which can be analyzed using the LDS. We detect HI gas with properties similar to those of the Lockman layer , although some of the parameters are more extreme. In every direction of the sky HI gas was found having a characteristic dispersion of 70 km s^{-1} and column densities in the range $1 - 2 \times 10^{19}$ cm^{-2}.

2 The Leiden/Dwingeloo Survey

The Leiden/Dwingeloo Survey (LDS) of the 21-cm line of Galactic neutral hydrogen has been carried out with the Dwingeloo 25-m telescope. The angular resolution is $36'$. The observations were made on a grid of $30'$ both in Galactic longitude and latitude. The spectral resolution is 1.03 km s^{-1}. Velocities between -450 km s^{-1} and $+400$ km s^{-1} (with respect to the Local Standard of Rest) are covered and allow the analysis of weak HI features exceeding the brightness-temperature rms limit of 0.07 K. Details of the observing and reduction procedures are given by Hartmann (1994) and by Hartmann & Burton (1996). The Hartmann & Burton publication includes an atlas of maps and a CD-ROM containing FITS files of the proper data.

It has been known since the demonstration by van Woerden et al. (1962) that the accuracy of Galactic HI observations, especially for weak lines at high Galactic latitudes, is hampered by sidelobe contamination of the antenna. 21-cm line profiles observed with parabolic reflector antennas suffer from stray radiation originating from the sidelobes of the antenna diagram which may account for more than 50% of the observed line radiation (Kalberla, 1978; Kalberla et al., 1980; Lockman et al., 1986). The LDS was expected to overcome these limitations. It provides high-quality line profiles of the entire sky north of $\delta = -30°$ and is the first large-scale survey corrected for stray radiation (Hartmann et al., 1996).

2.1   How reliable are the LDS profiles?

Compared to earlier large-scale HI surveys the LDS data are improved by an order of magnitude in either sensitivity, spatial and velocity resolution, or velocity coverage. Thus this survey should be ideal for investigations of HI in the outer parts of the Galaxy, including gas possibly associated with the Galactic halo. This motivated us to push the analysis of the LDS to the limits of its sensitivity. In the first place it is important to determine the reliability of the survey data. This was done in great detail. Since this cannot be the place to report on details we just outline what was done.

The analysis of halo gas is based on the high-latitude low-level HI emission in the extreme velocity regimes of the profiles. This emission can be significantly influenced by baseline effects and stray radiation. Both effects have previously been studied down to a level of 0.1 K (Hartmann, 1994; Hartmann et al., 1996). We averaged spectra over large areas on the sky and found problems with DC offsets in the autocorrelator. After Hanning smoothing we verified that the noise went down as expected. Next we found that the correction for ground reflections as suggested by Hartmann (1994) is limiting the accuracy of the averaged profiles causing negative amplitudes of up to 0.1 K in the most extreme cases. We applied the ground-reflection correction to the entire LDS and extracted 2700 spectra showing negative amplitudes which are not blended with the regular HI line emission. From this the parameters controlling the correction for ground reflections were re-adjusted. The new corrections were then applied to the survey (Kalberla et al., 1996).

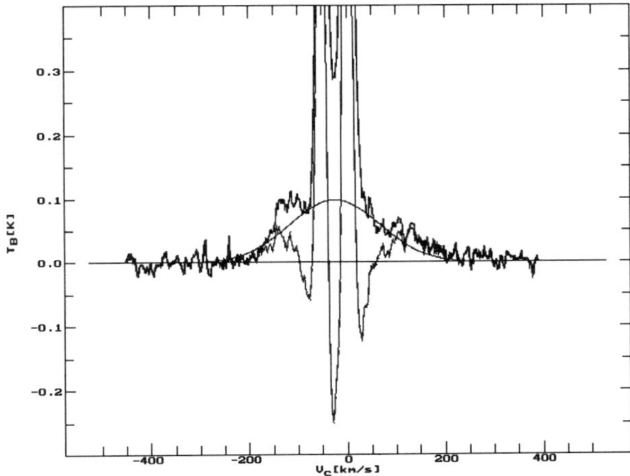

Figure 1: 21-cm line profile averaged over $5° \times 5°$ centered at $(l,b) = (142.5°, 42.5°)$. The upper line is the averaged profile after standard correction. A Gaussian fit to the Large-Velocity Dispersion (LVD) component is also shown. The lower thin line results from correcting the spectra with doubled sidelobe efficiencies. Such over-corrections cannot remove the wings beyond $|V_{lsr}| > 150$ km s^{-1}.

In section 4 we show that the averaged spectra display weak Hɪ emission components with large velocity dispersion, i.e. with extended profile wings. A key issue in the following analysis is whether or not these components are leftovers from the stray radiation correction or the baseline correction procedure.

To check this, we focused our attention on the extreme velocities of the profile wings. The stray radiation has been corrected for the sake of computational ease out to velocities $-300 < V_{lsr} < 300$ km s^{-1} only. This is sufficient to cover the necessary velocity range (see Fig. 4), except for parts of the Magellanic Stream and some very high velocity clouds, which need to be considered for future extensions of the Hɪ survey in the southern sky. It does not affect the present analysis of the LDS.

To find out how these wings are affected by stray radiation, we increased side- and backlobe levels such that the stray-radiation correction produced negative amplitudes in the corrected profiles. All such modifications did not reduce the profile wings at the most extreme velocities. In Fig. 1 the effect of a correction using doubled antenna parameter values is shown. We find that the far end of the wings is hardly affected by stray radiation correction. This can even be seen in profiles averaged over $360°$ in longitude. In such profiles stray radiation is extremely broad.

We found, that in general the extreme parts of the profile wings are not influenced by the stray radiation correction. Therefore we rejected all further modification of the antenna parameters. However, the parameters of the antenna pattern are known on average only. We cannot exclude that individual sidelobe structures may cause spurious features which could amount to $5 - 10$ mK. But these can not cause the wings at the most extreme velocities.

To check for leftovers of the baseline correction procedure we modified the baseline corrections. However we were unable to obtain improvements beyond a level of 5 mK. In other words the differences due to any improvements of the baseline correction were too small to be detected. The accuracy of the baseline corrections is ultimately limited by the accuracy in the determination of standing waves in the telescope as described by Hartmann (1994).

Thus the LDS (including the modified correction for reflections from ground) was found to the be the most reliable database of HI profiles available today. Still one should be cautious when drawing conclusions. Ultimately our results should be verified using other telescopes. The new Green Bank Telescope promises to be an ideal instrument in this respect.

In section 4 we present a comparison of the LDS with the Bell Labs Survey (hereafter BLS, Stark et al., 1992). For further details we refer to Kalberla et al. (1996). Details of the investigations described in this section will be published elsewhere.

3   Velocity Bridges between HV-, LV- and IV Gas

Low-level HI emission which appears to connect individual HVCs with emission at lower velocities was discovered in Effelsberg data by Meyerdierks (1988, 1992). Their data had been corrected for instrumental effects. A first systematic search for such connecting features, which we call velocity bridges (hereafter VB), has recently been carried out by Pietz et al. (1996) using the LDS. Analyzing HVCs in complex C they found eleven regions with significant VBs. Most of these have also been observed and confirmed using the Effelsberg telescope. Here we will illustrate the phenomenon by two examples.

Fig. 2 displays position-velocity plots at latitudes $b = 41°$ and $b = 43°$, between longitudes $82° < l < 98°$ obtained from the LDS. The profiles on the right were observed with the Effelsberg telescope at $(l, b) = (88.2°, 41°)$ (top) and $(l, b) = (93.6°, 42°)$ (bottom). These profiles were corrected for stray radiation (Kalberla et al., 1980). The shaded regions indicate the the range of the VBs.

Fig. 3 displays position-velocity plots observed with the Effelsberg telescope. The map at b = 55.5° (center) is in excellent agreement with a similar map extracted from the LDS (Pietz et al., 1996). Thus we conclude that the observed VB is real and not an artifact. The HVC and the VB in Fig. 3 are resolved by the Effelsberg telescope at a high level of significance.

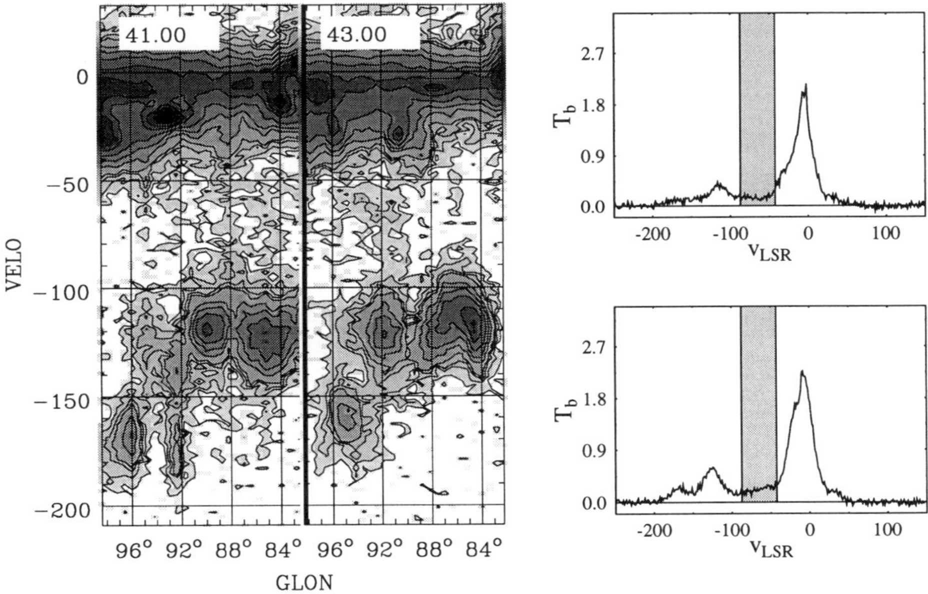

Figure 2: Position-velocity maps derived from Leiden/Dwingeloo Survey for $41° < b < 43°$ and $98° < l < 82°$. The lowest contours are drawn at brightness-temperature levels of 0.07, 0.15, 0.23, 0.3 K. The spectra on the right were observed with the Effelsberg telescope at $(l, b) = (88.2°, 41°)$ (top) and $(l, b) = (93.6°, 42°)$ (bottom). The shaded region indicates the velocity range used to display the velocity bridges (VBs) on the left.

Inspecting the LDS data set on an image-display system one is struck by how frequently low-level emission seems to connect individual HVCs with emission at lower velocities. VBs appear to be quite common. Because the astrophysical implications of these VBs may be far reaching, a careful analysis of the reliability of the data in the area of the bridges is mandatory. We analyzed the profiles for defects, either from the correction for stray radiation or from the baseline correction. Independent profiles at neighbour positions have been

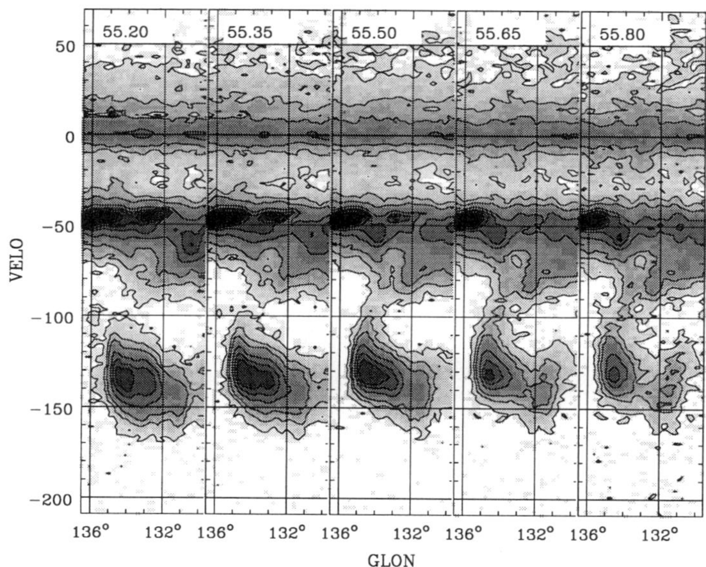

Figure 3: Position-velocity maps for a velocity bridge derived from Effelsberg data for $55.2° < b < 55.8°$ and $136.5° > l > 129.5°$. The lowest contours are at 0.1, 0.3, 0.5, 1.0 K.

compared. Also, observations with the Effelsberg telescope have been used systematically as a checking procedure.

We analysed the 3-D LDS data set using the AVS visualisation software package running on a Kubota display system. Slicing the 3-D data cube with arbitrary cuts at various aspect angles provides a useful tool for such visualisation. In searching for VBs we used criteria which select against possible instrumental effects. Without bias with respect to the origin of real features, we required continuity in the velocity-position space. Artifacts cause discontinuities in the data cube. These are correlated either with position or velocity and date of the observation depending on the origin of such errors. After visualisation we are convinced that the majority of VBs found in the LDS are real.

In summary: we find VBs connecting high-velocity gas and gas at lower velocities over velocity gaps of 20 to 100 km s^{-1}. We interpret this as indications for energetic interactions of HVCs under discussion here with gas in the Galactic halo or the disk-halo interface. It is difficult to understand an interaction between HVCs and IV gas in the context of current models which place these HVCs at z-distances of several kpc and the material at low and intermediate velocities much closer to the Galactic plane.

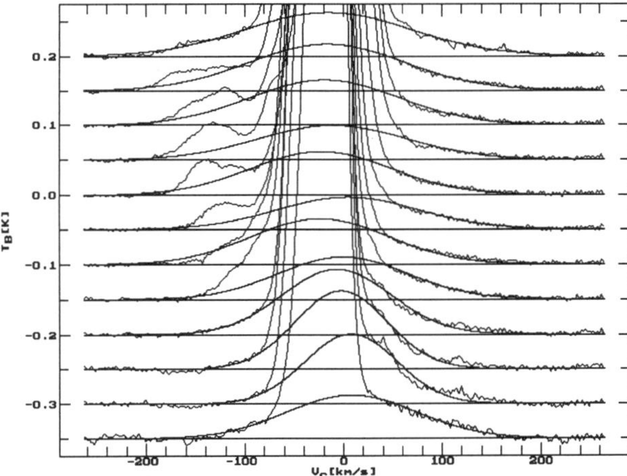

Figure 4: 21-cm line profiles averaged over all Galactic longitudes and a 5° of latitude interval with a rms of typically 3 mK. The thick line represents the Gaussian fit of the broad, large velocity dispersion component (LVD), which would be completely hidden in the noise of individual spectra. The top profile is at $30° < b < 35°$, the bottom one at $85° < b < 90°$. Radial velocity in kms^{-1} is along the abscissa and brightness temperature in K along the ordinate. The profiles are offset by 50 mK.

4   HI Gas with Large Velocity Dispersion

Our analysis of the LDS indicates the existence of extended profile wings in every direction on the sky. Fig. 4 displays the characteristics of the mean HI emission averaged over all longitudes and 5° wide strips in latitude from $b = 90°$ (bottom) to $b = 30°$ (top). The profile wings at positive velocities are smooth and approximately Gaussian in shape. At negative velocities emission from individual high velocity clouds is found to be superposed on the smooth profile wings. The HI gas giving rise to these smooth and very extended profile wings we call the large-velocity-dispersion component or LVD gas for short.

4.1   Extended profile wings

Observations of low-intensity features in HI profiles as discussed above depend critically on a reliable baseline determination. Such features may get lost if the observed velocity coverage is limited. A first comparison between observations of the Dwingeloo and the Bell Labs telescopes (Stark et al., 1992) indicated such problems with the BLS (see Kalberla et al., 1996). In Fig. 5 we display the averaged HI spectrum in the general direction of the south Galactic pole derived by averaging all HI spectra at $b < -45°$. The profile in the lower panel was derived from the BLS, the upper one from the LDS. The weak emission at $V_{lsr} \approx -120$ km s^{-1} observed with the Dwingeloo telescope is missing in the BLS data. In the positive-velocity range both profiles have slightly imperfect baselines.

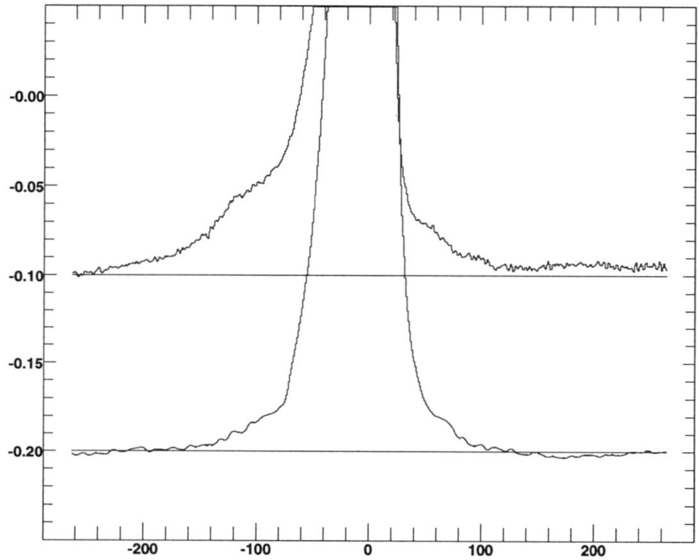

Figure 5: Averaged HI emission at $b < -45°$. The lower profile was derived from Bell Laboratory Survey (BLS), the upper one from the Leiden/Dwingeloo Survey (LDS). The profiles are offset by 0.1 K.

We examined the LDS observations carefully to explain the discrepant profile wing at $V_{lsr} \approx -120$ km s^{-1}. Investigating profiles from the LDS we found a large number of individual weak HI clouds at velocities -150 km s^{-1} $< V_{lsr} <$ -60 km s^{-1} which contribute to the high-velocity wing. The absence of these features in the BLS can be explained by the larger beam of the Crawford Hill antenna ($2° \times 2.8°$). Weak LVD components will be attenuated by beam dilution to amplitudes below the detection limit and will finally be removed by a baseline correction procedure.

Another example of the attenuation of extended wings is shown in Fig. 6. The

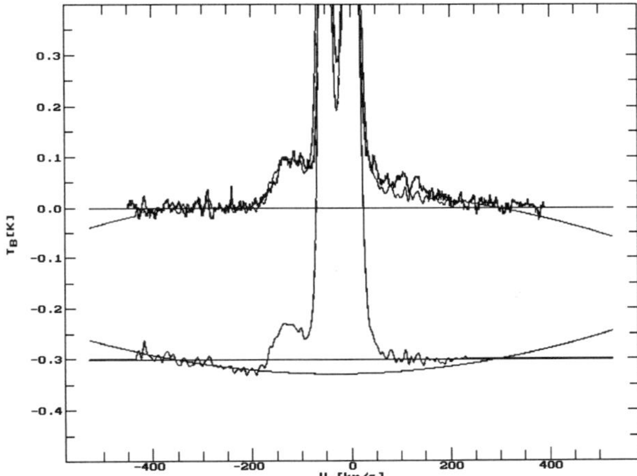

Figure 6: The profile as shown Fig. 1 was averaged over $5° \times 5°$ centered at $(l, b) = (142.5°, 42.5°)$. The lower panel was derived from the Bell Laboratory Survey (BLS). A second-order baseline readjustment is indicated. The upper panel is an overlay of the corresponding LDS profile (thick line) and the readjusted BLS profile (thin line).

lower panel shows an averaged profile derived from BLS while the top panel is an overlay of the corresponding LDS profile and the processed BLS profile. We readjusted the baseline for the BLS profile and scaled the profiles to the same peak temperature. After smoothing of the LDS profile to account for the different velocity resolutions, we scale the BLS profile by 1.33. This factor seems high, but according to Kuntz & Danly (1992) it is normal for the BLS data. Apart from small discrepancies around $v = 100$ kms^{-1} the two profiles appear identical. This is a typical example of the LVD HI gas observed in the LDS at low levels and missing in the BLS. We interpret the absence of such components in BLS as being due to an insufficient velocity coverage which does not allow a reliable determination of the instrumental baseline in the BLS.

A more detailed discussion of this finding and other instrumental effects is beyond the scope of this contribution. From the analysis of the LDS after correcting for contributions from ground reflection we are confident that these wings are not due to instrumental effects. Details will be published elsewhere.

4.2 Gaussian decomposition

Lockman & Gehman (1991, hereafter LG) tested the hypothesis of a hydrostatic HI distribution extending to large z distances by analyzing averaged HI profiles towards both Galactic poles. The systematic differences between the LDS and the BLS data gave reason to repeat this type of analysis. In addition, we searched for possible dependencies of these average profiles with respect to Galactic coordinates.

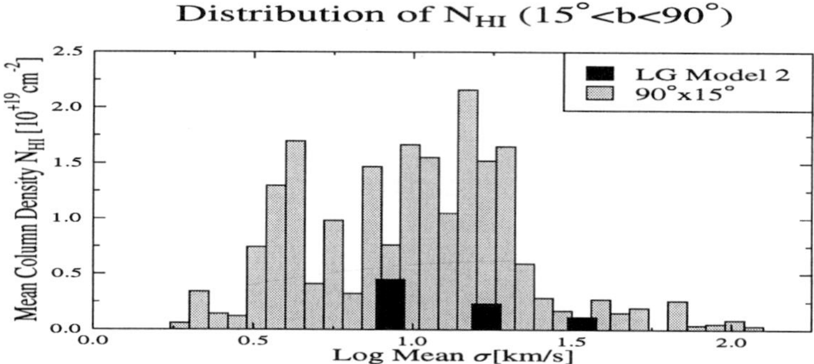

Figure 7: Distribution of the mean HI column densities weighted by $sin(|b|)$ as a function of the velocity dispersion. For comparison the parameters of model 2 from LG are plotted in black.

We calculated HI profiles from averages over 90° in longitude and 15° in latitude. To achieve equal weight to lines of sight at different latitudes for a plane-parallel distribution of the HI gas, the brightness temperature at each position was weighted by $sin(|b|)$. Each of these average profiles was decomposed into Gaussian components. Typically we found 5 components per profile. We took care that the extended profile wings were also included in the fit. However, we did not force a single component through both the positive and the negative velocity wing simultaneously.

In total we derived 209 components which were considered to characterize the observable HI distribution for $\delta > -30°$ (the observing limit of the LDS). The histogram in Fig. 7 shows the distribution of the mean column densities as a function of the velocity dispersion for all $b > 15°$. For comparison we included the parameters from model 2 of LG. We conclude that the extended profile wings visible in Fig. 1 & 4 can be characterized by Gaussian components with dispersions $\sigma > 30$ km s^{-1}, and up to 100 km s^{-1}.

Since extended profile wings are absent in the BLS, the broadest components seen in the LDS will not be present in the BLS. Interpreting the LVD components as emission from a static HI layer which reaches large z-distances, the z values derived from our data are considerably larger ($z \approx 3$ kpc) than the ones

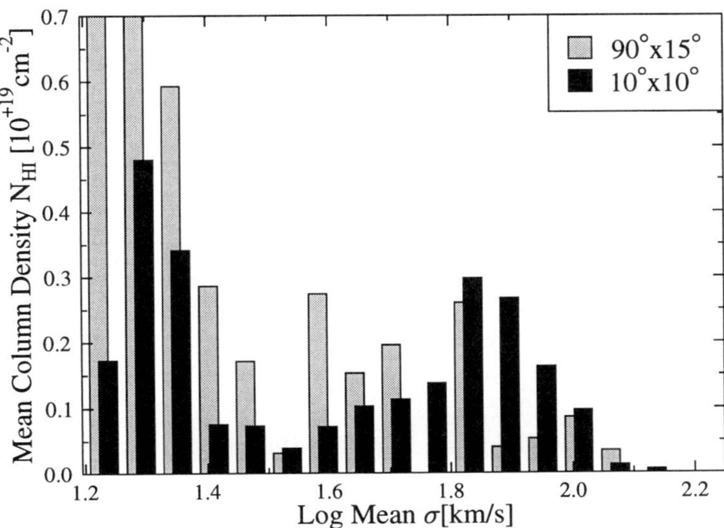

Figure 8: Distribution of the mean H I column densities as function of the velocity dispersion with $\sigma > 16$ kms^{-1}. The grey bars are the data from Fig. 7, the black bars represent $10° \times 10°$ averaged fields.

found by LG ($z \approx 1$ kpc).

The LVD components were studied by Westphalen et al. (1996) in more detail. These authors averaged the LDS data for $b > 20°$ over $10° \times 10°$ fields and decomposed the resulting spectra into Gaussians. They did force a single Gaussian to fit to the positive and the negative velocity wings of the profiles. The distribution of the H I column densities of all Gaussians found for $b > 20°$ as a function of the velocity dispersion is displayed in Fig. 8 for components with $\sigma > 16$ km s^{-1}. The distribution in Fig. 8 does not change significantly if the area on the sky for which the spectra are averaged is changed. The LVD component appears to be a ubiquitous phenomenon.

Fig. 8 shows that the LVD component separates clearly from the components with dispersions < 30 km s^{-1} ($\log \sigma = 1.48$). This suggests that this LVD gas represents a separate gas component and is not just a "tail" of previously known components. It may represent neutral gas in an otherwise ionised and highly turbulent plasma which fills large volumes above the Galactic plane.

References

Danly, L., Lockman, F.J., Meade, M.R. & Savage, B.D., 1992, ApJS, **81**, 125

Hartmann, D., 1994, PhD thesis, University of Leiden.

Hartmann, D. & Burton, W.B., 1996, Atlas of Galactic Neutral Hydrogen, Cambridge University Press, in press

Hartmann, D., Kalberla, P.M.W., Burton, W.B. & Mebold, U., 1996, A&AS, **119**, in press

Kalberla, P.M.W., 1978, PhD thesis, University of Bonn.

Kalberla, P.M.W., Mebold, U. & Reich, W., 1980, A&A, **82**, 275

Kalberla, P.M.W., Hartmann, D., Burton, W.B., Mebold, U. & Westphalen, G., 1996, High-sensitivity radio astronomy eds. N. Jackson & R. Davis, Cambridge University Press

Kuntz, K.D. & Danly, L., 1992, PASP, **104**, 1256

Lockman, F.J., Jahoda, K., McCammon, D., 1986, ApJ, **302**, 432

Lockman, F.J. & Gehman, C.S., 1991, ApJ, **382**, 182

Meyerdierks, H., 1988, PhD thesis, University of Bonn.

Meyerdierks, H., 1992, A&A, **253**, 515

Pietz, J., Kerp, J., Kalberla, P.M.W., Mebold, U., Burton, W.B., Hartmann, D., 1996, A&A, **308**, L37

Stark, A.A., Gammie, C.F., Wilson, R.W., Bally, J., Linke, R.A., Heiles, C. & Hurwitz, M., 1992, ApJS, **79**, 77

van Woerden, H., Takakubo, K. & Braes, L.L.E., 1962, BAN, **16**, 321

Westphalen, G., Kalberla, P.M.W., Mebold, U., Hartmann, D. & Burton, W.B., 1996, in Proc. of a Workshop held at Bad Honnef: "The Physics of Galactic Halos"

Addresses of the authors:

P.M.W. KALBERLA, G. WESTPHALEN, J. PIETZ, U. MEBOLD
Radioastronomisches Institut der Universität Bonn, Auf dem Hügel 71, D-53121 Bonn, Germany
DAP HARTMANN
Harvard-Smithsonian Center for Astrophysics, 60 Garden Street, Cambridge, MA 02138, U.S.A.
W.B. BURTON
Sterrewacht Leiden, Postbus 9513, 2300 RA, Leiden, The Netherlands

Two Observational Constraints on the Nature of HVCs and IVCs

W.B. Burton

1 Introduction

I discuss two observational constraints relevant to determining certain properties of both the high–velocity–cloud and the intermediate–velocity–cloud phenomena. The "scale–height constraint" follows from the observed vertical thickness of the galactic HI layer, which is well–measured along the subcentral–point locus to be $h_z \sim 100$ pc: either the HVCs and the IVCs do not significantly populate the subcentral–point locus, or else the vertical scale heights of HVCs and of IVCs are typically less than 100 pc. The "quiescent–terminal–velocity constraint" follows from the very modest scatter observed about the mean of the kinematic cutoffs used to determine the Galaxy's rotation curve: either the HVCs and the IVCs do not significantly populate the subcentral–point locus, or else the HVCs as well as IVCs typically have motions whose component parallel to the galactic equator is less than about 5 km/s.

I also briefly indicate two pressing questions, particularly relevant to consideration of the observational constraints, which are amenable to investigation using new HI and other survey data. The first of these involves the nature of the distinction between the HVC and IVC phenomena, which was originally arbitrarily invoked but which in fact seems to have a physical significance. The second observational challenge involves the continued search for indications of associations of HVCs with disturbances, either as "crash sites" or as "launching pads", in the conventional gaseous disk of the Milky Way.

Figure 1, reproduced from Norman & Ikeuchi (1989), is intended to set this discussion in an appropriate context. A range of different mechanisms has been proposed for the origin of HVCs. The review by Oort (1970) of the observational properties and plausible mechanisms is still significantly current; that of Wakker (1991) accounts for some of the more recent developments. Less attention has been given to the origin of IVCs, but the range of possibilities seems limited to consequences of energetic events common in the galactic disk.

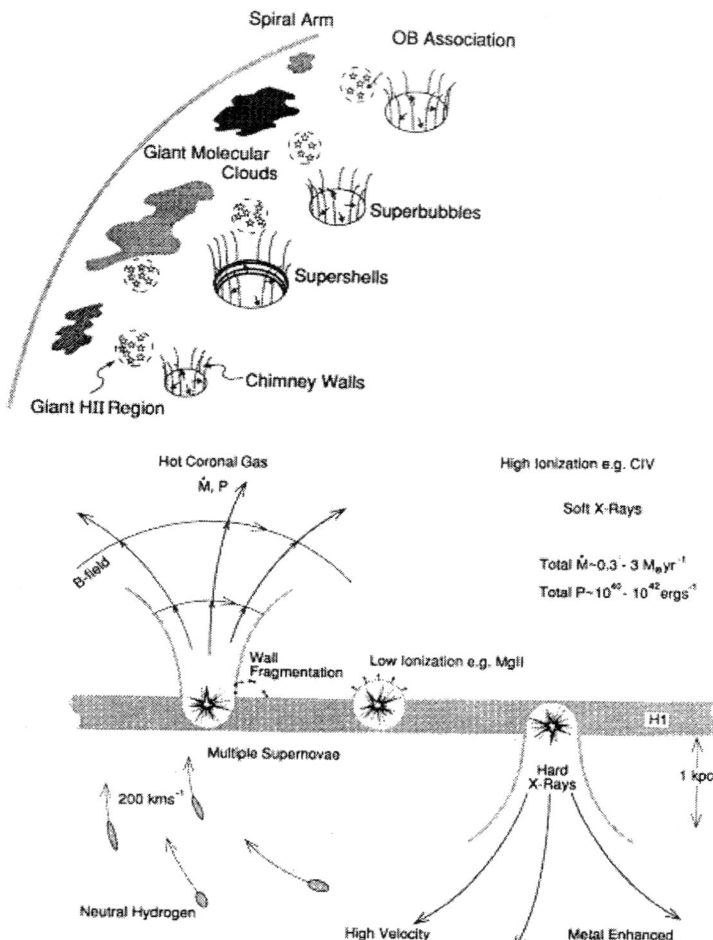

Figure 1: Two sketches, both from Norman and Ikeuchi (1989), showing aspects of the disk–halo interaction frequently invoked in discussions of HVCs and IVCs. The upper sketch indicates formation of supperbubbles and chimneys following massive star formation in giant molecular clouds located near the galactic equator. The lower sketch indicates the evolution and circulation of gas ejected from the equator to subsequently rain down on the disk from the lower halo as a galactic fountain. The two constraints discussed here limit the applicability of the scenarios depicted.

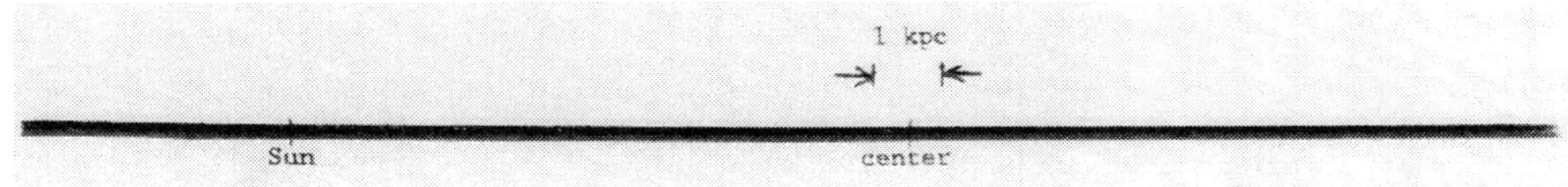

Figure 2: Simple, but in fact quite accurate, scale model of the conventional–velocity HI gaseous in the disk of the inner Milky Way. The model shows a cross–section cutting along the line of nodes of the warp. (The tilted gaseous disk in the inner few kpc is not indicated, because its line of nodes differs from that of the Galaxy at large; the gaseous disk in the Bulge region is in fact even thinner than the disk at larger radii indicated here.) At the scale indicated, the sketch is accurate regarding the degree of flatness ($\Delta z_0 < 20$pc) and the thickness (FWHM ~ 200 pc). Models for the HVC and IVC phenomena must not contradict the observations which support this simple scale model.

The sketches shown in Figure 1 incorporate several aspects of the galactic–fountain model (see Bregman 1980) which has been extensively invoked in discussions of HVCs, as well as indications of the sorts of energy inputs which seem plausible in the IVC context.

2 Scale–Height Constraint

The thickness of the conventional disk of interstellar material in the Milky Way is measured directly and reliably (see review by Burton 1993). Figure 2 shows a schematic scale model of the HI layer constituting the conventional galactic gaseous disk, drawn to scale. For tracers for which kinematics are available, notably HI and CO, the thickness at $R < R_o$ follows in a straightforward manner by converting the angular thickness observed near the terminal–velocity cutoff to a linear thickness using the geometrical distance appropriate to the subcentral region. At $R > R_o$, the galactic rotation curve is well enough known that kinematic distances may be relied upon to give quite plausible measures of the vertical thickness. For tracers for which kinematics are not available, notably dust, the techniques of radial unfolding suffice.

Figure 3 illustrates the sort of observations which have provided the HI vertical thickness. It is worth noting that the thickness measure samples a broad swath of the inner Galaxy, as the terminal–velocity cutoff refers to a length of path of several kpc extent because of the velocity crowding characteristic of the subcentral region; thus there is no reason to think that it is not a good general

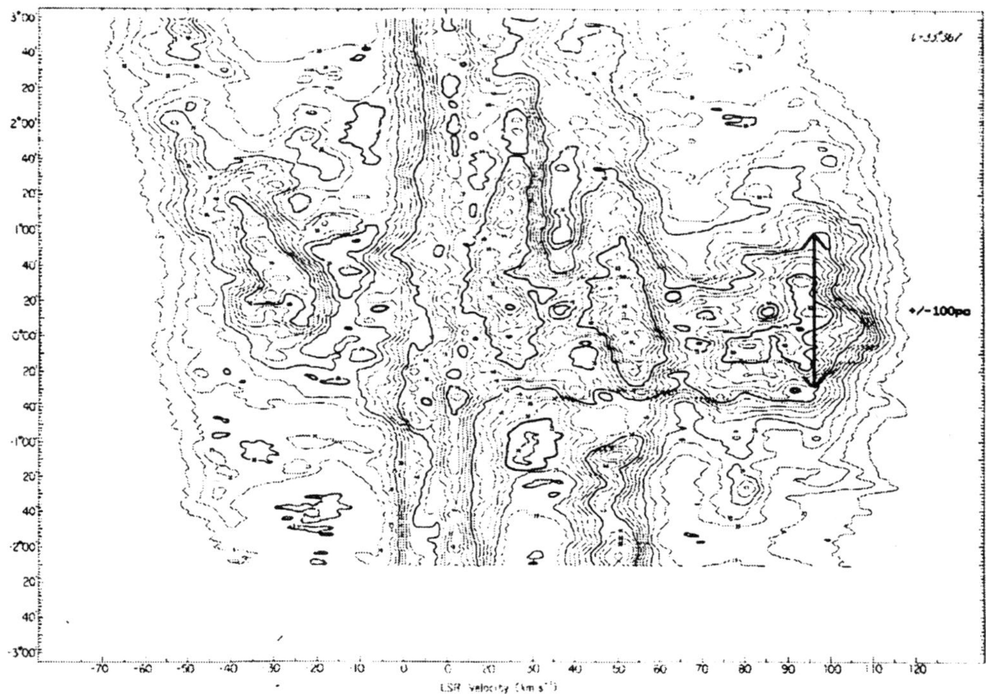

Figure 3: Latitude, velocity map of HI emission near the galactic disk at the representative longitude $l = 33°367$, from Bania & Lockman (1984). Observations such as these, which were made at the high angular resolution (4′) afforded by the Arecibo telescope, provide a firm measure of the thickness of gaseous disk of the inner Galaxy. The vertical bar drawn near the terminal velocity indicates two scale thicknesses, ±100 pc. Evidently neither HVCs nor IVCs, which together at high $|b|$ can contribute 10% to 50% of the total HI emission, significantly contaminate the low–latitude terminal–velocity regime.

measure. In the outer Galaxy, the thickness measure is a general one, not confined to a particular locus, and is as trustworthy as the kinematic distances; at $R \leq 1.5R_\circ$, 20% seems a conservative estimate of the plausible error.

The HI thickness measured at the terminal velocities gives a representative value of h_z of about 100 pc, or somewhat less at $R < R_\circ$ and somewhat more (but rapidly increasing at the largest R) at $R > R_\circ$. Radial unfolding using the total HI integral leads to similar results. Radial unfolding of the dust emissivities at 100 μm gives a thickness similar to that of the HI; the thickness derived from CO and most other molecular tracers is less. (The dust thickness may not be relevant to the HVC situation, because dust emission has not been measured from HVCs, but it is relevant to the IVCs, because of the frequently tight gas/dust correlation in these clouds; CO emission has not been found in HVCs, and only very rarely in IVCs.)

Both the IVC as well as the HVC phenomena carry a substantial portion of the total HI emissivity: IVCs are widely spread over essentially the entire sky, and contribute typically 10%, and in some directions substantially more, of the total HI column density; HVCs are concentrated in complexes covering about a quarter of the sky and contributing about 10% of the total emission in some directions. If the HVCs and IVCs were widespread phenomena – that is, not by some happenstance collimated solely above and below the Sun's location – located either in the galactic disk or in the lower regime of the disk/halo interface, then it would be expected that they would also pervade the terminal–velocity locus, and, generally, the outer–Galaxy disk. If that were the case, then the characteristic z height of the clouds would have to be consistent with the HI thickness measured; this would require that they be at characteristic $|z|$ heights less than about 100 pc.

The measured HI scale height constrains either the generality, or the distance scale, of HVCs and of IVCs. The scale–height constraint may be stated as follows. *Either* HVCs and IVCs do not populate the terminal–velocity locus, *or* the vertical scale height of HVCs and of IVCs is typically less than 100 pc. Figure 4 illustrates the possibilities. In the first possibility, HVCs and IVCs could be either collimated preferentially with respect to the Sun's location and at either large or small distances, or else they could be widespread but located well beyond the conventional galactic gaseous disk and the lower disk-halo interface. In the second possibility, HVCs and IVCs could be a general property of the galactic disk, but confined typically to $|z| < 100$ pc.

The scale–height argument applies to both HVCs and IVCs. But the distinction between HVCs and IVCs may well not be arbitrary, so that the same

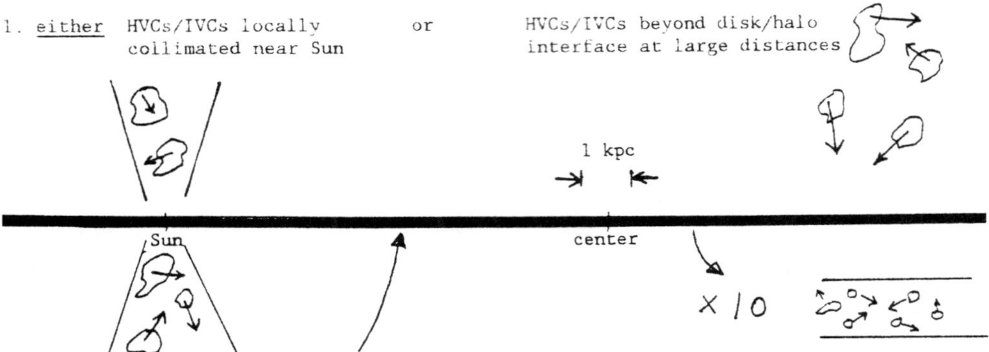

Figure 4: Schematic scale model of the gaseous disk of the Milky Way indicating the constraints imposed on the HVC and IVC morphologies by the measured thickness. Either HVCs and IVCs do not populate the subcentral–point region where h_z is measured to be ~ 100 pc, or the vertical scale height of the HVCs and IVCs is typically ≤ 100 pc. If the HVCs and IVCs do not populate the subcentral regions, the scale–height constraint does not distinguish between the anomalous–velocity clouds being locally confined and collimated near the Sun, or distributed generally, but well beyond the disk/halo interface.

possibilities to account for the restraint need not pertain for both class of objects. In particular, other arguments show that it is not likely that the HVCs are confined to $|z| < 100$ pc. If the HVCs were located in the lower halo, at typical distances from the Sun of several kpc (see e.g. Danly 1989 and Tamanaha 1996 and references in both papers), or even substantially more, and if the HVC's were a general phenomenon, that is, not confined to flows collimated on the Sun, then the problem is such that their $|z|$ heights must also be large enough that their emission does not contaminate the scale–height measurements made along the terminal–velocity locus. Thus, either the HVCs are in the lower disk/halo interface at $z = 1$ kpc but then found only in the vicinity of the Sun, and not generally distributed, or HVCs are well beyond the $|z| = 1$ kpc regime, where they may be widely distributed at much larger distances.

3 Quiescent–Terminal–Velocities Constraint

The kinematic cutoff measured on inner–Galaxy HI profiles is familiar from determinations of the Galaxy's rotation curve. HI measurements in the galactic equator of the sort illustrated in Figure 5 show that the scatter around the mean terminal velocity is very small. This scatter has been measured from HI data by Burton & Gordon (1978) to be about 5 km/s; Liszt et al. (1984) found the scatter in CO determinations to be 4.1 ± 1 km/s. These measurements along the terminal–velocity locus indicate that the kinematics of the neutral interstellar material dominating the galactic disk is remarkable quiescent. Excursions from the mean are rarely greater than 10 km/s, whereas the bulk motions of the HVC and IVC objects are typically an order of magnitude greater than this. (These remarks refer only to the line–of–sight component of the velocities; at the $|b| = 0°$ terminal–velocity locus, information on the z–component of the motions is not revealed.)

Anomalous–velocity gas of the IVC or HVC sort thus does not contaminate the terminal–velocity locus. The quiescent behavior of the HI terminal velocities constrains either the generality, or the distance scale, or the kinematics, of HVCs and of IVCs. The quiescent–terminal–velocity constraint may be stated as follows. *Either* HVCs and IVCs do not populate the terminal–velocity locus, *or* the HVCs and IVCs have low $< v_R >$ and low $< v_\Theta >$ but large $< v_z >$. Figure 6 illustrates the possibilities, which may pertain differently for the HVCs than for the IVCs. In the first possibility, HVCs and IVCs could be either collimated preferentially with respect to the Sun's location and at either large or small distances, with large mean velocities in the z direction and with the v_R and v_Θ components unconstrained, or else they could be widespread, and with velocities largely unconstrained, but typically located well beyond the conventional galactic gaseous disk and the lower disk-halo interface at z distances which lead to no contamination of the measured terminal velocities. In the second possibility, HVCs and IVCs could be a general property of the galactic gaseous disk, but with the anomalous velocities only in the z direction.

4 Conclusions from the Combined Consequences of the Two Constraints

Combining the consequences of the scale–height constraint and the quiescent–terminal–velocity constraint leads to some alternative conclusions (which need

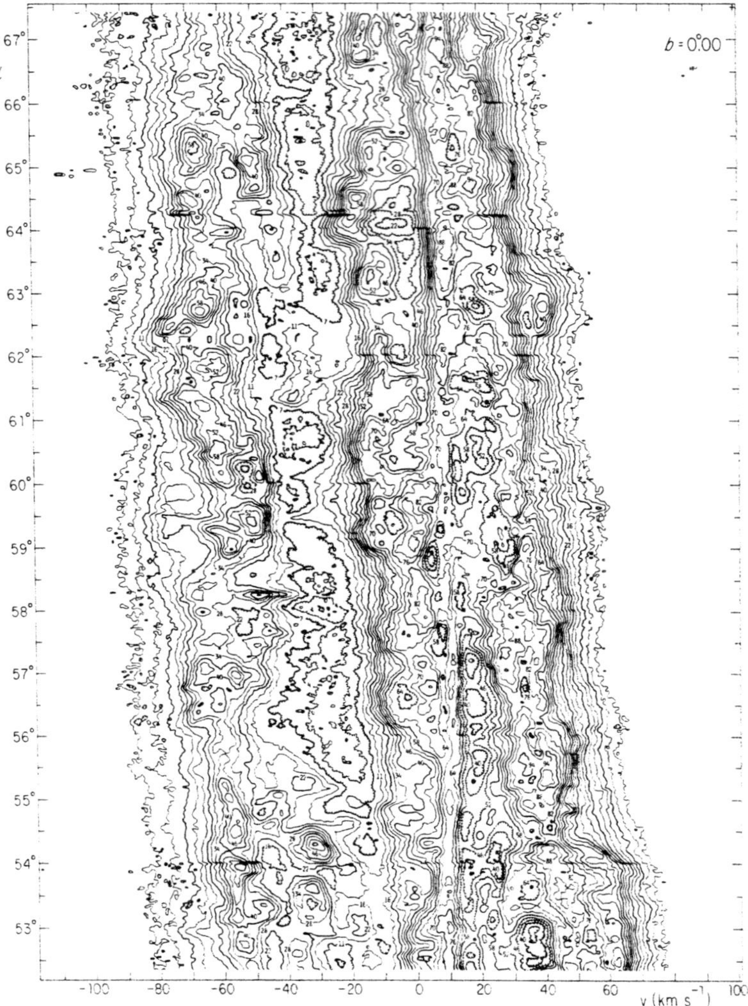

Figure 5: Longitude, velocity map of HI emission observed at $b = 0°$ over a representative portion of the inner Galaxy (from Baker & Burton, 1979), illustrating the quiescent behavior of the terminal velocities. The kinematic cutoff on the positive–velocity side (for the first longitude quadrant) provides the galactic rotation curve; the small scatter about the mean measures the line–of–sight component of the motions within the ensemble of galactic gas clouds; the dispersion of this component follows from the observations to be about $\delta_c = 5$ km/s. Neither the HVC nor the IVC phenomena are evident in the broad swath of disk sampled by the terminal–velocity measurements.

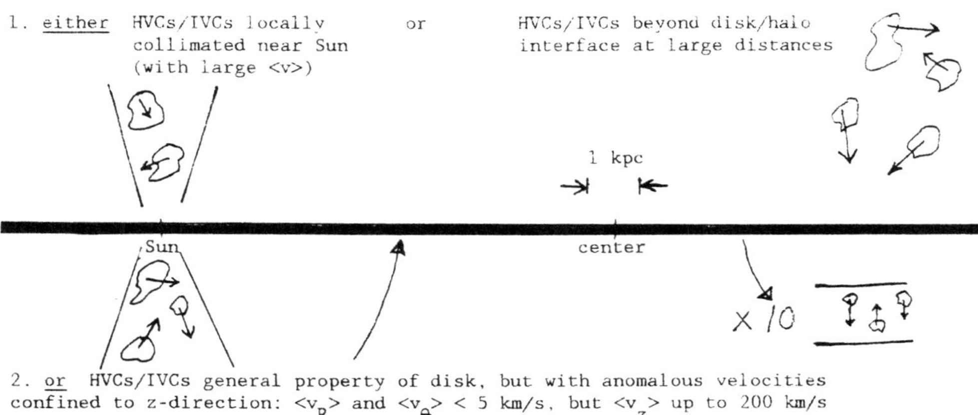

Figure 6: Schematic scale model of the gaseous disk of the Milky Way indicating the constraints imposed on the HVC and IVC properties by the quiescent behavior of the terminal velocities, whose scatter about the mean is measured to give the line–of–sight dispersion of the gas–cloud ensemble as $\delta_c = 5$ km/s. Either HVCs and IVCs do not populate the subcentral–point region where δ_c is measured, or both HVCs as well as IVCs have anomalous motions only in the z–direction. If the HVCs and IVCs do not populate the subcentral regions, the quiescent–behavior constraint does not distinguish between the anomalous-velocity clouds being locally confined and collimated near the Sun, or located well beyond the disk/halo interface at large distances.

not apply similarly to HVCs and IVCs): *either* HVCs and/or IVCs are collimated towards the Sun (with distances unconstrained by these arguments); *or* HVCs and/or IVCs are widely distributed but located at large z distances and thus not available to contaminate either the scale–height or the terminal-velocity measures; *or* HVCs and/or IVCs are widespread throughout the galactic disk but confined to $|z| < 100$ pc and with their anomalous motions only in the z direction.

Several remarks can be made regarding these alternatives, distinguishing on the basis of other information between the HVC and the IVC objects.

The first of the alternatives mentioned above, namely that the anomalous-velocity gas would be confined to regions above and below the Sun, seems implausible for the IVC objects because of the expectation that such features would be produced by events which are expected to be common throughout the galactic disk, at least throughout that portion of it experiencing strong star formation. It is more difficult to reject, on the basis of current knowledge, this alternative for the HVC objects; although it seems presumptuous to require confinement of the objects to cones oriented towards the Sun, observations of several external galaxies show high–velocity gas confined to single major complexes, clearly confined in extent.

The second of the alternatives mentioned above, namely that the anomalous–velocity gas would be located at large distances and thus not intimately associated with the conventional gaseous disk, seems implausible for the IVCs because of the other information on their distances, for example as derived from star counts (e.g. de Vries & Le Poole 1985), from their correlation with cirrus dust features, from their evident association with known OB associations and other galactic–plane phenomena. Distances of at least several kpc for the HVC objects seem plausible in view of the results of Danly (1989), Tamanaha (1996), and others.

The third of the alternatives allowed by these constraints, namely that the anomalous–velocity gas would be widespread in the galactic disk, but confined to $|z| < 100$ pc and with the anomalous motions only in the vertical direction, raises other questions. Such low z–heights seem contradicted by other data for the HVCs, but not for the IVCs. There is little information on the the space motions of either class of objects. Motions predominately in the z direction resulting from energetic events occurring in high density surroundings, with vertical escape favored by the lower densities away from the galactic disk, might well pertain for the IVC events. The IVC objects have largely been studied only at rather high $|b|$, because of confusion with conventional–velocity gas near the galactic disk. We note that the HVC objects have been followed over a wide range of latitude, so that it seems unlikely that their motions are strongly confined to the z direction.

5 Two Relevant Questions Amenable to Investigation with New HI Survey Data

There are two questions particularly relevant to the discussion here which have received attention in the past but which now could be considered anew exploiting aspects of the new Leiden/Dwingeloo survey of galactic HI (Hartmann 1994; Hartmann & Burton 1996).

Is the distinction between HVCs and IVCs arbitrary, or does it have a physical basis? The distinction was introduced as an arbitrary one. Studies of the anomalous–velocity gas dominated use of the Dwingeloo telescope during the 1960's; in order to divide the efforts between astronomers in Leiden and Groningen, a distinction was drawn at $v_{\mathrm{LSR}} = -70$ km/s: the more extreme velocities would be studied in Leiden, and the less extreme in Groningen. The distinction was useful for a practical reason too, because the number of spectral channels then, and until only recently, available has required independent observations with retuned receivers in order to cover the anomalous velocities adequately, or else the width of individual spectral channels has been necessarily so large as to preclude detailed investigation of the more intricate characteristics of the features. These restraints persisted through the large–scale HI surveys of the 1970's and 1980's, but have been ameliorated to an important degree by the new survey made with the Dwingeloo 25–m telescope.

It seems particularly important to investigate the degree of continuity across the demarcation velocity of -70 km/s, and to establish any other distinctions between the two regimes. Of course, some HVCs clearly transgress that demarcation: the Magellanic Stream and Complex C are the most obvious examples of bona fide HVCs with gradients in velocity which cause them to blend in velocity space with the IVCs. Yet distinctions between the two classes of anomalous–velocity objects seem clear in several regards. IVCs frequently have dust cirrus counterparts, HVCs do not; IVCs have distances evidently reliably determined from star counts, from association with energetic events at known distances, etc., which are of the order of 100 pc; HVC distances, although more difficult to determine, are probably much larger.

Are there definite cases of anomalous–velocity complexes interacting with the material in the in the conventional gaseous disk? This is also not a new question, but it is one which is particularly amenable to investigation with improved HI data as well as with other data now available, in particular the ROSAT X–ray observations, and various new material on ionized hydrogen. (Some related investigations are reported elsewhere in these proceedings.)

If the anomalous–velocity gas features are confined to the conventional galactic gas layer, or are produced in it, or collide from outside the Galaxy onto it, then it seems reasonable that there would be evidence of shocks and other disturbances stemming from supersonic interactions. There have been several investigations reporting such interactions; those for the HVC/disk interactions are particularly important but not yet utterly convincing. It seems important to apply all possible observational material to investigations of this question, because constraints following from a convincing association of HVCs with the conventional disk would be very robust.

References

Baker P.L., Burton W.B., 1979, A&AS 35, 129

Bania T.M., Lockman F.J., 1984, ApJS 54, 513

Bregman J.N., 1980, ApJ 236, 577

Burton W.B., 1993, "Distribution and Observational Properties of the ISM".
    In: Pfenniger D., Bartholdi P. (eds.) The Galactic Interstellar Medium:
    Saas-Fee Advanced Course N° 21. Springer–Verlag, Heidelberg, p. 1

Burton W.B., Gordon M.A., 1978, A&A 63, 7

Danly L., 1989, ApJ 342, 785

de Vries C.P., Le Poole R.S., 1985, A&A 145, L7

Hartmann Dap, 1994, Ph.D. Thesis, University of Leiden

Hartmann Dap, Burton W.B., 1996, "Atlas of Galactic Neutral Hydrogen".
    Cambridge University Press, in press

Liszt H.S., Burton W.B., Xiang D.L., 1984, A&A 140, 303

Norman C.A., Ikeuchi S., 1989, ApJ 345, 372

Oort J.H., 1970, A&A 7, 381

Tamanaha C.M., 1996, ApJS 104, 81

Wakker B.P., 1991, A&A 250, 499

Address of the author:

W.B. BURTON, Sterrewacht Leiden, P.O. Box 9513, 2300 RA, Leiden, The
Netherlands. (E-mail: burton@strw.leidenuniv.nl)

The Leiden/Dwingeloo Survey at its Limit: $\sigma \sim 70$ kms^{-1} HI emission at high latitudes

Gernot Westphalen, P.M.W. Kalberla, U. Mebold,
Dap Hartmann & W.B. Burton

1   Introduction

The existence of an extended gaseous halo of our Galaxy was already proposed
by Spitzer (1956). He argued that the clouds which had been seen in absorption far above the galactic plane could only persist there, if a surrounding
low-density gas shields them from rapid disintegration. This low-density gas
would have to be very hot to mount the necessary pressure ($T \sim 10^6$ K). Until
now, there have been considerable problems both with maintaining such a hot
halo (e.g., Schmutzler & Tscharnuter, 1993) and with the origin/condensation
of the cold absorbing clouds in it. Composition, density, temperature and extent of the galactic halo have all been subject to extensive debate.
As described by Kalberla et al. (1996a), the new Leiden/Dwingeloo 21-cm
line Survey (hereafter LDS) has improved sensitivity of low-level emission by
almost a factor of 10 compared with previous HI surveys. This stimulated us
to search for the signature of neutral hydrogen in the galactic halo, i.e. for
faint and very broad components in the HI emission profiles.
The basis for the present investigation is the LDS described by Hartmann
(1994) and Hartmann & Burton (1996). We improved the stray radiation
correction as described by Hartmann et al. (1996) by taking the ground reflection into account. Various methods have been used to isolate very low-level
HI emission with large velocity dispersions from systematic profile errors and
contamination features.

2   Data Processing & Analysis Method

Kalberla et al. (1996b, this volume) analysed the LDS by averaging spectra
over fields of $90° \times 15°$. The purpose of their analysis was to check the consis-
tency of low-level emission ($T_{\mathrm{B}} < 0.1$ K) with other data (e.g. the Bell Labs

Survey, hereafter BLS, Stark et al., 1992) and to deduce global properties for this emission. Their approach still allowed for the possibility that high-velocity clouds (HVCs) and other features blend together to form the broad line wings they found. Therefore one of us (Westphalen) has analysed the survey in more detail. He also spent additional effort to eliminate remaining contaminations. The spectra were averaged over fields of $10° \times 10°$, reducing the rms noise to ≈ 6 mK. To isolate the large velocity dispersion (LVD) component, the averaged profiles were decomposed into Gaussians. The aim of this decomposition was to get a simple and useful description of the very low-level emission presented by Kalberla et al. (1996b). Because of the very low peak temperature of the LVD component ($T_B \leq 0.1$ K) and the complexity of the decomposition, it proved impossible to automate the analysis. For control purpose we also decomposed the stray radiation profiles into Gaussians. We used about 1200 positions with a total of about 5000 components ($b > 20°$).

An efficient way to identify small-scale clouds or contaminations is a fluctuation analysis in the respective $10° \times 10°$ fields. We calculated the mean square deviation (or variance) of the spectra from the mean profile in a single $10° \times 10°$ field (cf. Mebold & Hills, 1975). This works as a high-pass filter for spatial frequencies: small-scale structures like HVCs are enhanced, while components distributed uniformly in the field are suppressed. Using this fluctuation analysis we determine the number of Gaussians for the decomposition and the channels which have to be excluded because of contamination.

Gaussian decomposition may give rise to large systematic errors (Takakubo & van Woerden, 1966; Kaper et al., 1966). Therefore the decompositions were carried out under various boundary conditions. We checked the stability of the decompositions against changing starting parameters, velocity range, and number of Gaussians used. Differences between the corresponding results were found to be stochastic and of the order of 20% (intensity), 10% (dispersion), 8% (column density) and ± 10 kms^{-1} (central velocity). Thus the decompositions give reproducible results and the scatter for one profile is a useful measure for the decomposition's intrinsic uncertainty. The details of this analysis and the results will be published elsewhere.

3 Existence of the LVD component

We confirm the existence of the ubiquitous LVD component ($\sigma \sim 70$ kms^{-1}) reported by Kalberla et al. (1996b). We are confident that this component is not merely an artefact. We investigated the band pass and baseline cor-

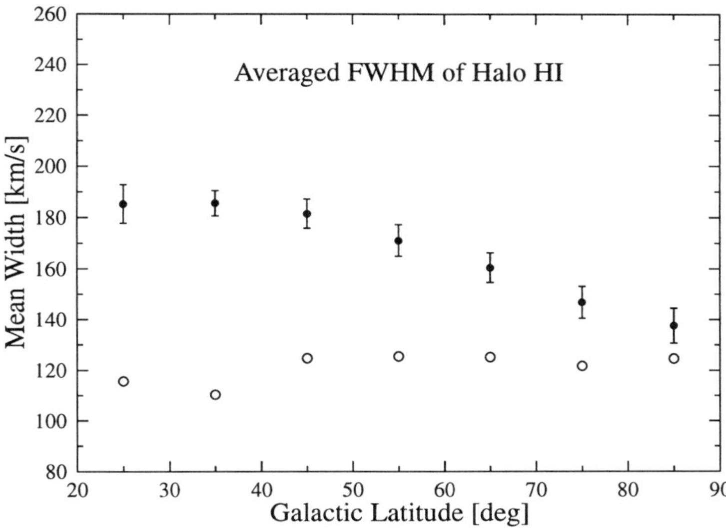

Figure 1: The halfwidth for the Large-Velocity Dispersion (LVD) HI component (•) is shown with the mean error. The circles (○) mark the halfwidth of the broadest stray radiation components (see text for details).

rections applied by Hartmann (1994). By averaging over large areas of the sky we detected a feature at ~ -300 kms^{-1} which could either be real or a residual baseline ripple ($T_B \sim 10$ mK). This feature occurs only in $\sim 10\%$ of the spectra and does not affect the LVD component. As we could not find any significant improvement (beyond removing a DC offset of the autocorrelator), we are confident that any residual uncertainties in the baseline are below our detection limit of 5 mK (Kalberla et al., 1996b).

Still, stray radiation, which is due to the diffraction pattern of the telescope, could be contributing to the LVD component. A non-negligible part of the stray radiation is ground reflection, i.e. radiation from the sky which is reflected by the ground into the far side-lobe pattern of the telescope. This ground reflection is of the same magnitude as the signal of the LVD component. A detailed analysis of the corrections applied previously revealed the necessity for an additional correction to the stray radiation correction (Kalberla et al., 1996b), in particular for the Dwingeloo site, where an extended heath in the South reflects radiation into the telescope. As a result the amplitude of the LVD component increased because the ground reflection had been overestimated initially.

The best argument for an astronomical origin of the LVD component is that stray radiation cannot account for the outermost velocity extent of this emission. For an example the reader is referred to Kalberla et al. (1996b). Fig. 1 shows the mean halfwidth of the Gaussians fitted to the LVD component and to the broadest wings of the stray radiation correction profile as a function of galactic latitude. The stray radiation correction profile includes the contributions of the near– and the far side lobes, as well as the ground reflection. This correction profile was decomposed into Gaussians in the same way as the corrected spectrum was. The large width, in particular at low latitudes, rules out that stray radiation is the source of the LVD component. Kalberla et al. (1996b) showed that any attempt to remove the LVD component by increasing the stray radiation correction leads to negative intensities, for lower velocities, clearly indicating an overcorrection. This leads us to conclude that the LVD component is very likely real and not an instrumental artefact.

4   Discussion

Assuming stratified layers of gas parallel to the galactic plane, the column density is expected to vary as $N_{\mathrm{HI}}(b)= \mathrm{A} \times cosec(b)$ as a function of galactic latitude b (e.g. Lockman & Gehman, 1991, hereafter LG). Fig. 2 shows the column density of the LVD component averaged over longitude. Analysis of high-latitude data indicates a deficit of HI gas in the solar neighbourhood. This is consistently interpreted as being due to the presence of the local hot bubble: an X–ray emitting region in our surroundings (Snowden et al., 1990). However, the deficit found from the LDS data is smaller than found by Lockman (1986). The reason is that there is *more* HI gas in our LVD component at high latitudes than expected from a stratified layer (dashed curve in Fig. 2). The HI column density in Fig. 2 can be described by $N_{\mathrm{HI}}(b) = \mathrm{A} \times csc(b) + \mathrm{B}$ (solid curve). This corresponds to a plane parallel layer with a column density of $\mathrm{A} = 4 \times 10^{18}$ cm^{-2} and a quasi spherical component with a column density of $\mathrm{B} = 1 \times 10^{19}$ cm^{-2}, possibly associated with the galactic halo. The quality of previous observations (e.g. the BLS) was not sufficient to derive this spherical halo component (Kalberla et al., 1996b). The halo component, ($N_{\mathrm{HI}}^{\mathrm{Halo}} = 1 \times 10^{19}$ cm^{-2}), is consistent with results of Danly et al. (1992) who found neutral hydrogen beyond 1 kpc with a column density of $N_{\mathrm{HI}}(z > 1 \text{ kpc}) = 0.5 - 2 \times 10^{19}$ cm^{-2}. The LVD component displays a complex behaviour in velocity space. It shows a systematic negative velocity of -9 ± 3 km/s when averaged over longitude. This fits well to results by de Boer (1996) and by LG for high galactic latitudes. It does not show a simple *sin l* or *sin 2l* variation with galactic longitude l, as would be expected if the gas were in a slowly rotating halo, or would be rather

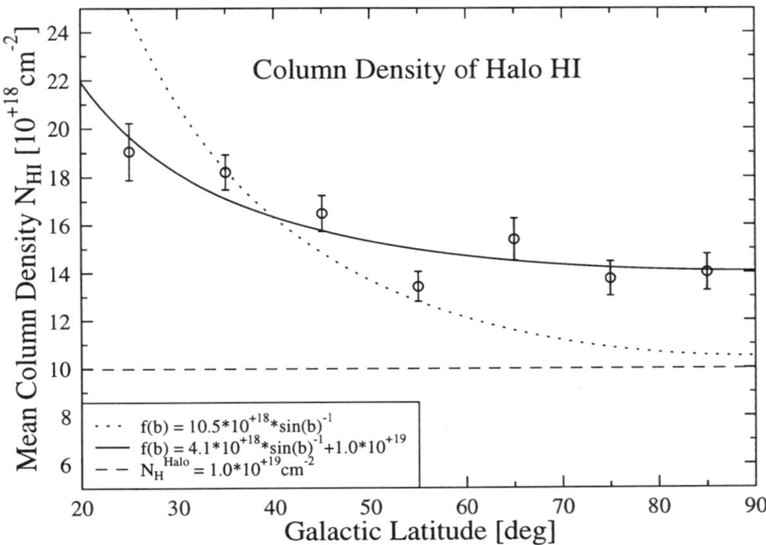

Figure 2: The longitude-averaged HI column density of the Large-Velocity Dispersion (LVD) component (o). The curved solid line indicates best fits using the least-square method assuming the plane parallel layer and constant offset originating from the halo (dashed). The dotted line represents a plane parallel layer only.

close by ($d < 1$ kpc), respectively.

The mean value of velocity dispersion ($\sigma = 71 \pm 1$ kms^{-1}) is significantly higher than previously reported for H$_{\rm I}$. However, it is not completely unexpected. While H$_{\rm I}$ emission lines with dispersions up to $\sigma \sim 35$ km/s are well established (LG and references therein), Danly (1989) finds strong low-ionisation absorption lines with dispersions up to $\sigma \sim 64$ km/s (towards 34 stars beyond 1 kpc). She argues that with 99.8% probability this high-dispersion gas is *not* an extension of the stratified disc gas. Danly et al. (1992) list 56 stars for which H$_{\rm I}$ emission spectra have been taken with the Green Bank 43-m telescope. Excluding 6 contaminated spectra, the H$_{\rm I}$ emission wings have a mean velocity spread of $\sigma_{+-} = 60 \pm 3$ kms^{-1} at the $T_{\rm B} = 0.05$ K level.

The above evidence suggests that a neutral, large velocity dispersion, low-density, extended halo does indeed exist. It is probably embedded in hot plasma ($T > 5 \times 10^5$ K) with an even lower density. Further investigations are necessary to determine the patchiness and the extent of this neutral halo. Analysis of the data for $b < -20°$ should reveal whether there is a global anisotropy between the northern and southern galactic hemispheres. Identical parameters for the LVD component would support an extended neutral galactic halo.

References

Danly, L., 1989, ApJ, **342**, 785.

Danly, L., Lockman, F.J., Meade, M.R. & Savage, B.D., 1992, ApJS, **81**, 125.

de Boer, K.S., 1996, in Proc. of a Workshop held at Bad Honnef: "The Physics of Galactic Halos".

Hartmann, Dap, 1994, PhD thesis, University of Leiden.

Hartmann, Dap, Kalberla, P.M.W, Burton, W.B. & Mebold, U., 1996, A&A, **119**, in press

Hartmann, Dap & Burton, W.B., 1996, Atlas of Galactic Neutral Hydrogen, Cambridge University Press, in press

Kalberla, P.M.W., Hartmann, D., Burton, W.B., Mebold, U. & Westphalen, G., 1996a, in High-Sensitivity Radio Astronomy, Cambridge University Press, Eds.: N. Jackson & R. Davis.

Kalberla, P.M.W, Mebold, U., Westphalen, G., Pietz, J., Hartmann, D. & Burton, W.B., 1996b, in Proc. of a Workshop held at Bad Honnef: "The Physics of Galactic Halos".

Kaper, H.G., Smits, D.W., Schwarz, U., Takakubo, K. & van Woerden, H., 1966, Bull. Astr. Inst. Neth., **18**, 465.

Lockman, F.J., 1986, in Proc. of a Workshop held at the NRAO Green Bank: "Gaseous Halos of Galaxies".

Lockman, F.J. & Gehman, C.S., 1991, ApJ, **382**, 182.

Mebold, U. & Hills, D.L., 1975, A&A, **42**, 187.

Schmutzler, T. & Tscharnuter, W.M., 1993, A&A, **273**, 318.

Snowden, S.L., Cox, D.P., McCammon, D., Sanders, W.T., 1990, ApJ, **354**, 211.

Spitzer, L., 1956, ApJ, **124**, 20.

Stark, A.A., Gammie, C.F., Wilson, R.W., Bally, J., Linke, R.A., Heiles, C., Hurwitz, M., 1992, ApJS, **79**, 77.

Takakubo, K. & van Woerden, H., 1966, Bull. Astr. Inst. Neth., **18**, 488.

Addresses of the authors:

GERNOT WESTPHALEN, P.M.W. KALBERLA, U. MEBOLD
Radioastronomisches Institut der Universität Bonn, Auf dem Hügel 71, D 53121 Bonn, FRG

DAP HARTMANN, Harvard-Smithsonian Center for Astrophysics, 60 Garden Street, Cambridge, MA 02138, U.S.A.

W.B. BURTON, Sterrewacht Leiden, Postbus 9513, 2300 RA, Leiden, The Netherlands

High Velocity Clouds: The Importance of Drag

Robert A. Benjamin

1 Motivation

High velocity clouds (HVCs) have remained an enigma since their discovery in 1963 (Muller et al 1963). A large part of their mystery is due to the fact that the highest velocity clouds, $|v_{LSR}| > 100 km \ s^{-1}$ are at unknown distances (c.f., de Boer et al 1994). The distances to several lower velocity clouds have been determined (Wesselius & Fejes 1973; Benjamin et al 1996, Kuntz & Danly 1996), suggesting that cloud velocities increase with z. Danly (1989) has also shown UV absorption line widths extend to increasingly negative velocity for increasing star distance. Figure 1 shows the velocity, v, and height, z, of clouds above $|z| \gtrsim 200 \ pc$. This figure shows, despite a paucity of accurate distances, a trend of velocity increasing with z.

Drag forces seem to have been largely neglected when considering the dynamics of HVCs falling towards the Galactic plane. Models of HVCs have tended to focus either on their ionization structure (Ferrara & Field 1994; Wolfire et al 1995) or have assumed that clouds fall ballistically in the gravitational field of the galaxy (Bregman 1980; Li & Ikeuchi 1992). In the following discussion, I demonstrate that drag forces associated with a cloud's motion through the gaseous galactic halo can decelerate the cloud to the observed velocities. I derive an empirical drag coefficient, and suggest that studies of individual cloud characteristics and morphologies together with numerical simulations of cloud dynamics may allow this technique to be used as a unique probe for the vertical density structure of the interstellar disk.

2 Deceleration of Halo Clouds

I start by considering the behavior of an idealized cloud. The cloud is assumed to be a flattened structure with a constant surface area, A, and mass, M. It is constrained to fall vertically toward the galactic plane with velocity, v, using the gravitational acceleration of Wolfire et al (1995). The pressure on the face of the cloud is the sum of ambient halo pressure plus the ram pressure, $p_h(z) + \rho_h(z)v^2$, where $\rho_h(z)$ is the halo mass density. The pressure on the back is assumed to be p_h. The pressure gradient across the cloud results in a drag force that decelerates the cloud. This force is given by $\frac{1}{2}C_D\rho_h v^2 A$, where C_D is the drag coefficient. For a perfectly streamlined cloud, $C_D = 0$; for completely efficient momentum transfer, $C_D = 2$. C_D can exceed 2 if the flow pattern produces an underpressurized region $p < p_h$ behind the cloud. The equation of motion is

$$\frac{dv}{dt} = \frac{1}{2}C_D\rho_h(z)v^2\frac{A}{M} - g(z) \tag{1}$$

I assume that $M = A\int_o^h \rho_c dz = A\mu N_{H\ I}/f$, where ρ_c is the cloud density, μ is the mass density divided by H atom density, and h is the vertical thickness of the cloud. The factor $f = N_{H\ I}/(N_{H\ I} + N_{H\ II})$ accounts for the fact that some fraction of the cloud mass is ionized. I assume an exponentially stratified halo, $n_h = n_{h,o}e^{-z/H}$, where $n_{h,o} = 0.025\ cm^{-3}$ and $H = 910\ pc$ (Reynolds 1993). The terminal velocity, where drag forces balance gravitational forces, is $v_T = \sqrt{2N_{H\ I}g(z)/C_D f n_{h,o}}\,e^{z/2H} \cong 26km\ s^{-1}\sqrt{N_{19}g_1/C_D f}\,e^{z/1.8\ kpc}$, where $N_{19} = N_{H\ I}/10^{19}\ cm^{-2}$ and $g_1 = g(1\ kpc) = 8.4 \times 10^{-9}cm\ s^{-2}$. This is plotted with a dotted line for $N_{H\ I} = 10^{19}\ cm^{-2}$, assuming $C_D f = 2$. This demonstrates that drag forces are potentially sufficient to decelerate infalling halo clouds to the observed velocities. I have also plotted the trajectories for a cloud with $N(H\ I) = 10^{19}\ cm^{-2}$ dropped from $z_i = 3, 5$, and $8\ kpc$, assuming $C_D f = 2$. The cloud initially falls ballistically. As it approaches the local terminal velocity, the acceleration decreases and will change sign for high enough drag coefficient or low enough cloud mass. In this limit, clouds will reach terminal velocity in less than a free-fall time scale, $t_{ff} \cong \sqrt{2z_i/g}$, and spend many subsequent free-fall times "locked" in at the local terminal velocity.

If this is the case in the halo, two simple predictions result: (1) For a fixed column density, more distant clouds fall faster than nearby clouds. (2) At a fixed distance, high column density clouds fall faster than low column density clouds. The first effect is amply demonstrated by Danly (1989); evidence for the second effect is also present. For instance, the ratio of absorption velocities observed for two stars at approximately equal height, HD121800 ($z = 1.7\ kpc; v = -92\ km\ s^{-1}$) and HD146813 ($z = 1.6\ kpc; v = -53\ km\ s^{-1}$), is 1.7, similar to the value 1.6 one would expect given the expression for terminal velocities and their estimated H I column densities of 4×10^{18} cm^{-2} and 1.5×10^{18} cm^{-2} (Kuntz & Danly 1996), respectively.

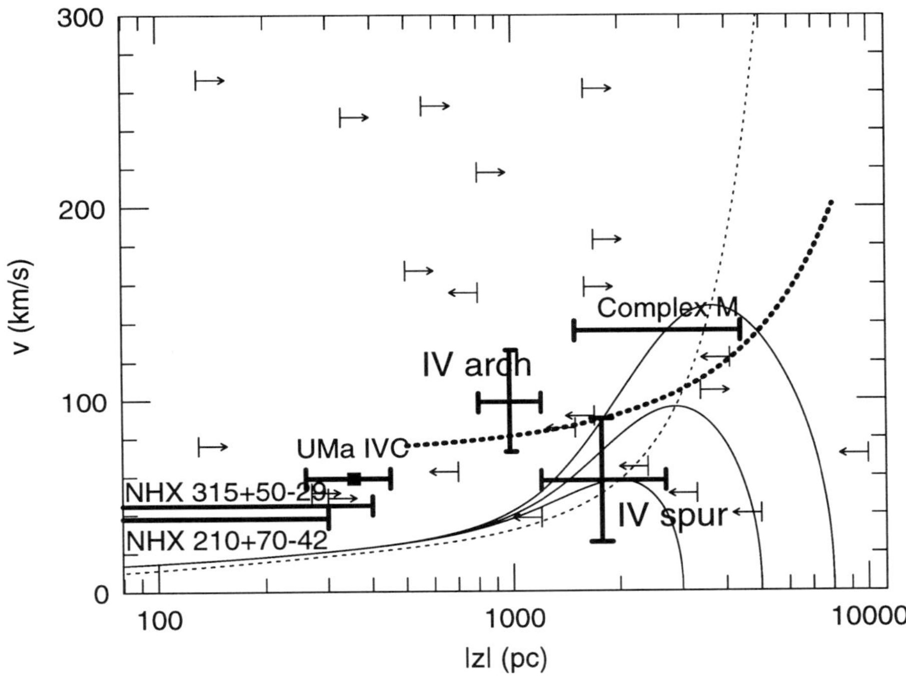

Figure 1: Velocity-distance relation for infalling halo clouds. $|v \csc b|$ is plotted for halo clouds with $|z| \gtrsim 200$ pc, $v < 0$, and $|b| > 30°$. The heavy dotted curve is a fit to the maximum negative velocity extent of UV absorption lines in Figure 4 of Danly (1989). Other curves are described in the text.

There are however several uncertainties in applying equation (1) to the observational data. First and foremost, $C_D f$ is unknown. Hydrodynamical simulations indicate that these clouds should be completely destroyed in much less than a free-fall timescale by Kelvin-Helmholtz instabilities (c.f., Klein et al 1994)! Given considerable theoretical uncertainties, I feel the proper way to proceed is to derive an empirical drag coefficient, and then determine whether any theoretical scenarios are consistent with this value. Taking z and v from Danly (1989), $N_{H\ I}$ from Kuntz & Danly (1989), and assuming the halo density structure of Reynolds(1993), I find that for $z < 500$ pc, $C_D f = 1.0 \pm 0.3$, while for $z > 500$ pc, $C_D f = 0.05 \pm 0.04$. This decrease is consistent with the expectation that clouds should be more ionized in the low pressure environment at high z, significantly lowering f. Morphological studies of these clouds and more sophisticated simulations of cloud deceleration will provide independent constraints on C_D.

If it becomes possible to constrain C_D sufficiently, the observed pattern of deceleration can be used to probe the density structure of the halo, put limits on the number of hot, presumably lower density, regions in the halo, and help estimate the density for $z >> H = 910pc$. The lowest column density features will be most useful in this regard as they are most easily locked to the local terminal velocity.

I would like to thank the organizers for allowing me to attend this stimulating workshop. The work presented here was supported by NASA LTSARP grant NAGW-3189 and the Minnesota Supercomputer Institute.

References

Benjamin, R.A., Venn, K. A., Hiltgen, D. D., & Sneden, C. 1996 ApJ, in press

Bregman, J.N. 1980 ApJ 236, 577

Danly, L. 1989 ApJ 342, 785

de Boer, K.S., Altan, A.Z, Bomans, D.J., Lilienthal, D, Moehler, S., van Woerden, H, Wakker, B.P., & Bregman, J.N. 1994 A & A 286, 925

Ferrara, A. & Field, G.B. 1994 ApJ 423, 665.

Klein, R.I., McKee, C.F., & Collela, P. 1994, ApJ 420, 213

Kuntz, K.D. & Danly, L. 1996 ApJ 457, 703

Li, F. & Ikeuchi, S. 1992 ApJ 390, 405

Muller, C.A., Oort, J.H., & Raimond, E. 1963 *C.R. Acad. Sci Paris* 257,1661

Reynolds, R.J. 1993 in *Back to the Galaxy: Proceedings of 3rd Annual Astrophysics Conference in Maryland: AIP Conf. Series No. 278* (eds. S.S. Holt & F. Verter) (AIP Press), p 156.

Wesselius, P.R. & Fejes, I. 1973 A & A 24, 15

Wolfire, M.G., McKee, C.F., Hollenbach, D., & Tielens, A.G.G.M. 1995 ApJ 453, 673

Address of the author:

ROBERT A. BENJAMIN, Department of Astronomy, University of Minnesota, 116 Church St, SE, Minneapolis, MN 55414, USA.

Magnetic Heating of High-Velocity Clouds

F. Zimmer, H. Lesch, G. Birk

1   Introduction

We present a numerical model for the magnetic heating of high-velocity clouds (HVC) falling onto the galactic disk with velocities up to 100 $\frac{\text{km}}{\text{s}}$. Observations with the X-ray satellite ROSAT (figure 1) show that the boundaries of these clouds are sources of strong X-ray emission revealing gas temperatures of a few million K.

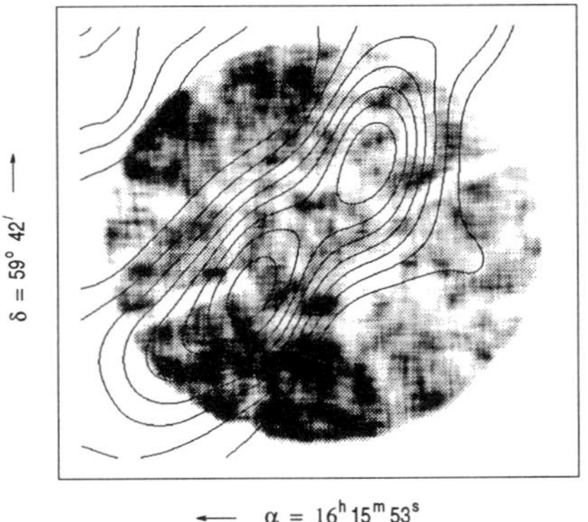

Figure 1: The $\frac{1}{4}$ keV ROSAT image of HVC90.5+42.5-130. The X-ray emission is shown as grey scale. Overlayed as contours is the HI column density distribution. Source: Kerp et al. (1994).

Assuming that a HVC colliding with the Reynolds layer excites a shock wave which propagates through this ionized gas layer the temperature ratio on both sides of the shock front can be calculated from the Rankine-Hugoniot jump relations for hydrodynamic shock waves. For cloud velocities up to 100 $\frac{\text{km}}{\text{s}}$ the gas temperature does not reach the observed value of a few million K (Kerp et al. 1996, Zimmer et al. 1996). In the presence of magnetic fields the sound velocity c_S has to be replaced by the generalized sound velocity $\sqrt{c_S^2 + v_A^2}$, where v_A denotes the Alfvén speed . Thus, if the magnetic field strength exceeds values of a few μG, it is very difficult to excite a magnetohydrodynamic shock wave - or in other words - a part of the kinetic energy of the clouds is transformed into magnetic energy instead of heat (Tidman & Krall 1971). Therefore it is not possible to heat up the gas to a few million K via shock waves. In this contribution an effective heating mechanism leading to the observed temperatures is presented.

2   Magnetic Heating

We know from observations that the galactic halo is a fully magnetized plasma which includes gas motions (Beuermann et al. 1985). High-velocity clouds are ionized as well due to cosmic rays and the soft X-ray background; magnetic fields are found in HVCs via Zeeman measurements with a strength of few 10 μG (Kerp & Güsten 1996). So, the interaction of two magnetized media with relative velocities up to 100 $\frac{\text{km}}{\text{s}}$ is considered. During the impact of a HVC into the Reynolds layer (Kerp et al. 1994) a boundary layer between the HVC and the ionized gas layer is formed, the plasma and the magnetic fields will be turbulently mixed and everywhere inside this interaction region magnetic field lines of different polarity approach each other. Thus, a large number of small scale reconnection sheets is formed, the gas is heated up, reaches finally a temperature of a few million K and emits the observed X-rays. To investigate such situations we perform resistive magnetohydrodynamic simulations. Here we present some results from $2\frac{1}{2}$D simulations.

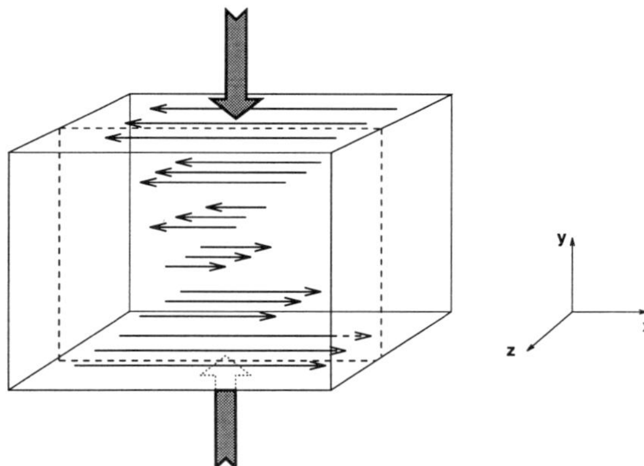

Figure 2: The MHD equations are integrated on the 2D box (dashed lines). The z-axis is invariant.

Magnetic flux is transported into a box and magnetic field lines of different polarity approach each other (figure 2). Due to the steepening of the gradient of the magnetic field the current density $\vec{j} \sim \nabla \times \vec{B}$ becomes larger and the drift velocity of the electrons relative to the protons exceeds some critical value. Plasma waves are excited and amplified and lead to an anomalous resistivity inside the current sheet region. So, noticeable heating rates $Q = \eta j^2$ and temperatures are expected. In our simulations the resistivity is produced by lower hybrid turbulences because of their low threshold for excitation (Papadopoulos 1979).

Figure 3 shows the magnetic field for a case with an initial temperature of about 900 000 K, the initial density is about 0.015 cm^{-3} and the magnetic field strength is roughly 17 μG: The magnetic field lines reconnect and the Lorentz-forces push the plasma out of the reconnection zone.

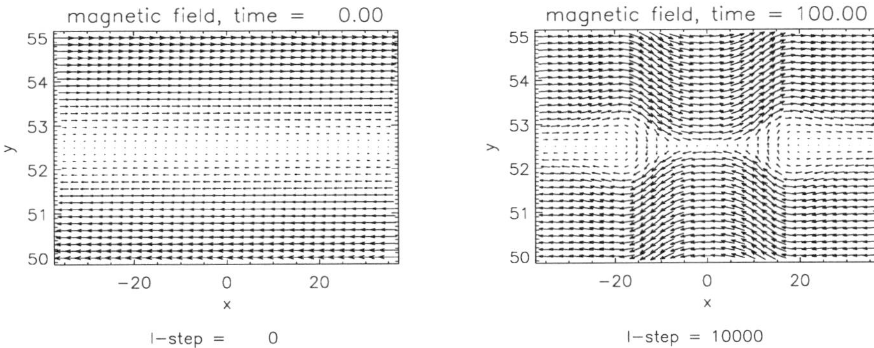

Figure 3: The time development of the magnetic field. The left figure shows the initial situation, the right figure the magnetic field after 100 Alfvén times ($\sim 5000\ s$).

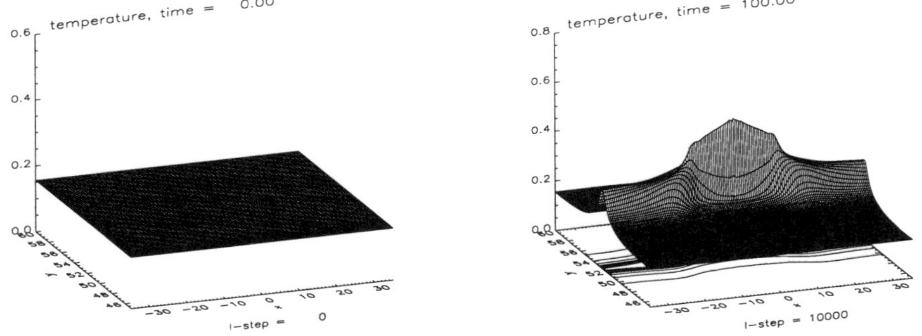

Figure 4: The time development of the corresponding temperatures.

The corresponding temperatures are shown in figure 4. Within 1.5 hours the temperature reaches a few million K via magnetic compression and magnetic reconnection. Simulations are also performed for other temperature regimes: gas with an initial temperature of 250 K, for example, is heated up to more than 2500 K. The typical cooling time of the gas given by $t_{cool} = \frac{nkT}{\Lambda} = \frac{nkT}{n^2\mathcal{L}}$, where Λ denotes the cooling rate and $\mathcal{L}$ the interstellar cooling function (Dalgarno & McCray 1972), is much less than the corresponding heating time $t_{heat} = \frac{nkT}{Q}$ derived from our simulations. Thus, the magnetic heating process will not be restricted by radiative losses, the magnetic fields heat the interstellar gas by magnetic compression and magnetic reconnection.

References

Beuermann K., Kanbach G., Berkhuijsen E.M., 1985, A&A 153, 17

Dalgarno A., McCray R.A., 1971, ARA&A 10, 375

Kerp J., Lesch H., Mack K.-H., 1994, A&A 286, L13

Kerp J., Mack K.-H., Egger R., Pietz J., Zimmer F., Mebold U., Burton W.B., Hartmann D., 1996, A&A (in press)

Kerp J., Güsten R., 1996, in prep.

Papadopoulos K., 1979, in Akasofu S.-I. (ed.), Dynamics of the Magnetosphere, D. Reidel, Dordrecht, 289-309

Tidman D.A., Krall N.A., 1971, in Shock Waves in Collisionless Plasmas, Wiley, New York

Zimmer F, Birk G., Lesch H., Kerp J., 1996, Astrophysical Letters & Communications (in press)

Addresses of the authors:

F. Zimmer, Radioastronomisches Institut der Universität Bonn, Auf dem Hügel 71, D-53121 Bonn

H. Lesch, Institut für Astronomie und Astrophysik der Universität München, Scheinerstraße 1, D-81679 München

G. Birk, Theoretische Physik IV, Ruhr-Universität Bochum, Universitätsstraße 150, D-44780 Bochum

High Velocity Gas in other Galaxies

J.M. van der Hulst

1 Introduction

The origin of the High Velocity Clouds (HVCs) in our Galaxy (Hulsbosch
& Raimond, Blaauw & Tolbert 1966, Oort 1970, Hulsbosch & Wakker 1988,
Wakker 1990, van Woerden et al. 1985) still is a puzzle. The general idea
of a galactic fountain (Bregman 1980) appears to work for a fair fraction of
the HVC population (Wakker 1990). Such a galactic fountain could be fed by
stellar winds and supernova explosions (Mac Low & McCray 1988, MacLow
et al. 1989, Elmegreen & Chiang 1982, Tomisaka 1992, Tomisaka & Ikeuchi
1986, Norman & Ikeuchi 1989, Tomisaka & Bregman 1993, Rosen & Bregman
1995). The numerous HI shells and filaments (Heiles 1979, 1984, 1990) in our
Galaxy, and the holes in the HI distributions of the Local Group galaxies M31
(Brinks & Bajaja 1986) and M33 (Deul & den Hartog 1990) suggest that this
is a viable possibility.

Other mechanisms may be important too, since the largest shells in the Galaxy
have energy requirements that surpass the output of stellar winds and supernovae in large associations. Such mechanisms could be infall of gas clouds
(Tenorio-Tagle et al. 1987, see also Tenorio-Tagle & Bodenheimer 1988). Evidence for collisions of gas clouds with the Galactic disk has been provided by
Heiles (1984) and Mirabel & Morras (1990). Such clouds could be primordial
or the result from tidal interaction between galaxies in a group (e.g. the Magellanic Stream and related HVCs in the case of our Galaxy, see also Wakker
1990).

The idea that supernovae deposit large amounts of energy into the interstellar
medium (ISM) is not new. Cox & Smith (1974) already discussed the hypothesis that the energy released by supernovae sets up tunnels of hot gas in
the ISM. This idea followed up on McKee & Ostriker (1977), who proposed a
three phase ISM with a hot ($T \approx 10^6$ K) phase powered by supernova events
in the Galactic disk, as a refinement of the original two phase model of Field,
Goldsmith & Habing (1969). This refinement was prompted by the discovery
of the $T \approx 10^6$ K component from X-ray observations and ubiquitous OVI

absorption lines (Gorenstein et al. 1974, Jenkins & Meloy 1974, Woodgate et al. 1974, York 1974).

Knowing how much kinetic energy is transferred into moving shells and filaments is of direct importance for assessing the energy balance in the ISM. It is important to track down the mechanisms which produce the HI shells and filaments since these are instrumental for determining the structure of the ISM and probably are also important for the heating and cooling balance in the ISM. Determining what fraction of the ISM is brought out into the halo of galaxies (and might eventually rain back onto the disk) is also of direct relevance to models which describe the chemical evolution of disk galaxies.

In our Galaxy we have the distinct disadvantage that it is difficult to determine the precise location of HI features in the disk. It is much more advantageous to try to find similar features in other, nearby galaxies and the advent of modern synthesis instruments such as the Westerbork Synthesis Radio telescope (WSRT) and the Very Large Array (VLA) has made this possible. In this paper I will describe some of the recent results which have become available and give some insight in the presence and nature of high velocity gas in other galaxies.

2   Observations

2.1   Face-on Galaxies

Face-on galaxies are the most ideal objects to search for high velocity HI, because one directly observes the radial velocity component perpendicular to the disk of a galaxy. In this way gas that is either moving out of the disk or falling into the disk can easily be observed through its difference in radial velocity with respect to the general rotation velocity of the gas in the disk.

The first specific search for high velocity gas in other galaxies was carried out by Wakker et al. (1989) using existing synthesis observations of the nearly face-on galaxies M51, M83 and NGC628. No clear high velocity component was found in these objects and mass limits to high velocity gas complexes are a few 10^6 $M_\odot$. There is high velocity gas in NGC628 but this is mostly associated with the warped outer parts (Kamphuis & Briggs 1992).

More recently Kamphuis (1993) studied the nearby, face-on galaxies M101 and NGC6946 in the 21-cm HI line using the WSRT. The HI column density distributions of these galaxies are shown in Figure 1. The HI extends farther than the Holmberg radius in both galaxies and shows spiral structure even outside the optical image. The resolution is $15''$ or 500 and 700 pc at the distances of M101 and NGC6946 respectively.

The percolated structure of the HI in the disks is quite obvious from figure 1 and tens of large holes with sizes of 1 to more than 5 kpc can easily be distinguished. Kamphuis (1993) found that the large, well defined holes occupy about 10% of the disk and are located throughout the whole HI disk, even in the outer regions where the spiral arms are barely visible in the optical.

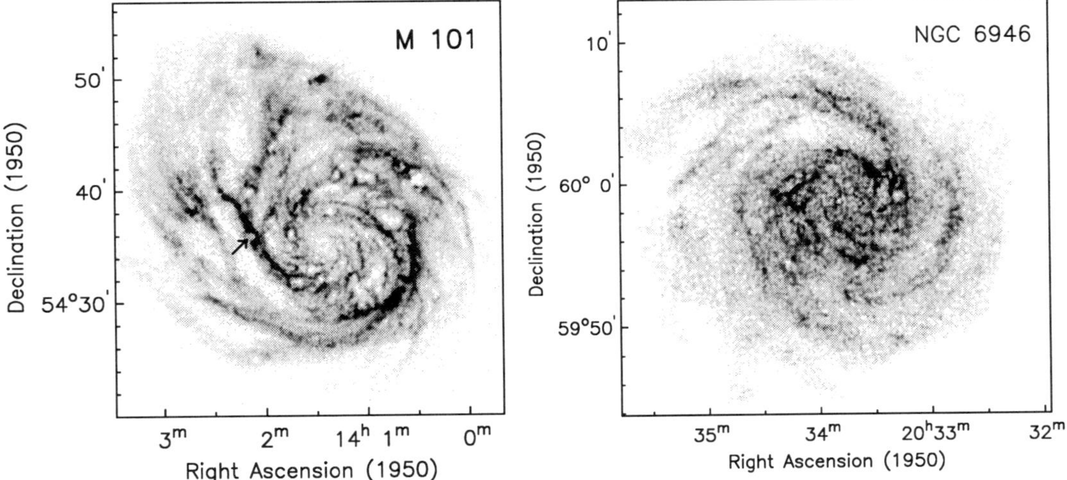

Figure 1: HI Distribution of the spiral galaxies M101 (left panel) and NGC6946 (right panel)

The first discovery of high velocity HI in M101 was made by Van der Hulst & Sancisi (1988, see also Kamphuis 1993). This high velocity HI complex is, however, exceptionally large and does not appear to bear much resemblance to the High Velocity Cloud complexes in our Galaxy. It is over 20 kpc in size, contains a few $10^8 M_\odot$ of HI and reaches velocities up to 160 km s^{-1} above the local rotation. The origin is not well understood. Most likely the complex is a result of a collision of M101 with a small galaxy or gas complex, perhaps even tidal debris from interaction with one of the nearby companion galaxies

NGC5477 or NGC5474. The fate of such a complex could be that it breaks up into smaller entities and eventually falls back into the disk. Viewed from the disk such an event could very well be similar to the High Velocity Cloud phenomenon in our Galaxy.

Kamphuis (1993) searched for high velocity gas in M101 and NGC6946 in a range of $\pm$ 100 km s^{-1} above and below the local rotation. Both M101 and NGC6946 have features in this range, sometimes connected to the holes in the HI distribution, mostly one-sided in velocity: either redshifted or blueshifted with respect to the local rotation. The features often are at the limit of the sensitivity and resolution.

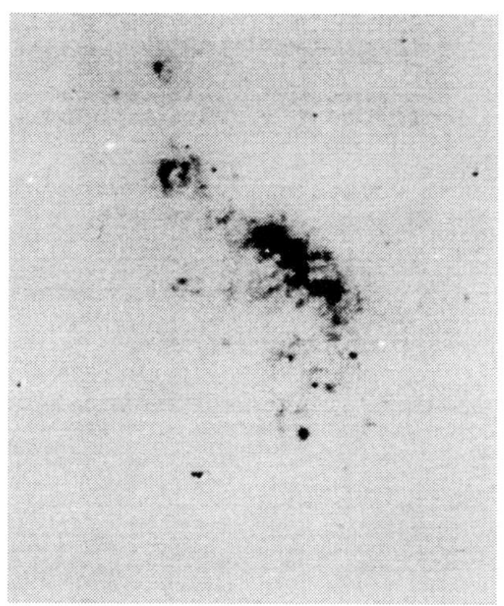
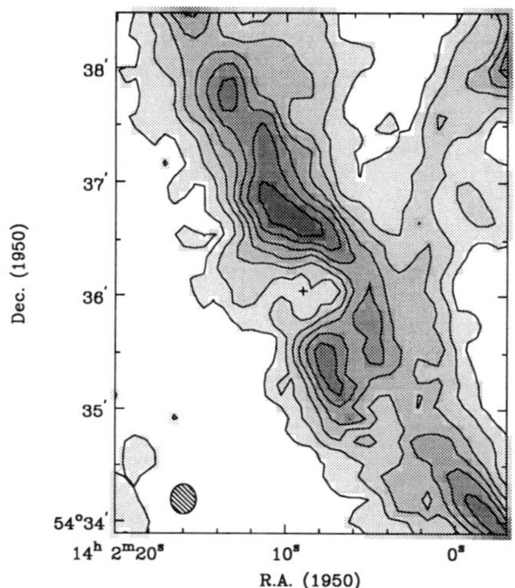

Figure 2: Distribution of Hα emission (left panel) and HI column density (right panel) of the giant HII region NGC5462 in M101.

One notable exception is the HII region NGC5462 at $\alpha = 14^h\ 2^m\ 9^s$, $\delta = 54° 36' 2.5''$ (indicated by the arrow in figure 1. This region is the best example of a well defined, symmetric HI bubble (Kamphuis et al. 1991) and deserves special attention. Figure 2 clearly shows the correspondence between the hole in the HI distribution and the HII region, in particular the faint filaments emerging from the HII region at the side of the HI hole. The Hα appears to follow the ridge of highest HI column density and curves around the HI hole. This HI hole was already noted by Allen et al. (1973) and Viallefond et al. (1981), but they did not have the sensitivity to detect faint HI around the hole. The observations of Kamphuis et al. (1991) are much more sensitive and reveal faint HI gas associated with the hole near NGC5462. It is the kinematic behavior of this faint gas which establishes beyond any doubt that this HI hole is a genuine superbubble.

Figure 3 best shows the kinematic behavior of the HI around the superbubble. This figure is a position − velocity map showing the HI profiles at each position along a line of position angle 30° centered on the superbubble. The high intensity ridge running from the SW to the NE along the position axis (centered around a velocity of ≈ 274 km s^{-1}) is the HI in the spiral arm in which NGC5462 is located. Close to the center of the bubble one clearly sees a slight increase in intensity and in velocity width of the HI profiles. At the center position the peak intensity drops and the profiles become very broad with a full width of ≈ 150 km s^{-1}. This signature is precisely what one expects to see in case of an expanding HI bubble. At the position of the center of the bubble one observes the full expansion velocity along the line of sight, both the blueshifted front and the redshifted backside of the bubble. As one moves away from the center the component along the line of sight gradually becomes smaller and one observes a gradually smaller velocity width.

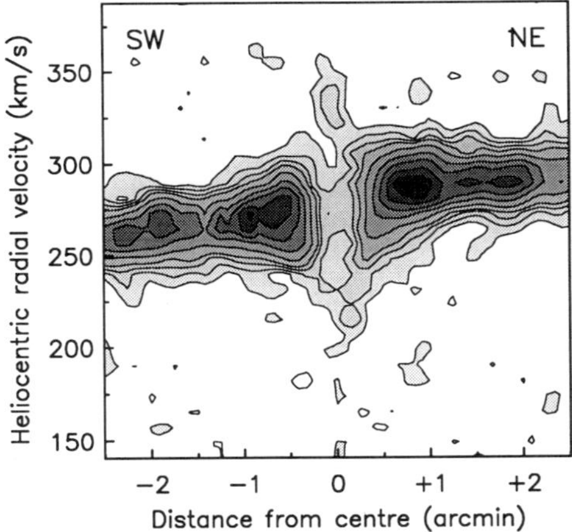

Figure 3: Position − velocity map at a position angle of 30° centered on the superbubble.

The total HI mass associated with the superbubble is estimated at 3×10^7 M$_\odot$. It is quite remarkable that the amount of HI required to fill the hole is 3.5×10^7 M$_\odot$, quite close to the estimated HI mass of the shell itself. The size of the bubble is about 1.5 kpc and the estimated age and kinetic energy are

1.5×10^7 yr and 2.5×10^{53} ergs.

The excellent kinematic symmetry of the HI and the good correspondence with a site of massive star formation provide very convincing evidence for such bubbles being the result of stellar winds and supernova explosions. The kinetic energy of the outflowing HI requires close to one thousand massive stars/super-novae. This may seem like a very large number, but one should keep in mind that such large HI regions as NGC5462 require many hundreds of O5 stars to produce the observed Hα luminosities.

As pointed out by Norman & Ikeuchi (1989) one expects the bubble to be filled with hot (a few 10^6 K) gas which should be visible in the X-ray regime. ROSAT observations of M101 (Williams & Chu 1995) show that one of the strongest X-ray sources in M101 is associated with NGC5462. The X-ray source is located right on top of the superbubble. Its luminosity is 1.4×10^{38} erg s^{-1}. Assuming a hot gas of temperature 5×10^6 K fills the 1.5 kpc hole, the estimated total gas mass is $\approx 2 \times 10^6$ M$_\odot$. This is remarkably close to the difference between the estimated HI mass and the amount of mass required to fill the hole.

The situation in NGC6946 is quite similar to that in M101, except that no fine example such as NGC5462 could be found. To get a better idea of how widespread high velocity gas features are, Kamphuis & Sancisi (1993) averaged position − velocity diagrams in a strip 6$'$ (18 kpc) wide along the major axis of the galaxy. This was only done for the receding half of the galaxy because the approaching half is partly confused with foreground Galactic HI emission. The averaging was performed on the smoothed 30$''$ resolution data. Figure 4 (left panel) shows the strip averaged position − velocity diagram. It clearly shows the presence of low intensity wings in the HI profiles (vertical slices through the position − velocity diagram), especially at radii from 0$'$ to 5$'$. These wings are indicated in figure 4 by the three levels of grey scale.

The faint profile wings are much more obvious if one corrects the data for the differential rotation of the galaxy by shifting the individual HI profiles to the same reference velocity. This procedure removes the rotation and other systematic motions but preserves profile shapes. The right panel in figure 4 shows the same data as shown in the left panel after such a de-rotation.

A very remarkable effect is that the low intensity extensions to the HI profiles are only present within the de Vaucouleurs radius (R$_{25}$). This strongly suggests a relation between the star formation activity in the disk and the presence of this fast moving, redshifted and blueshifted HI gas. The amount of HI in this excess velocity component is 5×10^8 M$_\odot$ or 2% of the total amount of HI in

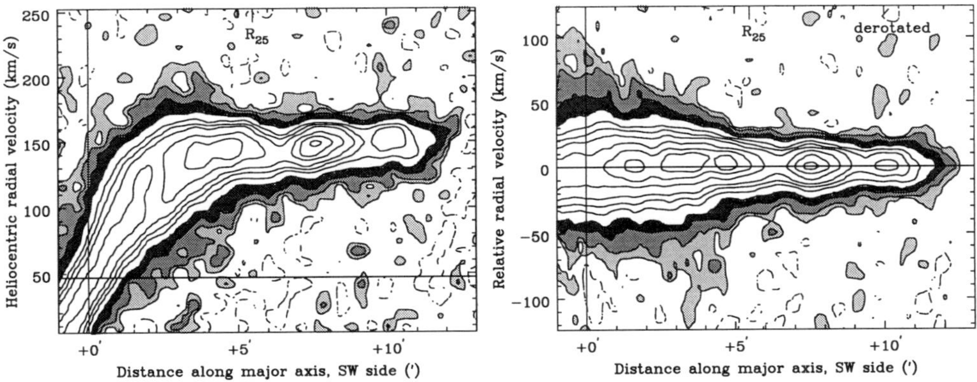

Figure 4: Left panel: Position − velocity diagram along the major axis of NGC6946 (position angle 60°), strip integrated over 6′ in the perpendicular direction. The de Vaucouleurs radius (R_{25}) is indicated in the top of the figure. Right panel: same data but derotated.

NGC6946. The total kinetic energy in this extra component is about 10^{55} ergs.

In this context it is of interest to look at the overall behaviour of the velocity dispersion in other face-on galaxies. Dickey et al. (1990) found a decrease in the velocity dispersion from about 12 km s^{-1} in the inner disk to 6 km s^{-1} in NGC1058. The velocity dispersion in M101 has a similar behavior (Kamphuis 1993) and decreases from 12 km s^{-1} in the inner parts to 8 km s^{-1} in the outer parts. Kamphuis (1993) compiled the available data on velocity dispersion measurements and the radial velocity dispersion profiles for 5 galaxies are shown in figure 5. The observations of NGC628 and NGC3938 are from Shostak & van der Kruit (1984) and van der Kruit & Shostak (1982). Most galaxies in figure 5 show a gradual decline in the velocity dispersion around the edge of the optical disk in support of the idea that star formation activity drives gas out of the disk into the halo. In none of these galaxies except NGC6946 was an extra broad component found. This can, however, be a result of lack of sensitivity since this broad component is a very faint feature, close to the detection limit of modern synthesis observations.

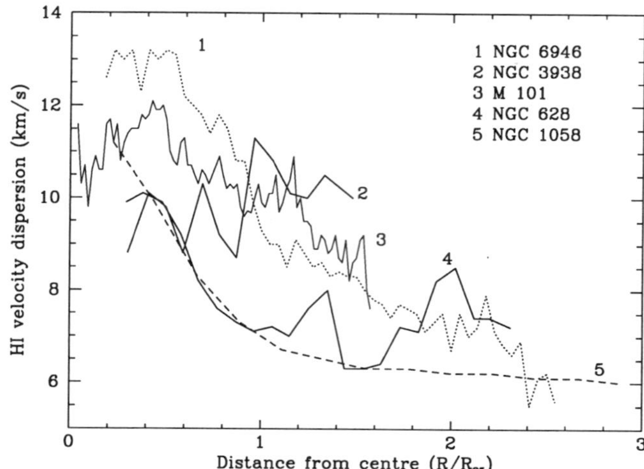

Figure 5: Azimuthally averaged HI velocity dispersion as a function of normalized galactocentric radius in 5 face-on galaxies.

2.2 Edge-on Galaxies

In the previous section I discussed the presence of high velocity gas features in face-on galaxies. All we know at present is that the gas is moving at velocities lower and higher than the general rotation in the disk. Since we have no information about whether the gas is in front or behind the disk, we do not know whether the gas is moving out of the disk or falling in. Most of the high and intermediate velocity gas in our Galaxy is moving toward the disk. In other galaxies indirect indications, such as the good global correlation with the star formation activity and the smooth connection in velocity with the gas in the disk, suggest that the gas is moving out. In either case we expect to see gas above and below the plane of a galaxy and the obvious way to confirm this picture is to study the HI in edge-on galaxies.

There are only a few suitable nearby, edge-on galaxies and not all have been observed with sufficient sensitivity. The best known nearby, edge-on galaxies are NGC3079 (Irwin & Seaquist 1990), NGC891 (Rupen 1991, Swaters et al. 1995, 1996), NGC4565 (Rupen 1991), and NGC4631 (Rand & van der Hulst 1993, Rand 1994).

NGC3079 is a galaxy with an active nucleus and has a peculiar distribution of non-thermal radio emission showing two radio lobes along the minor axis (Duric et al. 1983 and references therein). Irwin & Seaquist (1990) observed the galaxy in HI and found 20 extensions, which may be the equivalent of the worms and shells found in our Galaxy by Heiles. These extensions are randomly distributed over the HI disk and not concentrated toward the nuclear region, so that a relation with the non-thermal nuclear activity is unlikely. In addition they found 5 arcs, 2 to 6 kpc in size, in the outer regions of the galaxy. Here, however, the situation becomes confused because of the warp of the galaxy.

NGC4631 is a very complex system. It has several companion galaxies and the HI is very disturbed because of tidal interaction with these companion galaxies. This makes it more difficult to separate the effects of star formation from the effects of the interaction, including gas clouds which may have been separated from the disk by tidal effects and are now falling back into the disk and interacting with the gas in the disk. Rand & van der Hulst (1993) found two examples of superbubbles in the HI disk, one of which is associated with a bubble feature also seen in Hα. The other may be associated with an X-ray source. These are 'superbubbles' in the sense that the size and corresponding energy are too large to blame their existence simply on the local star formation activity in the disk. In some locations the HI extends to very large z $-$ distances, but this is most likely a result of tidal effects.

The VLA observations of NGC 4565 and NGC 891 by Rupen (1991) show extensions to moderate distances from the plane ($\approx$ a few kpc) at marginal column density levels. NGC 891 has been observed in Hα by Rand et al. (1990) and Dettmar (1990), and at very low levels they detect spurs of ionized gas extending to $\approx 2 - 4.5$ kpc in z. No clear correlation exists between the Hα filaments and extensions seen in the HI, though the HI resolution is not sufficient to be able to distinguish detailed correspondences between the HI and Hα emission.

NGC891 was reobserved with the WSRT to very low column density levels (Swaters et al. 1995, 1996). These observations give a very good view of faint HI at high z $-$ distances and have revealed some interesting features. From this new data it is clear that the HI extends to about 2$'$ or about 5 kpc from the plane. One should, however, keep in mind that this gas at high z $-$ distances could reside anywhere along the line of sight, for example only at the edge of and beyond the optical disk as is the case for a strongly flaring HI layer.

It is, however, possible to distinguish between the inner and the outer parts by using the velocity information. It will be clear that at the central position one cannot distinguish between the inner and the outer regions: all gas will be at the systemic velocity because all motion is perpendicular to the line of sight. If one moves away from the center of the galaxy, however, projection effects work such that one observes low velocities predominantly from regions in the outer parts and high velocities predominantly from regions in the inner part of the galaxy, close to the line of nodes. So by isolating the higher velocities from the lower velocities one has a way of distinguishing the inner from the outer parts.

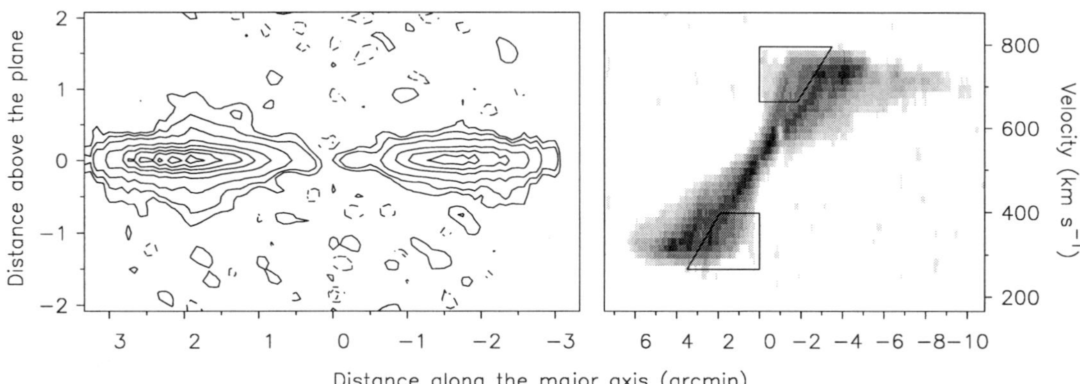

Figure 6: Major axis position − velocity diagram (right panel) of NGC891 and HI column density map (left panel) constructed by integrating over the velocities within the shaded area of the position − velocity diagram in order to isolate HI in the inner parts of the galaxy.

So it is possible to construct a column density map of HI in the inner parts by only integrating over those velocities which correspond to the inner parts. This has been done for NGC891 and the result is shown in figure 6. The left panel of figure 6 is such a column density map and has been constructed by integrating HI within the trapezoidal areas shown in the major axis position − velocity diagram (right panel of figure 6). If NGC891 exhibits normal differential rotation, the velocities within the trapezoidal areas can only arise from HI on orbits in the inner regions of the galaxy; gas in the outer regions would have radial velocities much closer to the systemic velocity of the galaxy. The HI column density dsitribution clearly shows that between $1'$ and $2.5'$ the NE part (left in the diagram) of the galaxy has HI extending up to z − distances of $1'$ or 2.5 kpc. The HI in the SW part of the galaxy does not extend as far from the plane as the HI in the NE. This is remarkable, and in good agreement with the Hα structure which also is much more extended in the NE than in the SW (see Rand et al. 1990). Also the non-thermal radio emission shows the same asymmetry (Dahlem et al. 1994). The distribution of star formation in the disk, as estimated from the Hα luminosity distribution is also asymmetric with much more vigorous star formation in the NE half than in the SW half of the galaxy. This correspondence is very suggestive of a direct relation between the star formation in the disk and the HI at high z distances.

The method outlined above is a quick and easy way of looking at the HI data. A more thorough approach is to compare the observations with a detailed model of the HI emission in the galaxy, in order to assess simultaneously the effects of a possible warping of the plane, a thickening of the HI disk with increasing galactocentric radius (flare), to discriminate between a thin and a thick disk, or possible combinations of the above. This has been done by Swaters et al. (1996). The modelling shows that the HI gas above and below the plane in the inner parts rotates more slowly than the gas in the disk by some 25 km s^{-1}. Swaters et al. (1996) consider two possibilites: a galactic fountain as proposed by Bregman (1980) for the High Velocity Clouds in the Milky Way or an overall asymmetric drift. It is not yet fully clear whether either of these are a fully satisfactory explanation for the observed rotation gradient.

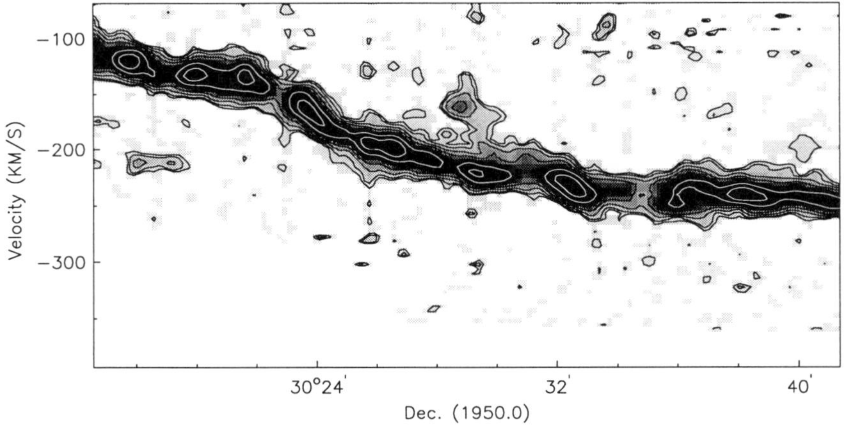

Figure 7: Example position $-$ velocity diagram in M33.

A detailed study of high velocity features on a smaller scale can at present only be done in very nearby galaxies. Such a study of M33 is underway (Kolkman & van der Hulst, in preparation), using the WSRT HI observations of Deul & van der Hulst (1987). An example of a high velocity feature in M33 is shown in figure 7. The general structure in figure 7 is very similar to the velocity structure seen in the hydrodynamical calculations of Rosen & Bregman (1995). The HI masses of such features are a few 10^5 M$_\odot$ and have kinetic energies of a few 10^{51} ergs, an amount that a few supernovae can easily produce.

3 Summary

The observations presented in the previous section demonstrate clearly that the small scale, spatial and velocity structure in the HI in nearby galaxies is rather complex, much like we are used to see in detailed images of the HI in our own Milky Way. Once sufficient angular resolution and sensitivity are available each galaxy examined reveals the presence of a rather frothy HI structure with evidence for HI shells, filaments and structures with deviations from the local differential rotation. The amounts of HI involved typically are a few percent of the total amount of HI in the disk. For large galaxies such as M101, NGC891 and NGC6946 this amounts to several 10^8 $M_\odot$ and represents a total bulk kinetic energy of 10^{54-55} erg s^{-1}.

The evidence for a direct relation to the star formation activity in the disk is circumstantial and global. No detailed correlation with the star forming regions in the disk of a galaxy has yet been established. The main results in this respect are the high-z gas found in the inner parts of NGC891 whose scale height so beautifully reflects the north-south asymmetry in the distribution of star formation in the disk and the declining velocity dispersion in face-on galaxies with as prime example the extra broad velocity component found within the optical disk of NGC6946. The decline of velocity dispersion with radius, measured in face-on galaxies, appears general, supporting the relation with the star forming disk.

Here one caveat should, however, be repeated: in face-on galaxies we only observe the radial velocity of the HI and have no way of knowing whether the gas we observe is in front or behind the main disk. We therefore cannot directly distinguish between gas falling in and gas moving out of the plane of a galaxy. The interpretation in terms of gas moving out of the disk into the halo is based on such nice and symmetric cases as NGC5462, on the connection in velocity with the gas in the disk and on the (subjective) notion that the current picture of the effects of stellar winds and supernovae suggests that outflows of gas are more readily expected. To really make sure one sees outflows one needs to perform an HI absorption measurement against a continuum source located in the disk of a galaxy. These sources exist but are too faint for performing an HI absorption experiment with the present instrumental capabilities.

Acknowledgments

I like to thank Olaf Kolkman, Rob Swaters and Jurjen Kamphuis for letting me use unpublished material. I thank them and also Renzo Sancisi for many valuable discussions regarding the broad subject of holes and bubbles in the ISM. I thank the organisers of the workshop *The Physics of Galactic Halos* for their invitation and hospitality.

References

Allen, R.J., Goss, W.M. & van Woerden, H. 1973, A&A, 29, 447

Blaauw, A. & Tolbert, C.R. 1966, BAN, 18, 405

Bregman, J.N. 1980, ApJ, 236, 577

Brinks, E. & Bajaja, E. 1986, A&A, 169, 14

Cox, D.P. & Smith, B.W. 1974 ApJ, 189, L105

Dahlem, M., Dettmar, R.J. & Hummel, E. 1994, A&A, 290, 384

Dettmar, R.J. 1990, A&A, 232, L15

Deul, E.R. & van der Hulst, J.M. 1987, A&AS, 67, 509

Deul, E.R. & den Hartog, R.H. 1990, A&A, 229, 362

Dickey, J.M., Murray Hanson, M. & Helou G. 1990, ApJ, 352, 522

Duric, N., Seaquist, E.R., Crane, P.C., Bignell, R.C. & Davis, L.E. 1983, ApJ, 273, 574

Elmegreen, B.G. & Chiang, W.H. 1982, ApJ, 253, 666

Field, G.B., Goldsmith, D.W. & Habing, H.J. 1969 ApJ, 155, L49

Gorenstein, P., Harnden, F.R. Jr. & Tucker, W.H. 1974, ApJ, 192, 661

Heiles, C. 1979, ApJ, 229, 533

Heiles, C. 1984, ApJS, 55, 585

Heiles, C. 1990, ApJ, 354, 483

Hulsbosch, A.N.M. & Wakker, B.P. 1988, A&AS, 75, 191

Hulsbosch, A.N.M. & Raimond, E. 1966 BAN, 18, 413

Irwin, J.A. & Seaquist, E.R. 1990, ApJ, 353, 469

Jenkins, E.B. & Meloy, D.A. 1974, ApJ, 193, L121

Kamphuis, J.J. 1993, Ph.D. Dissertation, University of Groningen

Kamphuis, J.J., Sancisi, R. & van der Hulst, J.M. 1991, A&A, 224, L29

Kamphuis, J.J., & Briggs, F. 1992, A&A, 253, 335

Kamphuis, J.J., & Sancisi, R. 1993, A&A, 273, L1

MacLow, M. & McCray, R. 1988, ApJ, 324, 776

MacLow, M., McCray, R. & Norman, M.L. 1989, ApJ, 337, 141

McKee, & Ostriker, J.P. 1977, ApJ, 218, 148

Mirabel, I.F. & Morras, R. 1990, ApJ, 356, 130

Norman, C.A. & Ikeuchi, S. 1989, ApJ, 345, 372

Oort, J.H. 1970, A&A, 7, 381

Rand, R.J. 1994, A&A, 285, 833

Rand, R. J., Kulkarni, S. R. & Hester J. J. 1990, ApJ, 352 , L1

Rand, R. J. & van der Hulst, J.M. 1993, AJ, 105, 2098

Rosen, A. & Bregman, J.N. 1995, ApJ, 440, 634

Rupen, M.P. 1991, AJ, 102, 48

Shostak, G.S. & van der Kruit, P.C. 1984, A&A, 132, 20

Swaters, R. Sancisi, R. & van der Hulst, J.M. 1995, Astro Lett. and Communications, Vol.31, 161

Swaters, R. Sancisi, R. & van der Hulst, J.M. 1996, ApJ, in preparation

Tenorio-Tagle, G. & Bodenheimer, P. 1988, ARA&A, 26, 145

Tenorio-Tagle, G., Franco, J., Bodenheimer, P. & Rozyczka, M. 1987, A&A, 179, 219

Tomisaka, K. 1992, PASJ, 44, 177

Tomisaka, K.& Ikeuchi, S. 1986, PASJ, 39, 109

Tomisaka, K. & Bregman, J. N. 1993, PASJ, 45, 513

van der Hulst, J.M. & Sancisi R. 1988, AJ, 95, 1354

van der Kruit, P.C. & Shostak, G.S. 1982, A&A, 105, 351

van Woerden, H., Schwarz, U.J. & Hulsbosch, A.N.M. 1985, IAU Symp. 106, eds. H. van Woerden, R.J. Allen, & W.B. Burton, p. 387

Viallefond, F., Allen, R.J. & Goss, W.M. 1981, A&A, 104, 107

Wakker, B.P. 1990, Ph.D. thesis Groningen University

Wakker B.P., Broeils, A.H., Tilanus, R.P.J. & Sancisi, R. 1989, A&A, 226, 57

Williams, R.M. & Chu, Y.H. 1995, ApJ, 439, 132

Woodgate, B.E., Stockman, H.S. Jr., Angel, J.R.P. & Kirshner, R.P. 1974, ApJ, 188, L79

York, D.G. 1974, ApJ, 193, L127

Address of the author:

J.M. van der Hulst, Kapteyn Astronomical Institute, Postbus 800, NL-9700 AV Groningen, The Netherlands

Warm and Hot Phases
in the Disk-Halo Interface

The Z-Distribution of the Ionized Interstellar Medium

R. J. Reynolds

1   Introduction

Diffuse ionized gas is a major, but little understood component of the interstellar medium, which consists of regions of warm (10^4 K), low density (10^{-1} cm^{-3}), nearly fully ionized hydrogen that occupy approximately 20% of the volume within a 2 kpc thick layer about the Galactic midplane. Hoyle & Ellis (1963) were the first to suggest the existence of an extensive layer of warm H II surrounding the Galactic disk, based on the spatial and spectral characteristics of the Galactic synchrotron background at very low radio frequencies (2-10 MHz). However, it was not until the discovery of pulsars a few years later that the existence of wide spread H II in the diffuse interstellar medium became generally accepted. This was soon followed by the detection of faint, diffuse optical line emission from the Galaxy (Reynolds, Roesler, & Scherb 1974), which provided information about the temperature, ionization state, kinematics, and spatial distribution of the gas within 2–3 kpc of the sun.

Only relatively recently has the full z-extent and mass surface density of this component been established. At the midplane, the space averaged density of H II is less than 5% that of the H I. However, because of its greater scale height, the total column density of interstellar H II along high Galactic latitude sight lines is relatively large, 1/4 to 1/2 that of the H I, and one kiloparsec above the midplane, warm H II may be the dominant state of the interstellar medium. The power required to maintain the ionization of this gas is large, 1×10^{-4} ergs s^{-1} per cm^2 of Galactic disk, which is approximately equal to the kinetic energy injected into the interstellar medium by supernovae. Of the known sources of power in the Galaxy, only supernovae and the Lyman continuum flux from O stars meet or exceed this power requirement.

Some of the evidence for the large scale height and mass surface density for the warm ionized medium are reviewed below.

2 The Thickness of the Galactic H II Layer

2.1 Pulsar Dispersion Measures

A direct method for estimating the vertical extent of the ionized hydrogen involves simply dividing the average H II column density N_p perpendicular to the Galactic plane by the mean electron (i.e., H II) density $< n_e >_0$ near the Galactic midplane. The result $H = N_p/ < n_e >_0$ is a measure of the half thickness of the ionized layer. If the mean density $< n_e >$ decreases exponentially with increasing distance $|z|$ from the midplane, then H is the scale height of the layer.

The value of $< n_e >_0$ is determined from the dispersion measures DM of pulsars within the Galactic disk that have known distances. Distances have been determined for approximately 15% of the more than 500 known pulsars through 21 cm absorption measurements (Weisberg, Rankin & Boriakoff 1980), VLBI parallax (Gwinn et al. 1986), and the fact that some pulsars reside in supernova remnants and globular clusters with known distances. These Galactic pulsars range in distance from 130 pc to 16 kpc, and thus sample a large portion of the disk (see, for example, Nordgren, Cordes, & Terzian 1992). The value of $< n_e >_0$ appears to be a function of galactocentric radius, with a value near 0.03 cm^{-3} within a few kpc of the sun (e.g., Frail & Weisberg 1990). The observed variations in the mean electron density from sightline to sightline suggest that the H II is distributed much more smoothly than the H I; this is seen if one compares a plot of DM $\sin |b|$ vs. $|z|$ in Reynolds (1991) with a similar plot for the column densities of the H I given by in Edgar & Savage (1989), for example. The pulsar data also show that the ionized gas extends more than 1 kpc from the midplane, with a total H II column density $N_p \simeq 7 \times 10^{19}$ cm^{-2}, and a column density above 800 pc of $2 - 3 \times 10^{19}$ cm^{-2} (Reynolds 1991a; Nordgren, Cordes, & Taylor 1992).

A measurement of the total column density N_p of the H II perpendicular to the disk is provided by pulsars in globular clusters far from the Galactic midplane. Unfortunately, the number of samples is small; only five clusters with detected pulsars are sufficiently far from the midplane (i.e., $|z| > 3$ kpc) to probe the entire H II column between the midplane and "infinity". For these five sightlines the maximum and minimum values for $N \sin |b|$ differ by less than a factor of two and provide a mean value for $N_p = 7.06 \times 10^{19}$ cm^{-2} (see Reynolds 1991a). The standard deviation for the five directions is 1.6×10^{19} cm^{-2}, which suggests that the uncertainty in this derived mean is about $\pm 0.8 \times 10^{19}$ cm^{-2}. Detections of pulsars in additional globular clusters are clearly needed to confirm the apparently low scatter in the values of $N \sin |b|$ and to probe the column density distribution of the gas at large distances from the midplane.

The derived value of N_p corresponds to a mass surface density for the ionized component of 1.6 $M_\odot$ pc^{-2} (including a factor of 1.4 for helium). Along these sightlines, the H II / H I column density ratios range from 0.23 to 0.63, indicating that the diffuse H II is a significant mass component of the interstellar medium. Power considerations suggest that nearly all of this H II column density, even at high $|z|$, can be attributed to warm ($\sim 10^4$ K) gas (Reynolds 1990b). For example, if the H II were at a temperature of 10^6 K or 10^5 K, then its cooling rate would exceed the supernovae power by factors of 30 and 5, respectively. This conclusion is supported by the optical emission line data, which reveal extensive regions of $\sim 10^4$ K ionized gas extending to $|z| > 1$ kpc (see section 2.2, below).

Before determining a value of H from the ratio $N_p/ < n_e >_0$, it is important to note that N_p is derived from high-latitude sight lines (i.e., $|b| > 27°$; see Fig. 1), which sample a region of the disk within about 2–3 kpc of the sun (anticipating a half thickness of about 1 kpc). Since the mean electron density near the midplane is known to vary with position in the Galaxy, it is therefore important that the value of $< n_e >_0$ used to determine H be derived only from the sample of Galactic disk pulsars with distances less than 3 kpc from the sun. In the the compilation of Nordgren, Cordes, & Terzian (1992) there are nine pulsars within 3 kpc that are uncontaminated by known classical H II regions. These pulsars give $< n_e >_0 = 0.024$ cm^{-3}. The standard deviation in the nine directions is 0.014 cm^{-3}, suggesting an uncertainty of ± 0.005 cm^{-3} in the derived mean.

The resulting mean value for H is 950 pc, with a formal uncertainty of ± 230 pc for the diffuse H II within 3 kpc of the sun. The location of the sightlines that were used to derive this value for H are shown in Figure 1. Their distribution suggests that the sky has been reasonably well sampled, given the small number of available sightlines.

2.2   Hα Emission

An independent and complementary probe of the properties of the warm ionized medium can be made through observations of faint optical emission lines at high spectral resolution. For example, the intensity of the interstellar Hα

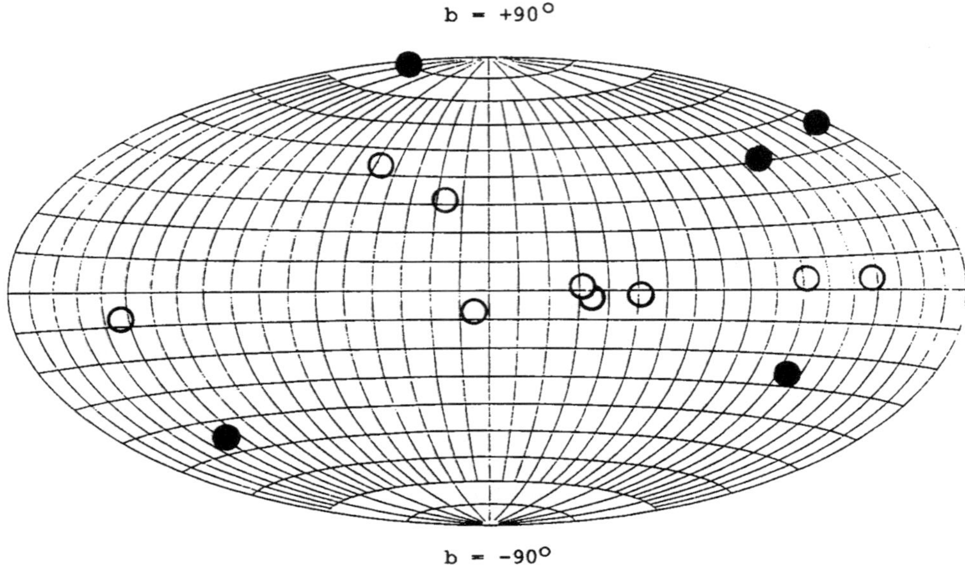

Figure 1: The locations of the pulsar sightlines used to derive the scale height H for the warm ionized medium. The filled circles represent the high-z globular cluster pulsars used to derive N_p, and the open circles represent the disk pulsars used to derive $< n_e >_0$. The two high latitude disk pulsars have distances less than 400 pc. The map is centered at $l = 180°$.

at high Galactic latitudes is a measure of the hydrogen recombination rate (Reynolds 1984). The existing high-latitude Hα data imply a rate of 5×10^6 recombinations s^{-1} per cm^2 of Galactic disk (Reynolds 1992), and with 13.6 eV per ionization, an ionizing power of 1×10^{-4} ergs s^{-1} per cm^2. The widths of the Hα, [N II], and [S II] emission lines imply a gas temperature near 8000 K (Reynolds 1985a), while the absence of detectable [O I] emission indicates that the hydrogen within the ionized regions is nearly fully ionized (Reynolds 1989).

This power requirement for the diffuse H II is approximately 100% of the kinetic energy injected into the interstellar medium by supernova (i.e., 5×10^{50} ergs per 35 yrs within 12 kpc of the Galactic center) or about 1/6 the total Lyman continuum photon production rate from hot stars, 90% of which is from O stars. The mechanism of ionization has not yet been identified; however, if the source is supernovae, the ionization mechanism must be very energy efficient, and if it is O stars, the ionizing photons must be able to travel hundreds of parsecs through the Galaxy's extended H I layer (e.g., Lockman, these proceedings). The ionizing photons must be able to travel large distances, not only away from the midplane, but also within the denser H I cloud layer close to the midplane (Reynolds 1990a). Furthermore, recent observations at both optical and radio wavelengths have found He I/Hα recombination line ratios in the diffuse interstellar medium that are significantly lower than ratios observed in traditional H II regions surrounding O stars (Reynolds & Tufte 1995; Heiles et al. 1996). This result suggests that the spectrum of the diffuse interstellar radiation field that ionizes the hydrogen is softer than that from the Galactic O star population. Therefore, the existence of the diffuse ionization appears to challenge our understanding of the principal energy sources within the interstellar medium as well as the morphology of the H I within the disk and halo. Alternative ionization mechanisms that have been proposed include Galactic microflares, powered perhaps ultimately by galactic rotation (Raymond 1992), and the decay of dark matter particles (Sciama 1990).

Hα observations provide independent evidence that warm ionized gas extends far above the disk of our Galaxy and others. For studies of other galaxies, see for example, the work of Dettmar (1992, and in these proceedings), Golla, Dettmar, & Domgörgen (1996), and Rand (1996), who found regions of Hα emitting gas extending a few kiloparsecs from the midplane of some edge-on spirals. In the Milky Way, a high z extent of the warm ionized medium is indicated by a number of Hα studies. A one degree angular resolution map of the Hα background within a $12° \times 11°$ region centered at $l = 144°$, $b = -21°$ (Reynolds 1980) shows Hα emitting "clouds" 1 kpc or more above the plane toward the Perseus spiral arm. These clouds are several degrees in extent, corresponding to sizes of order 100 pc and densities near 0.2 cm^{-3} (Reynolds et al. 1995), and they appear to be closely associated, both spatially and kinematically, with large H I shells, loops, and filamentary structures. Ionized gas at even larger distances is implied by the detection of extremely faint Hα emission from high velocity H I clouds (e.g., Kutyrev & Reynolds 1989; Songaila, Bryant, & Cowie 1989). Warm ionized gas associated with the cloud M I located at $z = 1.5$–4.4 kpc (Danly, Albert, & Kuntz 1993) has been clearly detected by the new WHAM spectrometer, which found both Hα and [S II] emission lines from this -110 km s^{-1} cloud (Tufte, Reynolds, and Haffner 1996).

In the Perseus spiral arm the thickness of the diffuse H II layer can be estimated from the observed decrease away from the Galactic plane in the intensity of the radial velocity component of the diffuse Hα emission identified with this spiral arm (Reynolds 1985b). The emission from this arm, located at a distance of approximately 2.5–3 kpc from the sun, appears as a well resolved radial velocity component near -40 km s^{-1} with respect to the local standard of rest in the longitude interval $100° < l < 150°$. Figure 2 shows the measured Hα intensities plotted against Galactic latitude for nine longitudes from $104°$ to $130°$. The data were obtained with the Wisconsin large aperture scanning Fabry-Perot spectrometer. The observations were confined only to negative latitudes because the spectrometer could not access declinations above $+70°$.

The scale height H of the electron density n_e (*i.e.*, twice the scale height of $n_e{}^2$) can be estimated from the slope of the intensity vs. latitude distribution in Figure 2; specifically,

$$H \simeq 2D \frac{\tan(|b_2|) - \tan(|b_1|)}{\ln(I_1) - \ln(I_2)}, \tag{1}$$

where D is the distance to the arm, and I_1 and I_2 are the Hα intensities at latitudes b_1 and b_2, respectively. The intensities presented in Figure 2 are all normalized to the intensity at $b = -10°$, and the data were not corrected for interstellar extinction. Extinction will tend to increase the derived value of H, but the effect should be significant only within 10 degrees of the plane, where the optical depth through the Galactic disk increases to unity and above. The three dotted lines indicate values for H of 0.5 kpc, 1.0 kpc, and 2.0 kpc, assuming a mean distance of 2.5 kpc to the spiral arm. A value of $H = 1$ kpc provides a good fit to the data at $-30° < b < -10°$ ($-1.4 < z < -0.4$ kpc), which is consistent with the pulsar dispersion measure result for the solar neighborhood presented above. Hα associated with the arm is detected down to a z distance of -1.4 kpc, where the mean intensity is approximately 0.14 R, which corresponds to an emission measure of 0.3 cm^{-6} pc. If the path length through the arm is 1 kpc, then the rms electron density at z = -1.4 kpc is 0.017 cm^{-3}.

3   Filling Fraction of the H II

Diffuse ionized gas is not distributed uniformly throughout the interstellar volume, but is clumped into regions that have an electron density significantly higher than the space averaged value of $0.024 \times \exp(-|z|/950)$ cm^{-3} derived in section 2.1. For example, the $1°$ angular resolution Hα map centered near $l = 144°$, $b = -21°$ shows that at least 30% of the H II up to a $|z| \sim 1$

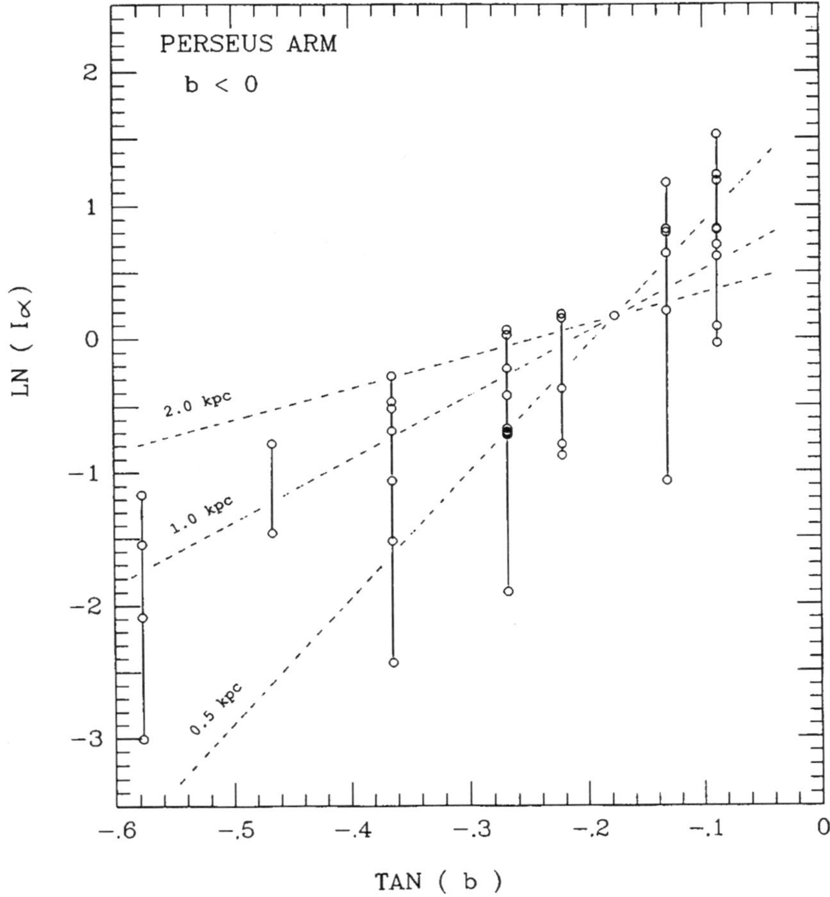

Figure 2: The natural log of the Hα intensity of the -40 km s^{-1} radial velocity component associated with the Perseus spiral arm plotted versus tan b. The observations were made at latitudes -5° (z= -200 pc) to -30° (-1400 pc) for nine longitude positions between 104° and 130°. The dashed lines represent the expected distribution for three different electron scale heights according to eq. 1. The intensities are in rayleighs (R), where 1 R = $10^6/4\pi$ photons cm^{-2} s^{-1} sr^{-1}; they have all been normalized to the value at $b = -10°$.

kpc is clumped into regions of density $n_e \simeq 0.2$ cm^{-3} (Reynolds et al. 1995). These prominent regions of enhanced emission are probably associated with the highest density H II clouds, while the remainder of the Hα background is from fainter, lower density regions that are not visible as discrete enhancements.

A more representative measurement of the amount of clumping within the medium is provided by comparisons between the emission measure (EM) and dispersion measure (DM) toward the four northern sky globular cluster pulsars with $|z| > 3$ kpc (the filled circles in Fig. 1, except for $l = 306°$, $b = -45°$); the fifth cluster, 47 Tuc, has too low a declination for the Wisconsin spectrometer (Reynolds 1991b). In these cases, characteristic densities n_c and line of sight occupation lengths L_c for the regions of H II can be derived from the ratios, EM/DM and DM^2/EM, respectively. The resulting values of n_c range from 0.05 cm^{-3} to 0.13 cm^{-3} and L_c from 230 pc to 410 pc (projected perpendicular to the disk). If the space averaged density $< n_e >$ decreases exponentially with $|z|$, then the average volume filling fraction f within the ionized layer is 0.18 with an electron density of 0.16 cm^{-3} within the ionized regions at $z = 0$ pc (see Reynolds 1991b). The warm ionized medium therefore seems to be distributed much more uniformly within the volume of the Galactic disk than is the ionized gas inside traditional H II regions, where $f \sim 0.03$ (Spitzer 1978; Osterbrock 1989).

An important question is whether the filling fraction varies with $|z|$. An analysis of the Hα and pulsar data by Kulkarni & Heiles (1987), for example, suggests that f increases from about 0.10 near the midplane to more than 0.40 at $|z| > 1$ kpc. Another examination of the subject, based essentially on the same data, reached a similar conclusion (Reynolds 1991b). Therefore, it is possible that warm ionized gas accounts for half or more of the interstellar volume at $|z| > 1$ kpc. However, given the uncertainties and the limited amount of data, such conclusions must be considered tentative.

4 Summary and Conclusions

Pulsar dispersion measures and diffuse Hα emission both confirm independently the existence of a 2 kpc thick layer of ionized hydrogen within a region of 3 kpc radius around the sun. The pulsar data imply a mass surface density for the H II that is approximately 1/3 that of the H I, and the Hα data imply an ionizing power that severely restricts the possible mechanisms of ionization. If the z-distribution of the H II, given by $< n_e(z) > = 0.024 \times \exp(-|z|/950)$

cm^{-3}, is compared to that of the H I given by Dickey & Lockman (1990), then it appears that the diffuse H II becomes the dominant mass component of the interstellar medium at $|z| > 700$ pc. The pulsar data also reveal that diffuse ionized hydrogen is present throughout a large portion of the Galactic disk (e.g., Taylor & Cordes 1993), although the large scale height and mass surface density can only be confirmed for the more local region. The scale height of the ionized hydrogen above the Perseus spiral arm derived from the $H\alpha$ intensities is equal, within the measurement uncertainties, to the scale height derived for the region within 2–3 kpc of the sun from the pulsar data. Finally, the combination of dispersion measure and emission measure data imply that the gas is clumped into "clouds" that have an electron density of 0.2 cm^{-3} near the midplane and that occupy on average about 20% of the volume within the 2 kpc layer.

This work was supported by a grant from the National Science Foundation (AST 91-22701).

References

Danly, L., Albert, C.E., and Kuntz, K.D. 1993, ApJ 416, L29

Dettmar, R.-J. 1992, Fund. Cosmic Physics, 15, 143

Dickey, J.M. & Lockman, F.J. 1990, ARA&A 28, 215

Edgar, R.J. & Savage B.D. 1989, ApJ 340, 762

Frail, D.A. & Weisberg, J.M. 1990, AJ 100, 743

Golla, G., Dettmar, R.-J., & Domgörgen, H. 1996, A&A 313, 439

Gwinn, C.R., Taylor, J.H., Weisberg, J.M., & Rawlings, L.A. 1986, AJ 91, 338

Heiles, C., Koo, B.-C., Levenson, N.A., & Reach, W.T. 1996, ApJ, 462, 326

Hoyle, F. & Ellis, G.R.A. 1963, Australian J. Phys., 16, 1

Kulkarni, S. & Heiles, C. 1987, in Interstellar Processes, ed. D.J. Hollenbach & H.A. Thronson, Jr. (Dordrecht:Reidel), p.87

Kutyrev, A.S. & Reynolds, R.J. 1989, ApJ 344, L9

Nordgren, T.E., Cordes, J.M., & Terzian, Y. 1992, AJ 104, 1465

Osterbrock, D.E. 1989, in Astrophysics of Gaseous Nebulae and Active Galactic Nuclei (Mill Valley:University Science Books)

Rand, R. 1996, ApJ 462, in press

Raymond, J.C. 1992, ApJ 384, 502

Reynolds, R.J. 1980, ApJ 236, 153

Reynolds, R.J. 1984, ApJ 282, 191

Reynolds, R.J. 1985a, ApJ 294, 256

Reynolds, R.J. 1985b, in Gaseous Halos of Galaxies, NRAO Workshop No. 12, eds. J.N. Bregman & F.J. Lockman (NRAO), p.53

Reynolds, R.J. 1989, ApJ 345, 811

Reynolds, R.J. 1990a, ApJ 348, 153

Reynolds, R.J. 1990b, ApJ 349, L17

Reynolds, R.J. 1991a, in IAU Symposium No. 144, The Interstellar Disk-Halo Connection in Galaxies, ed. H. Bloemen (Dordrecht:Kluwer), p.67

Reynolds, R.J. 1991b, ApJ 372, L17

Reynolds, R.J. 1992, ApJ 392, L35

Reynolds, R.J., Roesler, F.L., & Scherb, F. 1974, ApJ 193, L53

Reynolds, R.J. & Tufte, S.L. 1995, ApJ 439, L17

Reynolds, R.J., Tufte, S.L., Kung, D.T., McCullough, P.R., & Heiles, C. 1995, ApJ 448, 715

Sciama, D.W. 1990, ApJ 364, 549

Songaila, A., Bryant, W., & Cowie, L.L. 1989, ApJ 345, L71

Spitzer, L., Jr. 1978, in Physical Processes in the Interstellar Medium (New York:John Wiley), p.79

Taylor, J.H. & Cordes, J.M. 1993, ApJ 411, 674

Tufte, S.L., Reynolds, R.J., & Haffner, L.M. 1996, in preparation

Weisberg, J.M., Rankin, J.M., & Boriakoff, V. 1980, A&A, 88, 84

Address of the author:

R. J. REYNOLDS, University of Wisconsin, Department of Astronomy, 475 N. Charter St., Madison, WI 53706, USA.

The Ionization of the Diffuse Ionized Gas

Hilde Domgörgen

Abstract. A warm ionized gas is a highly important constituent of the interstellar medium in disks and halos of galaxies. Up to now only little is known about the physics of this gas phase, and not even its main ionization and heating processes are well understood yet. The knowledge of these processes, however, is crucial for an understanding of the complex interplay between stars, the interstellar gas and a possible large scale circulation of matter in galaxies. In this review I discuss mechanisms possibly responsible for the ionization of this warm ionized medium. In particular the possibility of photoionization by OB stars will be explored.

1 Introduction

Some of the most puzzling questions related to the physics of the ISM concern the existence of a warm ionized medium (WIM) or diffuse ionized gas (DIG). What we call the WIM is an Hα emitting gas phase with temperatures similar to those of HII regions ($T_e \sim 8000$ K), but lower densities ($n_e \sim 0.3$ cm^{-3}). High sensitivity Fabry-Perot measurements show that this gas phase is ubiquitous in the Galaxy (Reynolds, 1993). This finding is surprising since we are traditionally taught that the Hα emitting gas resides inside HII regions which are terminated by sharp Strömgren spheres (e.g. Osterbrock 1989; Spitzer 1978). The WIM, however, extends even to large distances above the galactic disk and has a scale height of h $\sim$ 900 pc in the Milky Way (Reynolds, this volume). This has to be compared with h $\sim$ 70 pc for HII regions (Reynolds 1990). Thus this phase of the ISM is an important constituent of disk *and* halo of galaxies. Being confronted with these observations one of the most obvious questions to ask is: "What is the mechanism responsible for the ionization of the WIM?"

In the following I will first briefly summarize the observational evidence for the existence of diffuse ionized gas in galaxies other than the Milky Way. Section 3 then presents the observational constraints on the ionizing sources. Finally, the possibility of photoionization by OB stars will be discussed.

2 Diffuse Ionized Gas in Galaxies

It is from deep Hα imaging of edge-on galaxies that we know about the existence of the DIG in halos of galaxies other than the Milky Way. The first galaxies for which widespread Hα halos were detected are two objects which had long been known for their thick radio continuum disks: NGC 891 and NGC 4631. In NGC 891 the observations show the presence of Hα emission extending up to 4 kpc from the midplane mostly concentrated above a region of intense star formation in the disk (Rand et al. 1990, Dettmar 1990). Hα maps of NGC 4631 have been published by Roy et al. (1991), Rand et al. (1992), Golla et al. (1996). Recently a very deep Hα image has been presented by Donahue et al. (1995). They used the Burrell Schmidt telescope on Kitt Peak to obtain a narrow band Hα map of the NGC 4631/NGC 4656 system down to an emission measure of 0.3 cm^{-6} pc. The sensitivity of these observations is thus comparable to those of the Fabry-Perot measurements of the Galactic DIG by Reynolds. It is found that gas with an emission measure of $\sim$ 1.1-2.2 cm^{-6} pc extends out to 16 kpc from the midplane of NGC 4631. A Hα survey of nine edge-on galaxies has recently been carried out by Rand (1996). With these galaxies and observations of single objects (Dettmar 1992; Rand, Kulkarni & Hester 1992; Pildis, Bregman, & Schombert 1994a) we now have Hα data with a fairly complete sample of nearby, edge-on galaxies on the northern hemisphere available.

It turns out that widespread layers of DIG such as present in NGC 891 and NGC 4631, are in fact not very common phenomena. Single plumes and filaments of ionized gas are more frequent and are likely related to star forming regions in the underlying disk (Rand 1996). Moreover, a correlation between the ratio of far infrared luminosity and the optical isophotal diameter at 25th magnitude (L_{FIR}/D_{25}^2), and extraplanar Hα emission indicates a dependence of the occurrence of DIG in halos of galaxies on the star formation rate per unit disk area (Dettmar 1995; Rand 1996; see also Dettmar, this volume). It is important to realize though that the data differ largely in their sensitivity. It is not yet clear what halos of galaxies look like on a surface brightness level comparable to that of the DIG at high Galactic latitude in the Milky Way.

The DIG also is an important constituent of the ISM in disks of galaxies. Hα imaging of M 31 (Walterbos & Braun 1994) and of the Magellanic Clouds (Kennicutt et al. 1995) shows that $\sim$ 30% of the Hα emission in these galaxies does not come from HII regions, but is diffuse instead. Moreover, large filaments of ionized gas are frequently observed in irregular galaxies (e.g., Hunter et al.

1993). However, the following discussion will mostly focus on the DIG in halos of galaxies.

3 Observational Constraints on the Ionizing Sources

A variety of energy sources such as the ionizing radiation of OB stars, supernovae and stellar winds, central stars of planetary nebulae and white dwarfs, the X-ray background or the extragalactic EUV background might be thought of as being responsible for the ionization of the DIG.

Most of these sources can be excluded as principal ionizing agents when considering the power requirement of the DIG. Moreover, the mechanism responsible for the ionization has to be able to account for the observed spectrum of the DIG.

3.1 Power requirement

From measurements of the Hα intensity at high Galactic latitude the hydrogen recombination rate of the DIG can be estimated as 5×10^6 s^{-1} cm^{-2} of galactic disk. This is equivalent to a power requirement of the DIG at 13.6 eV of 1×10^{-4} ergs s^{-1} per cm^2 of galactic disk. Assuming an exponential density distribution of the gas and a scale height between 500 and 1000 pc this value is valid in a region with a radius of 2 - 3 kpc around the sun (Reynolds, 1984).

This high power requirement clearly excludes most of the ionizing sources mentioned above. White dwarfs (Panagia & Terzian 1984), X-rays (McCammon et al. 1983), the EUV background (Vogel et al. 1995, Donahue et al. 1995), and cosmic rays (van Dishoeck & Black 1986) fall by far short in providing enough ionizing power. The power is equivalent to $\sim 100\%$ of the total energy input into the ISM by supernovae or $\sim 16\%$ of the OB star ionizing radiation (Abbott, 1982). Thus only the ionizing radiation of OB stars comfortably exceeds the power requirement of the DIG.

3.2 The Emission Line Spectrum of the DIG

Most information on the spectrum of the DIG we have comes from Fabry-Perot measurements of gas in the Galaxy (see R. Reynolds, 1993, and references therein). These observations show that the low ionization stages ([S II] and

Table 1: Typical Line Ratios

Object	[SII] $\lambda6717$/Hα	[NII] $\lambda6583$/Hα	References
Galaxy HII	$\sim$0.1	$\sim$0.3	Hawley (1978)
Galaxy DIG	0.3 - 0.5	0.3 - 0.5	Reynolds (1985)
NGC 891	$\leq$0.7	$\leq$1.1	Dettmar & Schulz (1992)
NGC 4631	$\leq$0.7	$\leq$0.7	Golla et al. (1996)

[N II]) are strong in the WIM, the [O III] is weak and [O I] is very weak, all considerably different from HII regions. Recently an upper limit has been set on the He I $\lambda5876$ emission (Reynolds & Tufte, 1995). Lines other than Hα have preferentially been observed at low Galactic latitude where its emission measure is comparatively high. For our understanding of the *extraplanar* DIG spectroscopy of edge-on galaxies is therefore of high importance.

Up to now the extraplanar DIG has been studied spectroscopically for only two edge-on spirals (NGC 891: Dettmar & Schulz 1992; NGC 4631: Golla, Dettmar & Domgörgen 1996). Both investigations have been restricted to the wavelength regime around Hα, i.e. the [NII] and [SII] lines, in order to also obtain velocity information from the data. Placing the slit positions perpendicular to the major axis of the galaxies, it was possible to investigate the variation of the line ratios with z, the vertical distance from the galactic plane.

Most importantly it was found that [NII] $\lambda6583$/Hα and [SII] $\lambda6717$/Hα do increase considerably with z. Both line ratios vary in a similar way. They rise significantly over what is typically measured for the DIG in the Galaxy. In NGC 891, for example, [NII] $\lambda6583$/Hα becomes as high as 1.1 at a z distance of 900 pc. The variation of the line ratios in the halo itself has been studied for NGC 4631. It is found that [NII] $\lambda6583$/Hα and [SII] $\lambda6717$/Hα change in a similar way and vary on scales $\sim$ 1 kpc. The results are summarized in Table 1. For comparison typical line ratios for the Galactic DIG and Galactic HII regions are also listed.

For NGC 891, additionally, an upper limit could be set on the strength of the [O I] $\lambda6300$ line implying [OI] $\lambda6300$/Hα $<$ 0.05 in agreement with the observations of DIG in the Milky Way.

4 Photoionization of the DIG

The discussion of the power requirement basically leaves us with OB stars as the only possible ionizing sources of the DIG. However, the spectrum of the WIM is considerably different from that of HII regions. Moreover, the optical depth for Lyman continuum photons is large and it is not obviously clear how OB stars located in the disk should ionize the DIG high above the galactic plane. In the following I will address these two problems.

The ionization of the DIG at high distances above the plane also motivated discussions about more 'exotic' ionization mechanisms for the WIM. An example is the decaying dark matter theory (DDM) which has been discussed by Sciama (1990). In the DDM a massive neutrino decays into another neutrino and a photon with an energy of ~ 13.7 eV (Sciama 1995). Thus the DDM cannot only account for the presence of dark matter in galaxies, but additionally delivers hydrogen ionizing photons *locally*. The fact that [NII] $\lambda6583/\text{H}\alpha$ as well as [SII] $\lambda6717/\text{H}\alpha$ change in a very similar way in all these galaxies is in contradiction with the DDM theory. This is because the 13.7 eV decay photon is able to ionize hydrogen and sulfur, but not nitrogen.

4.1 Photoionization Modeling

4.1.1 What Photoionization Models can explain

If both, HII regions and the WIM, are photoionized by OB stars, there is one main difference between these two morphological types of photoionized gas: whereas the classical HII region gas is close to the ionizing stellar cluster the DIG resides at a distance several hundred parsecs from its ionizing sources. Consequently, the DIG will see a radiation field that is considerably diluted with respect to the radiation field in a HII region. In terms of HII region physics this is equivalent to a low excitation parameter, $U = \frac{1}{3}\langle n_{ph}/n_e\rangle$. This parameter basically determines the ionization equilibrium, i.e. the distribution of the elements on different ionization stages.

We performed low-density low-excitation photoionization model calculations in order to model the line ratios of the DIG (Domgörgen & Mathis 1994). The excitation parameter enters the model calculation via the parameter $q = nf^2L_{50} \sim U^3$, where n is the nebular density, f the filling factor, and L_{50} the stellar luminosity in units of 10^{50} photons per sec. In the low density case considered here only this product, and neither n, f, nor L_{50} alone enter the calculations. Other free parameters of the model calculations are, of course, the chemical composition of the gas and the ionizing stellar spectrum. The outer edge of the model is determined by the fraction X_{edge} of neutral hydrogen there. i.e., depending on the choice of X_{edge} a photon bounded or a density bounded model is calculated.

It turns out that low-density low-excitation photoionization models are indeed able to reproduce some of the most important features of the DIG spectrum. Especially, the [SII] $\lambda6717/H\alpha$ and [NII] $\lambda6583/H\alpha$ line ratios increase and [OIII] $\lambda5007/H\alpha$ decreases with decreasing excitation parameter. The observed change of the line ratios with z can therefore naturally be explained by an increasingly diluted radiation field. The results of the model calculations are shown for [NII] $\lambda6583/H\alpha$ and [SII] $\lambda6717/H\alpha$ are shown in Figure 1. The calculations extend from the HII region regime ($log(q) \geq -1$) to excitation parameters comparable to those estimated for the Reynolds layer.

It is the low [OI] $\lambda6300$ upper limit that excludes the simplest photoionization models for the DIG in which H extends beyond the edge of the very dilute stellar radiation field. This is because low-excitation photoionization models do not have sharp Strömgren spheres, but contain a considerable fraction of neutral hydrogen. Since H^0 and O^0 are coupled via charge exchange, photon bounded models alone show [OI] $\lambda6300$ emission exceeding the observed value. Thus a mixture of two components, one accounting for the low [OI] emission, and one representing the edges of neutral clouds is needed to explain the observations.

4.1.2   Some Puzzles

Although energy considerations suggest that OB stars are the most viable sources for ionizing the DIG, and important spectral features of the DIG can be explained by photoionization models, there are some observational facts that speak against this picture.

The probably most puzzling problem is the non-detection of He I $\lambda5876$ emission on two lines of sight in the Galaxy (Reynolds & Tufte 1995). Photoionization models predict $He^+/H^+ \sim 0.6$ He/H. This number is deduced from the ratio of the helium ionizing photons to hydrogen ionizing photons of the mean spectral type ionizing the WIM. The upper limit set by the non-detection of He emission is $He^+/H^+ < 0.27$ He/H, implying that the DIG is ionized by a mean spectral type O8 and later. These late spectral types, although more abundant in number, contribute only $\sim 22\%$ to the ionizing flux (Abbott 1982), i.e. a fraction comparable to the flux needed to ionize the DIG.

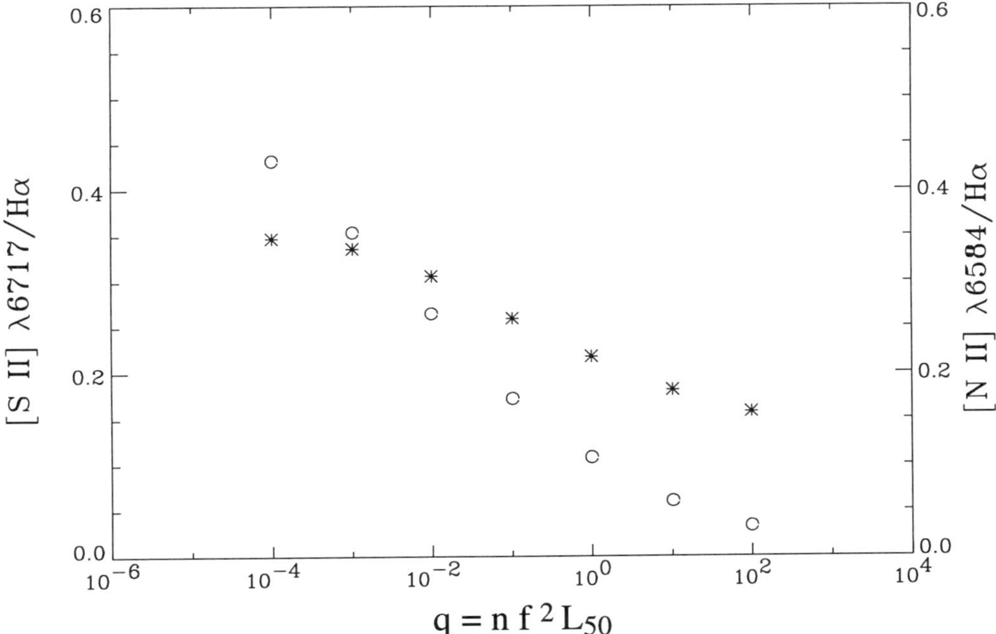

Figure 1: Line ratios [NII] $\lambda6583$/Hα (stars) and [SII] $\lambda6717$/Hα (circles) derived with the photoionization code of Mathis (1985). q is related to the commonly used excitation parameter U via: $U = 0.0013(T/10^4 K)^{-0.55}q^{1/3}$. Line ratios were calculated using a stellar model atmosphere from Kurucz (1979) with $log(g) = 4$ and $T_{eff} = 38{,}000$ K. The Orion abundances as determined by Peimbert (1992) were chosen for the chemical composition of the gas. These parameters were found to give the best fit for the Galactic DIG.

The optical observations by Reynolds & Tufte have recently been confirmed by Heiles et al. (1996). Heiles et al. observed radio recombination line emission at $\sim 1.4\,$GHz for many positions towards the Galactic interior. The recombination line ratios yield $He^+/H^+ < 0.13\,He/H$. The upper mass limit on the IMF estimated from the soft ionizing spectrum of the WIM is $m_u < 39\,M_\odot$, i.e. lower than the upper mass limit implied by the Galactic star formation rate and the observed thermal radio continuum flux, $m_u > 75\,M_\odot$. Although part of this difference can be explained by the observational uncertainties, it possibly suggests that a source other than O stars is responsible for the ionization of the WIM.

Secondly, photoionization model calculations cannot account for the most extreme values of [NII] $\lambda6583/H\alpha$ observed in NGC 891, and even the somewhat lower values found in NGC 4631 are hard to explain. This can best be seen in Figure 1 which shows [NII] $\lambda6583/H\alpha$ in dependence of the excitation parameter. The largest values for [NII] $\lambda6583/H\alpha$ are ~ 0.4, and somewhat larger ratios can be explained by a different chemical composition. However, there is no combination of input parameter that yields [NII] $\lambda6583/H\alpha \geq 1.0$. Consequently, if the extraplanar DIG in NGC 891 is photoionized there has to be an additional heating source. Additional heating could either be due to dust (Reynolds & Cox, 1992; Bakes & Tielens, 1994) or from a low energy cosmic ray population. Moreover, a hardening of the radiation field with distance from the plane would lead to a higher electron temperature in the ionized gas.

Finally, ultraviolet *HST* observations of interstellar absorption lines towards the Galactic halo star HD 93521 suggest that the WIM along this line of sight is not photoionized (Spitzer & Fitzpatrick 1993). Spitzer & Fitzpatrick determine the mean electron density $\langle n_e \rangle$, in each cloud on the line of sight from the column density of the CII*. Since the C^+ column density is not known, they assume that S^+/H is solar and that C^+ is depleted by a factor of 2, i.e. $N(C^+) = 10 \times N(S^+)$. For all clouds $\langle n_e \rangle$ is comparable to the $\langle n_e \rangle$ of the WIM determined from pulsar dispersion measurements. The similarity of the velocity structure of the CII* absorption to the HI emission line profile measured in the direction of HD 93521, and the fact that (S^+/H^0) is solar, both suggests that all the CII* absorption arises in a neutral medium. This means that the ionized gas and the neutral gas are completely intermixed. Only if it is a chance coincidence that (S^+/H^0) is solar, i.e. only if sulfur is partly depleted onto grains in the neutral clouds and some of the S^+ absorption is coming from a completely ionized medium, the high $\langle n_e \rangle$ gas might be due to a photoionized gas.

4.2 Geometry

The question of geometry has been addressed by Miller & Cox (1993) as well as Dove & Shull (1994).

The goal of the calculations presented by Miller & Cox (1993) was to model the distribution of the observed Hα emission measure and pulsar dispersion measure. To do so they used a model in which the structure of the interstellar medium was represented by dense opaque clouds attenuating the radiation field, and a smooth distribution of an intra-cloud medium with an exponential density distribution in z. The free parameter in the model is the mean free path between the opaque clouds $\lambda(z)$. The distribution of the OB stars resembles the observed distribution of OB stars in the 2-3 kpc vicinity of the sun.

Miller & Cox (1993) were able to reproduce the observed dispersion measure towards most pulsars within a factor of two. Even features of the observed electron density distribution like an observed asymmetry in z were successfully reproduced. The emission measure could be modeled with most of the emission coming from cloud surfaces rather than from the WIM itself.

The work by Dove & Shull (1994) was more focused on whether enough ionizing photons can penetrate the Lockman layer of neutral hydrogen in order to ionize the Reynolds layer. Their approach to the problem is a statistical one. Dove & Shull calculated the sizes of Strömgren spheres in dependence of the Lyman continuum production rate for the observed luminosity function of OB associations (Kennicutt, Edgar & Hodge 1989) normalized to the total Lyman continuum flux within the $D_{max} = 2.5$ kpc vicinity of the sun (Garmany, Conti & Chiosi 1982). The ISM is assumed to have a smooth distribution in z. Dove & Shull find that OB associations with a photon production rate $S \geq (3-4) \times 10^{49}$ s^{-1} produce ionization cones which allow the escape of ionizing photons into the halo. The fraction of ionizing photons penetrating into the Reynolds layer is in fact comparable to the power requirement of the WIM.

Not too surprisingly, considering the simplicity of the models compared to nature, both models have difficulties explaining some of the observations. The main difficulty of the Miller & Cox (1993) calculations is that they overestimate the midplane density and filling factor of the DIG by a factor of two. In the Dove & Shull models only $\sim 35\%$ of the ionizing photons that penetrate through the Lockman layer are actually absorbed by the Reynolds layer. 65% escape the Galaxy entirely. However, the authors do not take into account that the ionized parts of the HI layer might actually contribute to the Reynolds layer.

A realistic approach to the problem of geometry probably also has to take into account the combined effects of stellar winds and supernovae on the structure of the ISM. The holes blown into the ISM could increase the mean free path of ionizing photons in the immediate vicinity of the association. Additionally, their escape probability into the halo should be larger due to effects like super-bubble blow-out (MacLow, McCray & Norman 1989, Igumentshchev, Shustov & Tutokov 1990). On the other hand, the swept up shells are likely to reduce the transparency of the ISM to the ionizing radiation in the disk itself. The structure of the WIM might possibly be similar to a "worm" ionized medium where the WIM represents ionized chimney walls (e.g. Heiles 1993).

5   Summary and Conclusions

Since power estimates for the DIG in the Milky Way show that the Lyman continuum radiation of OB stars is the only viable ionizing source of the DIG (Reynolds 1984) the previous discussion has been focussed on the capability of OB star photoionization. It turns out that low-density low-excitation photoionization models are able to explain many spectral features of the DIG, in particular the high [NII] $\lambda6583$/Hα and [SII] $\lambda6717$/Hα and the low [OIII] $\lambda5007$/Hα line ratios observed in the Milky Way. A mixture of density bounded and photon bounded models is needed to account for the low [OI] $\lambda6300$/Hα, implying that a fraction of the OB star ionizing radiation will escape the Galaxy (Domgörgen & Mathis 1994). Additionally, it has been shown that the Strömgren spheres of OB star associations can become as large as several hundred parsecs. Consequently the ionizing radiation is able to travel to high distances above the plane (Miller & Cox 1993, Dove & Shull 1994).

But there are also problems with these models. (i) the low He$^+$/H$^+$ implies a too soft ionizing radiation field (Reynolds & Tufte 1995, Heiles et al. 1996). (ii) the fact that ionized and neutral gas seem to be well mixed on the line of sight to the Galactic halo star HD 93521, suggest a structure of the ISM that cannot be accounted for by photoionization (Spitzer & Fitzpatrick 1993). (iii) it is not possible to explain [NII] $\lambda6583$/H$\alpha \sim 1.1$ as observed in NGC 891 with photoionization models unless there is an additional heating source (Dettmar & Schulz 1992). (iv) it is hard to imagine that the large Hα halo that has recently been detected around NGC 4631 is photoionized (Donahue et al. 1995).

However, the power estimations for the Milky Way need not necessarily to be valid for the prominent Hα features seen in the halos of other galaxies. Starburst galaxies, for example, show widespread Hα emission in their halos. From multiwavelength studies, and the fact that kinematical evidence for outflows is present we know that this Hα emission can well be explained by a galactic wind (Heckman, Armus & Miley 1990). In this case shocks produced by the superwind are highly important for the ionization of the diffuse gas in the halo (Heckman, Armus, Miley 1990; Dahlem et al. 1996). The line ratios for the Hα emitting gas are in fact HII region like (i.e. explainable by photoionization) close to the plane and become LINER like (i.e. explainable by shock ionization) at large distances above the plane (Heckman, Armus, Miley 1990).

The correlation between star formation rate per disk area and DIG structures also found in galaxies with much lower star formation rates suggests that halo properties vary continuously in dependence of the energy input by massive young stars. The starburst galaxies then represent the most energetic end of the scale. At the low energy end there are galaxies with normal star formation rates. In these galaxies the DIG is a more quiescent and photoionization might actually be the important ionization mechanism (although the non-detection of the He I remains very puzzling).

The high [NII] $\lambda6583$/Hα in NGC 891 might then be due to an intermediate scenario, and it is likely that we see a local chimney (e.g. Norman & Ikeuchi 1989) rather then a galactic wind or a photoionized thick disk. This scenario is also indicated by the velocities of the Hα emitting gas above the plane (Pildis, Bregman, Schombert 1994b), as well as by the fact that NGC 891 has a radio continuum (Allen, Baldwin, Sancisi 1978) was well as an X-ray halo (Bregman & Pildis 1994).

Clearly, much work needs to be done before we can reach a deep understanding of the nature of the diffuse ionized gas in halos of non-starburst galaxies. In particular, our knowledge about the spectrum of the DIG is still very limited and needs to be improved. Due to the low surface brightness of DIG, lines other than [NII], [SII], and Hα are observed for only few lines of sight in the Milky Way ([OIII]: 3 lines of sight, [OI]: one sightline, He I: two sightlines). They are not observed in any other galaxies. The observations with the Wisconsin Hα mapper which are planned for the near future will clearly improve this situation.

References

Allen, R.J., Baldwin, J.E., Sancisi, R. 1978, A&A 62, 397

Abbott, D.C. 1982, ApJ 263, 723

Bregman, J.N., Pildis, R.A. 1994, ApJ 420, 570

Dahlem, M., Heckman, T.M., Fabbiano, G., Lehnert M.D., Gilmore, D. 1996, ApJ in press

Dettmar, R.-J. 1992, Fund. Cosmic Phys., 15, 143

Dettmar, R.J., Schulz, H. 1992, A&A 254, L25

Dettmar, R.J. 1995 in "The Physics of the interstellar medium and intergalactic medium", eds. A. Ferrara et al. ASP Conf. Ser. Vol. 80, p.389

Domgörgen, H., Mathis, J.M. 1994, ApJ 428, 647

Donahue, M., Aldering, G., Stocke, J.T. 1995, ApJ 450, L45

Golla, G., Dettmar, R.-J., Domgörgen, H. 1996, A&A 313, 439

Garmany, C.D., Conti, P.S., Chiosi, C. 1982, ApJ 263, 777

Hawley, S.A. 1978, ApJ 224, 417

Heckman, T.M., Armus, L., Miley, G.K. 1990, ApJS 74, 833

Hunter, D.A., Hawley, W.N., Gallagher III, J.S. 1993, AJ 106, 1797

Heiles, C. 1993, in "Star Formation, Galaxies and the Interstellar Medium", ed. J. Franco, F. Ferrini, and G. Tenorio-Tagle, Cambridge University Press, p. 245

Igumentshchev, I.V., Shustov, B.M., Tutukov, A.V. 1990, A&A 234, 396

Kennicutt, R.C., Edgar, B.K., Hodge, P.W. 1989, ApJ 337, 761

Kennicutt, R.C., Bresolin, F., Bomans, D.J., Bothun, G.D., Thompson, I.B., 1995, AJ 109, 594

Kurucz, R.L. 1979, ApJS 40, 1

Mac Low, M-M., McCray, R., Norman, M.L. 1989, ApJ 337, 141

Mathis, J.S. 1985, ApJ 291, 247

McCammon, D., Burrows, D.N., Sanders, W.T., Kraushaar, W.L. 1983, ApJ 269, 107

Norman, C.A., Ikeuchi, S. 1989, ApJ 345, 372

Osterbrock, D.E. 1989, "Astrophysics of Gaseous Nebulae and Active Galactic Nuclei", p.13

Panagia, N., Terzian, Y. 1984, ApJ 287, 315

Peimbert, M., Torres-Peimbert, S., Ruiz, M.T. 1992, Rev. Mexicana, Astron. Af. 24 155

Pildis, R.A., Bregman, J.N., Schombert, J.M. 1994a, ApJ 427, 160

Pildis, R.A., Bregman, J.N., Schombert, J.M. 1994b, ApJ 423, 190

Rand, R.J. 1996, ApJ in press

Rand, R.J., Kulkarni, S.R., Hester, J.J. 1992, ApJ 396, 97

Reynolds, R.J. 1984, ApJ 282, 191

Reynolds, R. J. 1985, ApJ 298, L27

Reynolds, R.J. 1990, in "The Interstellar Disk-Halo Connection in Galaxies",
     ed. H. Bloemen, IAU Symposium No. 144
Reynolds, R.J. 1993 in *Back to the Galaxy*, eds. Holt et al. APS Conf. Ser.
     Vol. 80, p. 178
Reynolds, R.J., Tufte, S.L. 1995, ApJ 439, L17
Roy R.-J., Wang, J., Arsenault, R. 1991, AJ 101, 825
Sciama, D.W. 1990, ApJ 364, 549
Sciama, D.W. 1995, ApJ 448, 667
Spitzer, L. 1978, "Physical Processes in the Interstellar Medium",
Spitzer, L., Fitzpatrick, E.L. 1993, ApJ 409, 299
van Dishoeck, E.F., Black, J.N. 1986, ApJS 62, 109
Vogel, S., Weymann, R., Rauch, M., Hamilton, T. 1995, ApJ 441, 162
Walterbos, R.A.M., Braun, R. 1995, ApJ 431, 156

Address of the author:

H. DOMGÖRGEN, Sternwarte der Universität Bonn, Auf dem Hügel 71, D-
53121 Bonn, Germany.

Diffuse ionized gas and star formation in spiral galaxies

R.-J. Dettmar

The detection of a thick layer of DIG in NGC891 which was found to be similar to the Reynolds-layer of the Milky Way (Rand et al. 1990, Dettmar 1990) immediately raised the question how typical such a component of the ISM would be for a 'normal' spiral galaxy. About two dozen 'normal' (i.e. excluding nuclear starbursts) edge-on galaxies are meanwhile searched for extraplanar Hα emission (Walterbos 1991, Dettmar 1992, Rand et al. 1992, Pildis et al. 1994a, Rand 1996). Only few of them are showing evidence of a wide spread DIG in the halo comparable to NGC891, the Milky Way twin and prototype for studies of the interstellar disk-halo connection. In NGC891 the DIG is distributed in long filaments (Rand et al. 1990, Dettmar 1990) and bubbles (Pildis et al. 1994b) of ionized gas embedded in a smooth background. The origin and the source of ionization for DIG in the disk-halo interface of spiral galaxies is generally discussed in the framework of a large scale circulation of matter between actively starforming areas in the disk and the halo such as *galactic fountains* (Shapiro and Field 1976), *chimneys* (Norman and Ikeuchi 1989), or *winds* (Habe and Ikeuchi 1985).

Ionization and excitation conditions for extraplanar DIG in edge-on galaxies are discussed in detail in the previous contribution by Domgörgen. In the following, a brief summary of evidence for a correlation of extraplanar DIG with starformation in the underlying disk is given.

Correlation with starforming regions

In the case of NGC891 a correlation of extraplanar DIG and star formation in the underlying disk has been studied in detail (Dettmar 1992, Dahlem et al. 1994). Unfortunately, absorption by dust strongly affects the observed distribution of the HII regions in the disk of edge-on systems. However, star-formation can be traced by emission at wavelengths for which galactic disks are optically thin such as FIR radiation.

The thermal emission of dust in galaxies observable in the FIR is known to be a good tracer of star formation (e.g., Thronson and Telesco 1986). In the

Tab. 1: DIG properties and SFR

Object(Ref.)	$L_{FIR}/D_{25}^2 \times 10^{40\,(a)}$ (erg/sec kpc^2)	DIG distrib.&morphol.	radio halo	X-ray halo
NGC 4666 (1)	14.5	bright, diffuse filaments	X	X
NGC 3079 (2,3)	8.9	bright, diffuse filaments	X	
NGC 5775 (4,5)	8.1	bright, diffuse+ filament	X	
NGC 253 (6-8)	8.1	bright, diffuse+ filamen	X	X
NGC 3044 (5,9)	4.0	bright, diffuse	X	
NGC 891 (10-13)	3.3	bright, diffuse+ filamen	X	X
Galaxy (14-16)	3.0	diffuse	X	X
NGC 4013 (5,9)	2.6	faint diffuse	?	
NGC 4631 (17-19)	1.8	bright diffuse	X	X
NGC 4302 (5,10,20)	<2.3	faint diffuse	?	
NGC 973 (5,20)	1.0	plumes	—	
UGC 3326 (5,20)	0.8	—		
NGC 5907 (9)	0.8			
UGC 12281 (20)	0.5	plumes		
UGC 2092 (5,20)	0.4	plumes	—	
UGC 10288 (5,9)	0.4	3 or 4 plumes	—	
NGC 4565 (10,21)	03	—	—	X
NGC 5746 (9)	0.2			
NGC 4244 (22,23)	<0.04	—	—	—
NGC 5023 (9)	<0.09	1 or 2 plumes		
NGC 4217 (5,9)	<0.12	2 faint patches	X	
UGC 4278 (9)	<0.04			
NGC 4762 (9)	<0.15			

Notes: *(a)* distances from Tully 1988; FIR fluxes from Rice et al. 1989 or Fullmer and Lonsdale 1989; FIR for Galaxy from Cox and Mezger 1989. (1) Dahlem et al. (in prep.) (2) Veilleux et al. 1995 (3) Duric et al. 1983 (4) Dettmar 1992 (5) Hummel et al. 1991 (6) Carilli et al. 1992 (7) Pietsch et al. (in prep.) (8) Golla et al. (in prep.) (9) Rand 1996 (10) Rand et al. 1990 (11) Dettmar 1990 (12) Dahlem et al. 1994 (13) Bregman and Pildis 1994 (14) Reynolds 1990 (15) Beuermann et al. 1985 (16) Kerp, these proceedings (17) Rand et al. 1992 (18) Hummel and Dettmar 1990 (19) Wang et al. 1995 (20) Pildis et al. 1994a (21) Sukumar and Allen 1991 (22) Walterbos 1991 (23) Schlickeiser et al. 1986 (current) lack of FIR observations with sufficient angular resolution for most galaxies only integral fluxes are known. These can be used to derive, at least relative, global star formation rates, one of the quantities that controls the physical state of the ISM.

In Tab. 1 we have compiled the DIG morphology and its brightness together with IRAS FIR luminosity normalized to the optical diameter, as well as evidence for thick radio and X-ray halos. FIR luminosities of these galaxies differ by almost two orders of magnitude and from this sample the earlier impression, that extraplanar DIG is related to star formation, emerges again. Bright and diffuse DIG in the halo is observed in galaxies with higher star formation rate per unit area. More active objects preferentially show in addition radio (see also Golla, these proceedings) and X-ray halos. Only the observed X-ray halo in NGC4565 and the radio thick disk of NGC4217 do not fit into a scenario that relates extraplanar DIG, radio, and X-ray halos to the SFR in the disk. It is worth mentioning that the Milky Way actually fits into this picture. A more quantitative analyses of the correlation is currently difficult since the compilation presented in Tab. 1 is based on very inhomogeneous data, in particular with regard to detection limits.

A more detailed discussion is given in Dettmar 1995.

References

Beuermann, K., Kanbach, G., Berkhuijsen, E. M. 1985, A&A 153, 17

Bregman, J. N., Pildis, R. A. 1994, ApJ 420, 570

Carilli, C. L., Holdaway, M. A., Ho, P. T. P., de Pree, C. G. 1992, ApJ 399, L59

Cox, P., Mezger, P. G. 1989, A&AR 1, 49

Dahlem, M., Dettmar, R.-J., Hummel, E. 1994, A&A 290, 284

Dettmar, R.-J. 1990, A&A 232, L15

Dettmar, R.-J. 1992, Fund. Cosm. Phys. 15, 143

Dettmar, R.-J. 1995, in *The Physics of the Interstellar Medium and Intergalactic Medium* eds. A. Ferrara, C. F. McKee, C. Heiles, and P. R. Shapiro, ASP Conf. Ser. Vol. 80, p. 398

Duric, N., Seaquist, E. R., Crane, P. C., Bignell, R. C., Davies, L. E. 1983, ApJ 273, L11

Fullmer, L., Lonsdale, C. 1989, Cataloged Galaxies and Quasars observed in the IRAS Survey, JPL

Habe, A., Ikeuchi, S. 1980, Prog. Theor. Phys., 64, 1995
Hummel, E., Dettmar, R.-J. 1990, A&A 236, 33
Hummel, E., Beck, R., Dettmar, R.-J. 1991, A&AS 87, 309
Norman, C. A., Ikeuchi, S. 1989, ApJ 345, 372
Pildis, R. A., Bregman, J. N., Schombert, J. M. 1994a, ApJ 427, 160
Pildis, R. A., Bregman, J. N., Schombert, J. M. 1994b, ApJ 423, 190
Rand, R.-J. 1996, ApJ 462, 712
Rand, R. J., Kulkarni, S. R., Hester, J. J. 1990, ApJ 352, L1 (Err. 362, L35)
Rand, R. J., Kulkarni, S. R., Hester, J. J. 1992, ApJ 396, 97
Reynolds, R. J. 1990, in *IAU Symp. 139 Galactic and Extragalactic Background Radiation* eds. S. Bowyer and C. Leinert, Kluwer, p. 157
Rice, W., Lonsdale, C. J., Soifer, B. T. et al. 1988, ApJS 68, 91
Schlickeiser, R., Werner, W., Wielebinski, R. 1984, A&A 140, 277
Shapiro, P. R., Field, G. B. 1976, ApJ 205, 762
Sukumar, S., Allen, R. J. 1991, ApJ 382, 100
Thronson, H. A., Telesco, C. M. 1986, ApJ 311, 98
Tully, R. B. 1988, *Nearby Galaxies Catalog*, Cambridge Univ. Press
Veilleux, S., Cecil, G., Bland-Hawthorn, J., 1995, ApJ, 445, 112
Wainscoat, R. J., de Jong, T., Wesselius, P. R., 1987, A&A 181, 225
Walterbos, R. A. M. 1991, in *IAU Symp. 144 The interstellar disk-halo connection in galaxies* ed. H. Bloemen, Kluwer, p. 223
Wang, D., Walterbos, R., Steakley, M., Norman, C. N., Braun, R. 1995, ApJ 439, 176

Address of the Author:

R.-J. DETTMAR
Ruhr-University Bochum, Astronomical Institute, Universitätsstrasse 150/NA 7, D-44780 BOCHUM, GERMANY

On the Connection between Star Formation and Radio Halos of Galaxies

Götz Golla

Scenarios of star formation induced disk–halo interaction (DHI) naturally follow from a dynamical multi-phase picture of the ISM, where the hot phases are heated by shock waves of violent stellar winds and supernova explosions (cf. Field 1986, and references therein). As soon as the dynamical time for the hot gas to reach one scale height of the disk is shorter than the cooling time, outflow will occur (Norman & Ikeuchi 1989). The outflowing gas will facilitate the transport of cosmic rays and magnetic fields into the halo, leading to the formation of a radio halo. Such radio halos have been observed in several galaxies, the most prominent being NGC253 (Carilli et al. 1992) and NGC4631 (Hummel & Dettmar 1990; Golla & Hummel 1994).

Dahlem, Lisenfeld & Golla (1995; hereafter DLG95) addressed the question what minimum energy input into the ISM by star formation is required to facilitate the formation of radio halos. In this short contribution I will ask the same questions as in DLG95, but use a slightly different approach and sample of galaxies.

For our purpose parameters are required which describe star formation and DHI in a galaxy. An adequate, physical approach is the concept of a threshold in star formation above which DHI occurs, corresponding to a break-out condition for a hot superbubble into the halo. In this approach, as a first approximation, one only needs to know whether DHI is occuring or not. Star formation, on the other hand, may be conveniently described by the supernova rate of a galaxy, being the best indicator for the energy input by massive stars into the ISM. In contrast to DLG95, I derive the supernova rates from the FIR luminosities assuming an "extended" Miller-Scalo IMF (Kennicutt 1983). Using relations in Sauvage & Thuan (1992) and Condon (1992) I find

$$\nu_{SN}[yr^{-1}] = 9.2 \cdot 10^{-12} L_{FIR}[\mathrm{L}_\odot]. \tag{1}$$

In Table 1 a list of (nearly) edge-on galaxies is presented, ranging from the quiescent NGC4244 to the prototype starburst galaxies NGC253 and M82. The sample is identical to the one of DLG95, except for the latter starburst galaxies. Table 1 of Dettmar (page 83, these proceedings) and Table 1 of Beck (page 137) list the properties of the ionized gas and radio emission in the halos of these galaxies. I present in column (4) of Table 1 the supernova rates as derived from the FIR luminosites of the galaxies, assuming the distances listed in column (3) (see DLG95 for the references).

	radio halo emission	D [Mpc]	ν_{SNR} [yr^{-1}]	r_{SF}	r_{inner} [kpc]	$\dot{E}_{tot}^A$	$\dot{E}_{tot}^{inner}$ $[10^{-3}\,\mathrm{erg\,s^{-1}\,cm^{-2}}]$
NGC4244	no	3.1	0.0024	4.5		0.05	
NGC4565	marginal	9.7	0.14	16.9		0.09	
NGC5907	marginal	14.9	0.06	18.1		0.18	
NGC3044	filaments	20.6	0.075	10.0		0.80	
NGC891	filaments	9.5	0.32	9.6		1.4	
NGC4666	yes	14.1	0.45	13.0		0.8	
NGC4631	yes	10.0	0.23	17.3	1.8	0.55	164
NGC5775	filaments	26.7	0.23	9.6		2.7	
M82	yes	3.6	0.21		0.5		920
NGC253	yes	4.2	0.28	20.0	5.0	0.75	12

Table 1: Parameters of the Galaxies in our Sample

Figure 1 shows a histogram of the supernova rates for the galaxies in Table 1. From this Figure the general trend is confirmed that only galaxies with high supernova rates exhibit eDIG and radio halos. In particular, NGC253 hosts the largest synchrotron halo of a galaxy known to date (Carilli et al. 1992), and all the galaxies with $\nu_{snr} > 0.1\mathrm{yr}^{-1}$ are known to exhibit large radio halos (NGC891: Dahlem et al. 1994, NGC4631: Golla & Hummel 1994, NGC4666: Dahlem et al. 1996, NGC5775:, M82: Seaquist & Odegard 1991).

As has been pointed out by DLG95, to quantify the threshold condition for DHI it is necessary to normalize the supernova rates to the areas of the disks where star formation occurs. In Table 1 the radii of the star forming disks are listed as derived from IRAS CPC and Hα data (see DLG95). The areas πr_{SF}^2 usually comprise only one-half to two-thirds of the optically visible disks.

I used equation (8) of DLG95 to derive the energy input by SN into the ISM normalized to those star forming disks, and list the values of $\dot{E}_{tot}^A$ in Table 1. In Figure 1 the resulting histogram of $\dot{E}_{tot}^A$ is presented. Despite the huge radio halo of NGC4631, $\dot{E}_{tot}^A$ for this galaxy is even lower than $\dot{E}_{tot}^A$ of NGC3044. In general the sequence in $\dot{E}_{tot}^A$ does not correspond very well to the intensity of DHI observed.

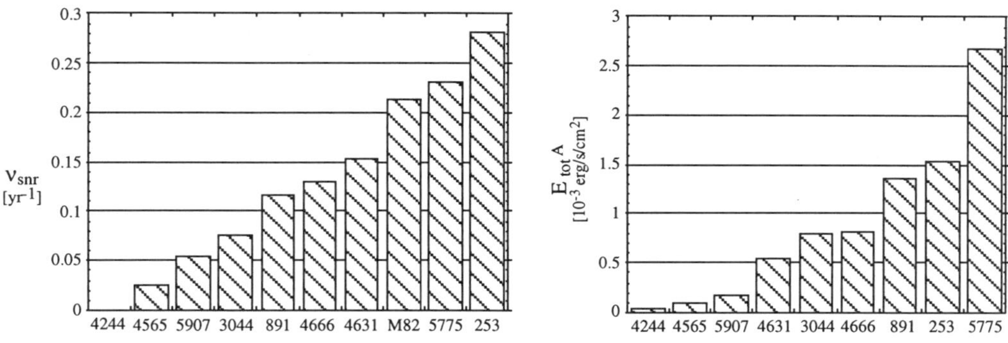

Figure 1: Histogram of the estimated SN and energy input rates

These inconsistencies may be resolved by examining the distribution of star forming regions in more detail. In many galaxies star formation is not distributed smoothly over the area described by r_{SF}, even on scales of a few kpc. In particular, in NGC253 and NGC4631 *most* of the star formation is limited to the inner rigidly rotating region, and indications for outflow are found *only* above these regions. So, as a first approximation, the whole FIR luminosity can be regarded as coming from that inner disk. The radii of these regions r_{inner} have to be used instead of r_{SF} to give meaningful values for $\dot{E}_{tot}^{A}$.

The resulting $\dot{E}_{tot}^{inner}$ for NGC253 and NGC4631 are listed in Table 1. For M82 only $\dot{E}_{tot}^{inner}$ is given because no star formation in the outer parts of the galaxy occurs. For the other galaxies, either no high resolution data of the distribution of star formation with radius are available, or star formation is not as concentrated to the inner disk, so that a simple approximation as for NGC253 and NGC4631 cannot be applied. The energy input rates for NGC253 and NGC4631 over the inner disks $\dot{E}_{tot}^{inner}$ are substantially higher than the overall $\dot{E}_{tot}^{A}$. For NGC4631 the difference is two orders of magnitude. Still, there is another order of magnitude to reach the extreme energy input rates of M82 of almost $1\,\mathrm{erg\,s^{-1}\,cm^{-2}}$.

The energy input rates $\dot{E}_{tot}^{A}$ are lower limits to the energy deposition into the ISM which might happen locally. DLG95 estimated that, based on typical numbers of giant HII regions in a galaxy, local energy input rates may be two orders of magnitude higher. This might occur in galaxies with star formation distributed over the whole disk, and mainly concentrated in giant HII regions. The 'chimney' model for DHI (Norman & Ikeuchi 1989), involving highly correlated supernova explosions in OB associations, and the formation of superbubbles of hot gas, is an adequate description for the DHI occuring in such galaxies. Such 'chimneys' may be operating in NGC891 and NGC5775.

Our simple calculations show that in NGC4631 and M82 extreme energy input rates are reached over the entire *inner* disks. Thus, for these galaxies the chimney model does not apply. Rather, an overall wind blows from their inner disks. NGC253, even though a classical starburst galaxy like M82, exhibits an $\dot{E}_{tot}^{inner}$ two orders of magnitude lower than M82. This means that the starbursts in these two galaxies are fundamentally different.

In the outer disks of NGC4631 and NGC253 the chimney mode may be operating like in NGC891 and NGC5775. Detailed spatially resolved data of the star formation and SN rates, as well as the disk halo connection at radio continuum frequencies in theses galaxies will be necessary to deduce observationally the threshold values needed for local outflows to occur. I expect these thresholds to be in the range of $0.01 - 0.1\,\mathrm{erg\,s^{-1}\,cm^{-2}}$ (see also DLG95).

It is my impression from the available multi-waveband data of edge-on galaxies that different thresholds might be relevant for the formation of radio thick disks, extended radio halos with aligned & perpendicular magnetic fields, extraplanar DIG and extraplanar X-ray emission. Thick radio disks probably have the lowest threshold, i.e. they are most easily triggered, whereas an overall wind with $\dot{E}_{tot} > 0.1\,\mathrm{erg\,s^{-1}\,cm^{-2}}$ may be necessary to produce aligned magnetic fields perpendicular to a galaxy's plane.

References

Carilli, C.L., Holdaway, M.A., Ho, P.T.P., de Pree, C.G., 1992, ApJ **399**, L59

Dahlem, M., Dettmar, R.-J., Hummel, E., 1994, A&A **290**, 384

Dahlem, M., Lisenfeld, U., Golla, G., 1995, ApJ **444**, 119

Dahlem, M., Petr, M.G., Lehnert, M. et al. 1996, in prep.

Field, G.B., 1986, in Highlights of Modern Astrophysics, eds. Shaprio, S.L., Teukolsky, S.A., Wiley Interscience (New York), 235

Golla, G., Hummel, E., 1994, A&A **284**, 777

Hummel, E., & Dettmar, R.-J., 1990, A&A **236**, 33

Norman, C.A., Ikeuchi, S., 1989, ApJ **345**, 372

Seaquist, E.R., & Odegard, N., 1991, ApJ **369**, 320

Author:

G. GOLLA

Universität Bochum, Astronomisches Institut, D-44780 Bochum, Germany.

Absorption lines towards halo objects

Klaas S. de Boer

Abstract: In this review, I will attempt to discuss the observational data relevant for the understanding of halo clouds. First I will address the principal observational constraints in the quest of detecting interstellar absorption against halo objects, as well as on the instrumentation available for the halo gas studies (Practicalities). Observing extragalactic objects provides the best estimates of how halo gas may look like in absorption while questions about the spatial structure of clouds are addressed briefly (Properties). Finally, I will review what really has been seen in terms of absorption (Pertinent Data), consisting of a discussion of general data for the halo direction, a presentation of distance limits to individual clouds, and a brief discussion of results for highly ionized gas, as well as a note on two peculiar extragalactic lines of sight.

1 Prelude

The detection of interstellar gas in absorption provides the only way to get information about the distance of such gas. In order to make sure that a detection (or non-detection for that matter) is meaningful, we should know that the absorbing species (atom, ion, molecule) is available in free form in sufficient abundance in that cloud of gas, as well as that the cloud looked for is well spread (on the spatial scale of interest) on the sky.

Detecting absorption by such species relies on knowledge of the wavelength at which a resonance electronic transition of the species is found, as well as on knowledge of its transition probability. The absorption lines in the visual include the well known Ca II H+K and the Na I D_1 and D_2 doublets, as well as weak transitions from K I, Ca I, Fe I, and Ti II, and of the molecules CH^+ and CN, all tracing neutral gas. Then the satellite UV instruments opened up a spectral domain full of interesting and very useful and sensitive interstellar absorption lines for neutral gas (Mg II, Si II, C II, Fe II, etc.) as well as those representative for ionized gas (Si III, Si IV, C IV, N V, and O VI).

Several of the technical aspects have to be discussed in order to really appreciate what has been measured, what has been detected and what not, and what we can reliably conclude from the data.

2 Practicalities

2.1 Spectral resolution, instruments

To detect an interstellar cloud in an absorption line, the spectrograph must provide adequate spectral resolution. Given the natural width of interstellar lines due to the characteristic temperatures of and the turbulence in gas clouds, and knowing that normally several clouds exist on a line of sight with different radial velocities, a spectral resolution of the order of 10 $\mathrm{km\,s}^{-1}$ is an optimal value. Such spectrographs are available or can be purpose built for the visual. The two UV instruments relevant for the present topic are the International Ultraviolet Explorer (IUE) with $\approx$20 $\mathrm{km\,s}^{-1}$ resolution, and the Hubble Space Telescope (HST) where the GHRS offers a choice of similar or even better spectral resolution. The magnitude limit for the IUE is $\approx$ 14 mag, so that the brightest stars of the Magellanic Clouds and several extragalactic objects are within reach. The HST surpasses these possibilities in principle, but in practice spectra are obtained of similar resolution as those from the IUE albeit with much better signal-to-noise and from somewhat fainter sources.

2.2 Background light sources

Objects at apropriate distances should be avialable. Any type of star is in principle suitable to detect interstellar absorption in the spectrum, provided the spectrum is clean near the line of interest. This means that either the star must be of early spectral type or is a fast rotator (thus widening the stellar lines to shallow depressions in the spectra). Early type stars are rare at high galactic latitudes. However, subdwarf stars are available and after determining their distances (e.g. Moehler, Heber & de Boer 1990, Schmidt et al. 1996) they are very useful (see de Boer et al. 1994, Centurión et al. 1994; Schwarz, Wakker & van Woerden 1995). RR Lyr stars, however, pose serious problems because they are too cool and have to many stellar lines (see Lilienthal, Meyerdierks & de Boer 1990). Fast rotating A stars work well too (Lilienthal & de Boer 1991).

2.3 Atomic data

At the start of the ultraviolet spectroscopy era atomic data were known for a fair number of transitions. However, for many absorption lines the transition

Table 1: Expected *relative* optical depth for common absorption lines[a]

Species (X)	Transition (Å)	$\log N = [X/H]^b$	$\log f_\lambda\lambda^b$	$\log Nf\lambda$
	ultraviolet			
Mg II	2795.53	7.59	3.23	10.82
C II	1334.53	8.56	2.23	10.79
O I	1302.17	8.93	1.80	10.73
Si II	1260.42[f]	7.55	3.10	10.65
Mg II	2802.70	7.59	2.93	10.52
Fe II	2600.17	7.51	2.76	10.27
Si II	1526.70	7.55	2.54	10.09
Fe II	2585.87	7.51	2.23	9.74
Si II	1304.37[f]	7.55	2.28	9.83
Fe II	1608.45	7.51	2.00	9.51
Mg I	2852.13	5.59[c]	3.72	9.31
	visual			
Ca II	3933.66	6.34[d]	3.40	9.74
Ti II[e]	3383.79	4.93	3.06	7.99
Na I	5889.95	4.31[c]	3.57	7.88
Fe I[e]	3239.04	5.51[c]	2.18	7.79
K I[e]	7698.97	3.13[c]	3.42	6.55

[a] Wavelengths smaller than 1200Å are not commonly accessible at present.

[b] Solar abundance $[X/H]$ from compilation by Morton (1991), in scale $\log N(\mathrm{H}) = 12.0$; $\log f_\lambda\lambda$ also from Morton (1991).

[c] For the lower ionization stages (Mg I, Na I, Fe I and K I) an abundance of 1/100 of that of the element was assumed.

[d] A (possibly substantial) fraction of all Ca may exist as Ca III in neutral gas.

[e] The Ti II, Fe I, and K I lines lie in difficult parts of the atmospheric window.

[f] Line blends with other disturbing spectral structures.

probabilities had still to be determined and this was done in part from the UV data themselves. The newest listing by Morton (1991) contains all the information needed for interstellar absorption studies.

The product of the gas column density of a species X, $N(X)$, for which we in first instance take values equal to the solar abundance $A = \log N(X)/N(\mathrm{H})$ (or $[X/H]$) of the species, and the $f_\lambda\lambda$-value of the transition can be calculated. Those transitions having the highest optical depth τ ($\tau \sim Nf_\lambda\lambda$) are the most promising ones to detect gas. For neutral gas, the Mg II lines near 2800 Å, the C II line near 1334 Å, and the O I line near 1302 Å are at the top of the list. An order of magnitude inferior are the Ca II H+K lines while the Na I D_1+D_2 lines have 3 orders of magnitude less optical depth than the ones at the top of the full list. However, in the visual one has the advantage of possibly using large telescopes and long integration times (and Na I is present as stellar line only in cool stars). If a species were depleted or has a smaller intrinsic abundance the ranking in such a list is changed accordingly (Ca and Ti?). If the gas has a general deficit of metals in comparison with the solar abundances, then the ranking stays the same, only the chance to detect gas becomes smaller accordingly. Table 1 contains such data for the most relevant absorption lines (see also de Boer et al. 1994).

The best accessible lines for highly ionized gas are the C IV lines near 1548 and 1550 Å. Other possible lines are the Si IV doublet near 1400 Å, but they may also be produced in the stellar H II-region, and the N V doublet near 1248 Å, but both the intinsic abundance of N as well as the f-value of its transitions are smaller than the respective values of C (the product being a factor of 10 smaller). Therefore, the lines studied are mostly the C IV lines. Column densities for these species can be derived with curve-of-growth methods. However, in many cases the absorption is due to gas with high velocity dispersion such that a transformation of absorption depth to optical depth leads to reliable column density profiles $N_{\rm C\,IV}(v)$, as shown by Sembach & Savage (1992).

2.4   Detection limits

The strength of the absorption line looked for should be large enough to be detectable. This means that the amount of gas must be large enough for the optical depth τ to produce a measurable signal. Adopting a spectral resolution of 10 to 20 $\rm km\,s^{-1}$ as adequate, one can define the actual limits for the instruments currently available.

IUE. The equivalent width detection limit for absorption due to a cloud isolated in velocity space is $\log W_\lambda/\lambda \approx -5.0$, which is reached in a 6 hr exposure for a $V \approx 13$ mag star of $T \approx 20\,000$ K (see de Boer et al. 1994). This limit represents for Mg II a column density limit of $\approx 10^{12}$ cm^{-2} and the equivalent H I column density is $N({\rm H(Mg)}) \approx 3 \times 10^{16}$ cm^{-2}.

visual. In the visual a similar detection limit is reached for the Ca II and Na I lines with a 2.5m telescope for the same star with a 30 min exposure (see Centurión et al. 1994). The Ca II column density limit is here $\approx 3\ 10^{11}$ cm^{-2}, and the equivalent gas column density (assuming solar abundance) is $N({\rm H(Ca)}) \approx 10^{17}$ cm^{-2}.

HST. Here the signal to noise is much better than with the IUE so that shorter exposures result in better limits than the ones obtained with the IUE.

It is clear that the halo gas detected in 21-cm down to $N(\mathrm{H}) \approx 10^{18}$ cm^{-2} should produce, if it were in front of the star and if interstellar gaseous abundances are solar, absorption by the ions of Table 1. With the IUE, whose operations will definitely end Dec. 1997 the latest, no new results are to be expected. The UV domain will continue to be accessible with the HST, which will require exposure times of the order of a few hours per spectral range for the indicated star. In the visual, due to the lower abundance of the species detectable there, a larger effort must go into the measurements, by using large telescopes and long or repeated exposures of the spectral range of interest.

Visual spectra of numerous stars have been investigated for the presence of interstellar absorption lines. A comprehensive list of all data (complete up to 1988) can be found in the compilation by Garcia (1991), also available at the *Centre des Données Stellaires* in Strasbourg.

3   Properties of Halo Absorption

The best way to know the nature of absorptions by halo gas in the spectra of halo objects is to look for such information in the spectra of *extragalactic* objects. Contradictory as this may seem, the detections in such spectra demonstrate what really can be expected if gas clouds are present on the line of sight to (and in front of) halo objects (see the schematics given by Schwarz, Wakker & van Woerden 1995).

The first UV spectra of extragalactic objects obtained with high spectral resolution were those for stars of the Large Magellanic Cloud (Savage & de Boer 1979). In these, absorption by the species normally detectable in the UV was found in two clouds, at $+60$ and $+120$ km s^{-1}. These gas clouds were recognised as residing in the halo of the Milky Way and the abundance of the free metals was found to be between 1/3 and 1/10 of Solar (see also discussion by Blades et al., 1988, and by de Boer, Morras & Bajaja, 1990). Since then spectra with absorption of gas at velocities pertaining to the halo of the Milky Way have been obtained toward QSO's, extragalactic supernovae (see papers cited in Vladilo et al., 1994, or in Ho & Filippenko, 1995), and high brightness knots in galaxies (see e.g., Morton & Blades 1986; Robertson et al. 1991, Bowen, Blades & Pettini 1995, and many others).

All the extragalactic cases show that the metal content of the clouds in the Milky Way halo is large enough to lead to clear detections in their absorption lines. Although the exact number for the metal content is not well defined yet, one can safely say that $[M/H] \approx -0.3$ to -1.0. In this no allowance is made for the possibility that metals are locked into dust. However, since the extinction on the lines of sight is normally small and since little evidence is found for the presence of large amounts of dust, one can use the given range as a good working hypothesis for further investigations. Yet, Ca may be severely depleted (Wakker et al. 1996).

One limitation of halo gas absorption studies has been uncovered in radio synthesis investigations of isolated halo clouds. When looking at the overall sky maps of high-velocity halo gas at high galactic latitudes (see Wakker 1991) it is clear that gas complexes have a rather patchy and filamentary structure, on scales of $0.5°$. In addition, Wakker & Schwarz (1991) looked at some isolated halo clouds and found their structure on arcmin scale to be of the same nature as that found for large scale clouds in the surveys: they are quite filamentary.

It was then pointed out, that there is the risk of selecting a target for absorption line studies in the exact direction of an unrecognised gap in the cloud and thus that a non-detection of absorption need not mean that the cloud is behind the star. On the other hand, the gaps in clouds were found in radio synthesis maps in which large scale structures are suppressed. Single beam observations of the isolated clouds indicated that indeed smoothly distributed H I might be missed and thus that 'gaps' cannot be truly empty and that the column densities in those are normally still large enough to produce detectable absorption lines. At any rate, the extragalactic objects observed (see Sect. 4) prove that clouds do show absorption indeed.

A proper comparison of absorption with emission spectra is essential in the investigation of halo clouds. It is clear that the narrowest 21-cm radio telescope beam is needed. The best single dish instrument for the purpose is the Effelsberg 100-m telescope with a $9'$ beam, otherwise radiosynthesis pencil beams give the required information.

4   Pertinent Data

Due to slow observational progress data on absorption toward halo objects is rather scanty and scattered (a few spectra of a few stars at the time; the unattractiveness of publishing just non-detections). We will exclude here details of nearby molecular clouds at high galactic latitudes and low velocities such as discussed by Penprase (1992).

Elise Albert and collaborators used mostly the Ti II and Ca II lines (Albert 1983; Albert et al. 1993) toward high latitude stars in combination with closely alligned extragalctic objects. The stars are scattered over the sky and spectra show in several cases absorption with velocities of $3v3 \approx 50$ km s^{-1}.

A substantial body of IUE-data was presented by Danly (1989), further expanded by Danly et al. (1992). Here the stars are scattered over the sky, mostly selected for their UV brightness and possibly large distance. Absorptions are seen with velocities down to $v \approx -120$ km s^{-1}.

Several studies are available investigating the distances of individual high-velocity clouds. They are presented in Sect. 4.2.

Galactic globular clusters stars may provide a dense grid of background light sources. The group in Belfast pursued the investigation of interstellar Ca II and Na I absorption lines toward such stars and to foreground stars in the same direction. One particular result from the visual data is that the foreground galactic gas has a very fine structure on sub parsec scale, a result confirming earlier works (e.g. Molaro et al. 1993 on galactic gas using stars around SN 1987A as background light sources). Because of the relevance for the study of fine structure in clouds a listing of the cluster papers is given in Table 2.

Table 2: Results from studies toward globular clusters

Cluster	l, b	z	data	$v(\,\text{km s}^{-1})$	z-limit (kpc)	ref.
M 3	42, +79	8.7	IUE	≈ -80	<8.7	dBS84
M 13	59, +41	4.1	IUE	≈ -80	<4.1	dBS83
			vis.	≈ -80	>0.2	B+
M 10	15, +23	1.7	vis.		only local gas	KBK
M 4	351, +16	0.6	vis.		only local gas	L+
M 55	9, −23	1.8	vis.	−21, +33, +55	$0.7 < z < 1.8$	LBK

B+= Bates et al. (1995); BC= Bates & Catney (1991); dBS83= de Boer & Savage (1983); dBS84= de Boer & Savage (1984); KBK= Kennedy, Bates & Kemp (1996); L+= Lyons et al. (1995); LBK= Lyons, Bates & Kemp (1994).

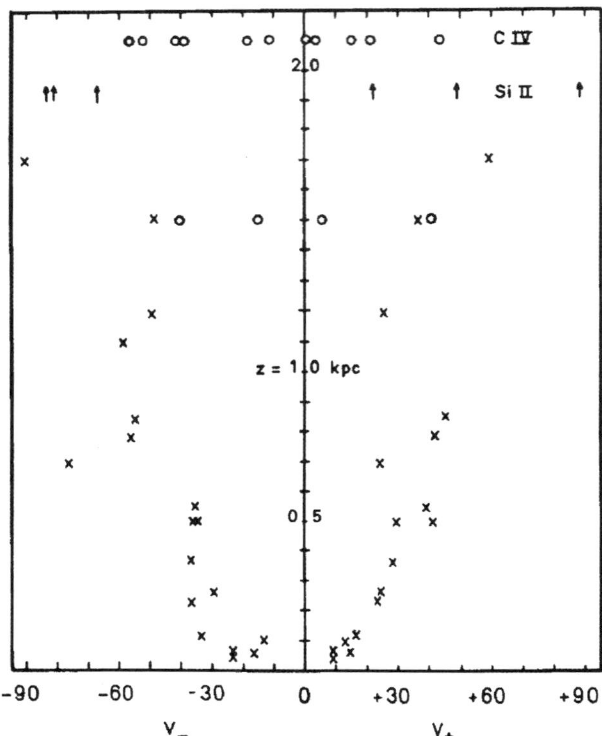

Figure 1: For stars with $|b| > 50°$ from Danly et al. (1992) the velocity limits v_- and v_+ (in $\mathrm{km\,s^{-1}}$) of Si II absorption are plotted ($\times$) against the z-distance in kpc of the bakground star. For stars beyond 2 kpc the limits are given with up-arrows. Because of the latitude limit, these data are essentially free of effects of galactic rotation. Note that close to the disk the velocity limits are small, that they increase with z, but approach limiting values of -60 and $+50$ $\mathrm{km\,s^{-1}}$ near $z= 1$ kpc. The velocity limits for detections of C IV are added ($\circ$) and data for stars beyond 2 kpc are collected at the top. No detections of C IV are present at $z < 1.2$ kpc, even in spectra where Si II is seen. Apparently, C IV exists only beyond $z= 1$ kpc.

4.1 Overall gas

Looking into the most complete sample of UV spectra of Danly et al. (1992) combined with the data of Albert et al. (1993) one can make an analysis of the systematics of the absorption. Danly et al. show that in most directions broad H I 21-cm emission is present of which no absorption complement can be found. Also, the absorption is normally at moderate velocities, only in a few cases exceeding 50 $\mathrm{km\,s^{-1}}$ (higher velocities in their Table are understood as either due to stellar lines or due to low latitude rotating disk gas).

There are 56 stars in the Danly et al. sample (most of the Albert et al. stars were also included by Danly et al.) of which 12 have $z < 1$ kpc and 14 are at $z > 1$ kpc. However, since lines of sight at low latitudes go through substantial portions of disturbing disk gas, the stars at $|b| > 20°$ are most relevant; there are 22 of these. Also, effects of galactic rotation may substantially influence the actual velocities and Danly et al. have investigated this to some extent.

To quantify how the gas behaves in the first 1 kpc perpendicular to the disk I have taken only the stars in a cone with $|b| > 50°$ from Danly et al. (1992). In this way one samples the gas on the line of sight into the nearby halo without rotation effects. That sub-sample contains 15 stars (11 to the north, 4 to the south). Fig. 1 shows the velocity range of the Si II absorption. For nearby stars the limiting velocities are small (≈ 20 $\mathrm{km\,s^{-1}}$), for more distant stars they are larger. With increasing z they approach limits of -60 and $+50$ $\mathrm{km\,s^{-1}}$. This width may be compared with Doppler widths of hot gas: in 10^6 K gas Si atoms would have a velocity distribution with width of 31 $\mathrm{km\,s^{-1}}$, for 10^5 K it is 10 $\mathrm{km\,s^{-1}}$. The observed width may therefore point at high kinetic temperatures in the gas although bulk gas motions certainly will have contributed to the width, too. Alternatively, the lack of increase in velocity width and thus absorption beyond about 0.5 kpc may indicate a limit to the presence of Si II with z.

4.2 Individual Clouds

On individual clouds a number of papers has been published, as discussed below. For the names of the various complexes I refer to the maps of and discussion by Wakker & van Woerden (1991). Note that the clouds normally are somewhat fragmented, are in the form of 'chains' or even form 'complexes' and this structure normally also means differences in velocity. Therfore, for gas clouds appearing in the literature with the same general name distances may differ as well.

Table 3: Distance limits to well known clouds

Cloud		l, b^a	$v(\mathrm{km\,s^{-1}})$	z (kpc)	ref.
	A 0	133, +23	−110	>0.2	LMdB
	A I	159, +29	−180	>0.2	SCW, LMdB
	A III	148, +35	−150	>1.6	SWvW, WCR
	A IV	153, +39	−150	>0.2	SCW, LMdB
	A V	160, +40	−150	>0.3	WCR
	M I	157, +65	−94	>3.1	SWvW
	M II-III^b	183, +62	−120	$1.2 < z < 4.4$	DAK, K+95
	C	59, +41	−80	$0.2 < z < 4.1$	dBS, B+
	C I B	81, +40	−100	<0.8	C+, dB+
	C I B	81, +40	−140	>0.8	C+, dB+
	C I B	86, +44	−110	>1.6	D+
	C I A	100, +46	−120	>0.5	C+
	C I A	100, +46	−70	<0.5	C+
	C I A	100, +47	−110	>2.4	SWvW, dB+, C+
	C	113, +50	−80	<1.7	D+
	C	113, +50	−140	>1.7	D+
	C III	112, +55	−124	>0.4	SWvW
Complex L		347, +35	−100	<1.4	A+
Complex L		347, +35	−128	<1.4	A+
Draco cloud					
	HVC	90, +40	≈ -140	>0.2	L+
	mol.cld.	90, +40	−30	>0.2	L+
ACC, CS					
		167, −45	−120	>0.3	K+94, C+
MCF		270, −32	+60	$1 < z < 25$	SdB, M+
MCF		270, −32	+120	$1 < z < 25$	SdB, M+

^{a} The coordinates given are in most cases averages for several stars in that general direction; for further details see the papers cited.

^{b} Stars investigated are in the space between M II and M III.

A+= Albert et al. (1989); B+= Bates et al. (1995); C+= Centurión et al. (1994); D+= Danly et al. (1992); dB+= de Boer et al. (1994); DAK= Danly, Albert & Kuntz (1993); K+94= Kemp et al. (1994); K+95= Keenan et al. (1995); L+= Lilienthal et al. (1991); LMdB= Lilienthal, Meyerdierks & de Boer (1990); M+= Molaro et al. (1993); SCW= Songaila, Cowie & Weaver (1988); SdB= Savage & de Boer (1981); SWvW= Schwarz, Wakker & van Woerden (1995); WCR= Welsh, Craig & Roberts (1996).

Chain A: The first report on a high-velocity cloud distance came from Songaila, Cowie & Weaver (1988). However, in an in-depth study of stars in this direction, Lilienthal, Meyerdierks & de Boer (1990) showed that in part stellar absorptions had been mistaken for interstellar ones. Yet, distance limits were determined. Further data are available from Schwarz, Wakker & van Woerden (1995). For data see Table 3.

Complex C: A large complex with spatial and velocity structure covering a fair fraction of the sky between $60< l <160$ and $+40< b <+70$ has been investigated on several occasions. Visual data are available from Songaila, Cowie & Weaver (1988) [but see also Lilienthal, Meyerdierks & de Boer (1990) on stellar lines], Centurión et al. (1994), and Schwarz, Wakker & van Woerden (1995). IUE-spectra of stars in the direction of Complex C were investigated by de Boer et al. (1994) and they found that in the direction of the star PG 1648+536 gas with $v = -100$ kms^{-1} is in front, and gas with $v = -140$ kms^{-1} is behind the star. Further IUE data are available from Danly et al. (1992). Note that also the globular cluster M 13 (see above) lies in this direction.

In this large and complex structure C the distance limits refer always only to the direction of the stars observed (see Table 3).

Draco Cloud: This cloud (named D by Wakker & van Woerden) has been investigated intensively by the group of Mebold. A comprehensive discussion is given by Lilienthal et al. (1991).

Clouds M: Albert (1983) and Danly, Albert & Kuntz (1993) find a distance limit for gas in a weak bridge between M II and M III. Further data are given by Keenan et al. (1995). For data see Table 3.

Complex L: This northern cloud has $v \approx -115$ kms^{-1}. A distance limit for this cloud, $z < 1.4$ kpc, is given by Albert et al. (1989).

Complex WA: Wakker & van Woerden (1991) explain that parts WB to WD have velocities close to permitted galactic velocities. Component WA has $v \approx +147$ kms^{-1}. Some optical data in this general direction are available from Danly et al. (1992) and Albert et al. (1993), but there is no good positional match so that no distances can be given.

Complex H: This cloud is at $3b3 < 10°$ and is therefore not in the halo. Its velocity is near -200 kms^{-1}. No absorption was found in spectra of stars closer than 2 kpc (Centurión et al. 1994). This result was confirmed by Schwarz, Wakker & van Woerden (1995).

Anti-Center Clouds, Cohen Stream: Absorption due to a stream of gas near $l = 160°$, $b = -45°$ with $v \approx -120$ km s^{-1} was investigated by Kemp et al. (1994). They found no absorption in Mg II toward their most distant star at 450 pc. A little off the stream gas with $v \approx -40$ km s^{-1} was found in absorption; the distance of this gas is $z < -890$ pc. Also Centurión et al. (1994) have data in this direction. For data see Table 3.

Magellanic Cloud foreground clouds: In the spectra of almost all LMC stars absorption by halo clouds is detected. In contrast, galactic foreground stars in the same direction with distances up to 1 kpc do not show these clouds in absorption. The first UV detections (Savage & de Boer 1979) are analysed in detail by Savage & de Boer (1981) who also summarise earlier optical data. Further detailed optical investigations are availble from Molaro et al. (1993) and I refer to that paper for further literature on this topic. The spatial extent of the clouds on the sky is relatively large, so these clouds are not part of the LMC, nor stripped off the LMC (de Boer, Morras & Bajaja 1990).

Magellanic Stream: Gas in this stream is supposedly associated with the passage of the Magellanic Clouds through the galactic halo. Its distance is therefore expected to be large. Indeed, none of the invesitgations towards galactic stars showed this gas in absorption (Albert 1983, Danly et al. 1992, Centurión et al. 1994).

4.3   Highly ionized gas

Ionized gas has been investigated in spectra of halo stars by Savage & de Boer (1981), Pettini & West (1982), de Boer & Savage (1982, 1983), Savage & Massa (1985), and Danly (1989). The newest data (including the earlier results) and analyses are given by Sembach & Savage (1992) and Danly et al. (1992). Detections of C IV are few and only to distant objects.

Over the long sightlines skimming the galactic disk C IV is present but it has velocities clearly related to galactic rotation. The long sightlines make it virtually impossible to get a handle on the distances of the absorbing gas complexes. Likely the C IV found here pertains to hot gas intimitely related with Milky Way disk stuctures.

On some sight lines C IV is present as discrete absorptions (clouds), e.g. on the lines of sight to the LMC (Savage & de Boer 1981, Savage et al. 1989). Here the C IV is present at the same velocity as absorption by neutral gas, possibly indicating transitions zones between neutral clouds and yet hotter gas in the general field. No further detection related with high-velocity clouds has been reported from spectra of halo objects. However, in the spectra of SN 1993J in M 81 very clear C IV absorption is seen from individual halo gas clouds (see below).

Inspection of the Danly et al. and Sembach & Savage data shows that stars with $|b| > 50°$ show C IV only when the stars are at $z > 1.2$ kpc, see Fig. 1. Stars closer by do not exhibit C IV in absorption. In a way, this behaviour is complementary to the one of Si II shown in Fig. 1: C IV starts showing up at z distances where the Si II absorption becomes constant. And since absorptions detected are cumulative with distance we may conclude that on average Si II and C IV exist in a consecutive gas layer when looking out of the Milky Way disk.

5 Peculiar extragalactic sightlines

5.1 Line of sight to Mrk509

Sembach et al. (1995) observed Mrk509 ($l = 36°$, $b = -29°$) with the GHRS of HST. The sightline passes at 0.5° by known HVCs having $v = -260$ km s^{-1}, but there is no high-velocity gas present in H I in the sight line itself. The HST spectra show *no* absorption in Si II (1526 Å) at that velocity. However, C IV is present with strong absorption *at the velocity of the HVC*. This demonstrates the association of diffuse ionized gas with a HVC.

5.2 Line of sight to SN 1993J in M81

The bright supernova in M81 allowed to obtain very good visual (Vladilo et al. 1993) and IUE (de Boer et al. 1993) spectra. This was supplemented with H I 21-cm spectra at Effelsberg. A full account of the data is given by Vladilo et al. (1994). The spectra show absorption at galactic velocities, at galactic halo velocities of -80 km s^{-1} (also in C IV), and at velocities pertaining to M 81 near -130 km s^{-1} (*but no C IV*).

Very special is broad and strong absorption by C IV as well as by Ca II, Na I, and even Ca I and K I at $v = +110$ to $+ 150$ km s^{-1}. There is no H I 21-cm detected at that velocity. The gas seen in absorption can be explained only by a HVC so distant that it does not fill the Effelsberg beam by a large factor. It may be a HVC falling onto M 81 but it can also be an intergalactic cloud somewhere on the 3.25 Mpc line of sight.

6 Conclusions

All observational data can be summarized in the following way.

On a journey from the Sun out of the Milky Way disk we encounter over up to 1 or 1.5 kpc diffuse neutral gas rotating along with the disk but at slightly smaller velocities.

At larger distances, little extra neutral gas is encountered but on occasion we will meet neutral clouds heading toward us at great speed. These clouds are mostly of isolated nature, sometimes exist as 'chains' or as large fragmented structures. The true space velocity of these clouds is unknown, given the fact that proper motions are basically impossible to determine. However, if those clouds are seen at lower galactic latitude they clearly also partake in the rotation of the Milky Way, albeit at a lesser pace. Most of these clouds fit in the galactic fountain scenario (Shapiro & Field 1976, Bregman 1980, Kaelble, de Boer & Grewing 1985, Wakker 1990).

Well ionised gas, such as that showing $C\,IV$, is seen in several directions, but generally not toward stars with $z < 1$ kpc. Sometimes it is found in discrete absorption features but the associated neutral gas possibly indicates it is gas in transition zones towards hotter gas in the general field. Diffuse $C\,IV$ is present mostly beyond the layer showing $Si\,II$.

References

Albert C.E., 1983, ApJ **272**, 509

Albert C.E., Blades J.C., Morton D.C., Proulx M., Lockman F.J., 1989, in IAU Coll. 120, eds. G. Tenorio-Tagle et al., Lec. Notes in Physics, Springer, p. 442

Albert C.E., Blades J.C., Morton D.C., Lockman F.J., Proulx M., Ferrarese L., 1993, ApJS **88**, 81

Bates B., Shaw C.R., Kemp S.N., Keenan F.P., Davies R.D., 1995, ApJ **444**, 672

Blades J.C., Wheatley J.M., Panagia N., Grewing M., Pettini M., Wamsteker A., 1988, ApJ **334**, 308

Bowen D.V., Blades J.C., Pettini M., 1995, ApJ **448**, 634

Bregman J.N., 1980, ApJ **236**, 577

Centurión M., Vladilo G., de Boer K.S., Herbstmeier U., Schwarz U.J., 1994, A&A **292**, 261

Danly L., 1989, ApJ **342**, 785

Danly L., Albert C.E., Kuntz K.D., 1993, ApJ **416**, L29

Danly L., Lockman F.J., Meade M.R., Savage B.D., 1992, ApJS **81**, 125

de Boer K.S., Savage B.D., 1983, ApJ **265**, 210

de Boer K.S., Savage B.D., 1984, A&A **136**, L7

de Boer K.S., Morras R., Bajaja E., 1990, A&A **233**, 523

de Boer K.S., Rodriguez Pascual P., Wamsteker W., Sonneborn G., Fransson C., Bomans D.J., Kirshner R.P., 1993, A&A **280**, L15

de Boer K.S., Altan A.Z., Bomans D.J., Lilienthal D., Moehler S., van Woerden H., Schwarz U.J., Bregman J.N., 1994, A&A **286**, 925

Garcia B., 1991, A&AS **89**, 469

Ho L.C., Filippenko A.V., 1995, ApJ **444**, 165

Kaelble A., de Boer K.S., Grewing M., 1985, A&A **143**, 408

Keenan F.P., Shaw C.R., Bates B., Dufton P.L., Kemp S.N., 1995, MNRAS **272**, 599

Kennedy D.C., Bates B., Kemp S.N., 1996, A&A in press

Kemp S.N., Bates B., Dufton P.L., Keenan F.P., Montgomery A.S., 1994, MNRAS **270**, 597

Lilienthal D., de Boer K.S., 1991, A&AS **87**, 471

Lilienthal D., Meyerdierks H., de Boer K.S., 1990, A&A **240**, 487

Lilienthal D., Wennmacher A., Herbstmeier U., Mebold U., 1991, A&A **250**, 150

Lyons M.A., Bates B., Kemp S.N., 1994, A&A **286**, 535

Lyons M.A., Bates B., Kemp S.N., Davies R.D., 1995, MNRAS **277**, 113

Moehler S., Heber U., de Boer K.S., 1990, A&A **239**, 265

Molaro P., Vladilo G., Monai S., D'Odorico S., Ferler R., Vidal-Madjar A., Dennefeld M., 1993, A&A **274**, 505

Morton D.C., 1991, ApJS **77**, 119

Morton D.C., Blades J.C., 1986, MNRAS **220**, 927

Penprase B.E., 1992 ApJS **83**, 273

Pettini M., West K.M., 1982, ApJ **260**, 561

Robertson J.G., Schwarz U.J., van Woerden H., Murray J.D., Morton D.C., Hulsbosch A.N.M., 1991, MNRAS **248**, 508

Savage B.D., de Boer K.S., 1979, ApJ **230**, L77

Savage B.D., de Boer K.S., 1981, ApJ **243**, 460

Savage B.D., Massa D., 1985, ApJ **295**, L9

Savage B.D., Jenkins E.B., Joseph C.L., de Boer K.S., 1989, ApJ **345**, 393

Schmidt J.H.K., de Boer K.S., Heber U., Moehler S., 1996, A&A in prep.

Schwarz U.J., Wakker B.P., van Woerden H., 1995, A&A **302**, 364

Sembach K.R., Savage B.D., 1992, ApJS **83**, 147

Sembach K.R., Savage B.D., Lu L., Murphy E.M., 1995, ApJ **451**, 616

Shapiro P.R., Field G.B., 1976, ApJ **205**, 762

Songaila A., Cowie L.L., Weaver H., 1988, ApJ **329**, 580

Vladilo G., Centurión M., de Boer K.S., King D.L., Lipman K., Stegert J., Unger S.W., Walton N., 1993, A&A **280**, L11

Vladilo G., Centurión M., de Boer K.S., King D.L., Lipman K., Stegert J., Unger S.W., Walton N., 1994, A&A **291**, 425

Wakker B.P., 1990, Doctoral Thesis, Univ. Groningen, Ch. 5

Wakker B.P., 1991, A&A **250**, 499

Wakker B.P., Schwarz U.J., 1991, A&A **250**, 484

Wakker B.P., van Woerden H., 1991, A&A **250**, 509

Wakker B.P., van Woerden H., Schwarz U.J., Peletier R.F., Douglas N.G., 1996, A&A **306**, L 25

Welsh B.Y., Craig N., Roberts B., 1996, A&A, in press

Address of the author:

K. S. DE BOER

Sternwarte, Univ. Bonn, Auf dem Hügel 71, D-53121 Bonn, Germany

Diffuse galactic soft X-ray radiation

Jürgen Kerp

The discovery of the soft X-ray background (SXRB) radiation by Bowyer et al. (1968) was the starting point of an extensive debate about its origin, whether it belongs to the local interstellar medium or to the galactic halo. Now we can give an answer, it belongs to both! A recent review on the Local Hot Bubble (LHB) and its X-ray emission can be found in Snowden (1996). Therefore this topic will not further be discussed in this paper. I would like to focus on the diffuse soft X-ray emission originating beyond the bulk of the galactic neutral gas, which is attributed to the hot galactic corona.

1   The LHB and Draco's shadow

The analysis of the very first SXRB observations revealed an apparent negative correlation between the soft X-ray intensity, $E \leq 0.28\,\text{keV}$, and the galactic neutral hydrogen distribution (Bowyer et al. 1968). This observational result suggested that the X-rays have to originate beyond the galactic hydrogen layers. The negative correlation was interpreted in terms of photoelectric absorption. Moreover, the observation seemed to confirm the prediction of Spitzer (1956) of a hot galactic corona which confines the neutral clouds high above the galactic disk.

Serious quantitative problems associated with the evaluation of the soft X-ray radiation transfer equation suggested that this simple galactic X-ray halo model is only a rough approximation of the "real" SXRB. Moreover, other observations indicate that the very soft X-rays (Be-band $E \in [0.078; 0.111]\,\text{keV}$, B-band $E \in [0.13; 0.188]\,\text{keV}$) reveal an apparently constant intensity ratio across large areas of the sky (Juda et al. 1991). This is remarkable because both band-averaged photoelectric absorption cross sections differ by a factor of five. A thin neutral cloud of only $N_{\text{HI}} \simeq 1 \cdot 10^{19}\,\text{cm}^{-2}$, located on the line of sight, would significantly change this intensity ratio. These facts led to an increasing popularity of the LHB model (Snowden et al. 1990). It offered the best approximation to the SXRB observations with only the mean free path length through the X-ray plasma as parameter.

The discovery of an "X-ray shadow" associated with the Draco cloud by Snowden et al. (1991) demonstrated two points: The power of the fast imaging

X-ray optics of *ROSAT* and, more important, the existence of an extended X-ray emitting component beyond the Draco nebula. Follow-up *ROSAT* X-ray observations towards the Draco nebula revealed that the entire cloud is detectable as an absorption feature against a distant soft X-ray source. Burrows & Mendenhall (1991, 1993) suggested that this distant X-ray source is a galactic hot spot and the LHB is still the dominant source of diffuse soft X-ray radiation. From these first detections of a distant diffuse soft X-ray source beyond the Draco cloud it is still a long way to an X-ray emitting galactic corona.

2   The two-component SXRB model

Burrows & Mendenhall (1993) mentioned the fact that the "ON" Draco positions, which denote the regions of the strongest attenuation of the SXRB radiation by the neutral gas of the Draco cloud, appear to be "softer" than the "OFF" positions. This result is a key to understand most of the present view of the SXRB.

Kerp (1994) analysed two pointed *ROSAT* PSPC observations towards the Draco cloud which have a slight positional overlap (Fig. 1). One PSPC pointing is centred on the "nose" of the Draco cloud. The other pointed observation covers the HI filament LVC 86+38−2. It is important to recall that the Draco cloud is located at a distance greater than 300 pc (Lilienthal et al. 1991), while LVC 86+38−2 is embedded within the LHB gas (Wennmacher et al. 1992, Kerp et al. 1993). This very special constellation gives us the opportunity to overcome all calibration uncertainties of the soft X-ray observations by the alignment of the absolute fluxes in the overlapping *ROSAT* fields of view. Moreover, we can solve the soft X-ray radiation transfer equation along a path length of several 100 pc through the galactic interstellar medium. The main results are: First, the theoretical photoelectric absorption cross section is in agreement with the amount of absorbing neutral matter derived from stray-radiation corrected HI 21-cm line measurements. Secondly, spectral investigations suggested that the LHB plasma is significantly cooler ($T_{\mathrm{LHB}} \simeq 10^{5.8}$ K) than the distant X-ray plasma ($T_{\mathrm{distant}} \simeq 10^{6.3}$ K). This temperature difference is of utmost importance for the understanding of the SXRB. Due to the strong energy dependence of the photoelectric absorption cross section (Morrison & McCammon 1983) a soft X-ray plasma radiation is becoming "harder" if it penetrates through an X-ray absorbing medium. This suggests that towards the areas of the highest X-ray absorbing column density of the Draco cloud the X-ray spectrum has

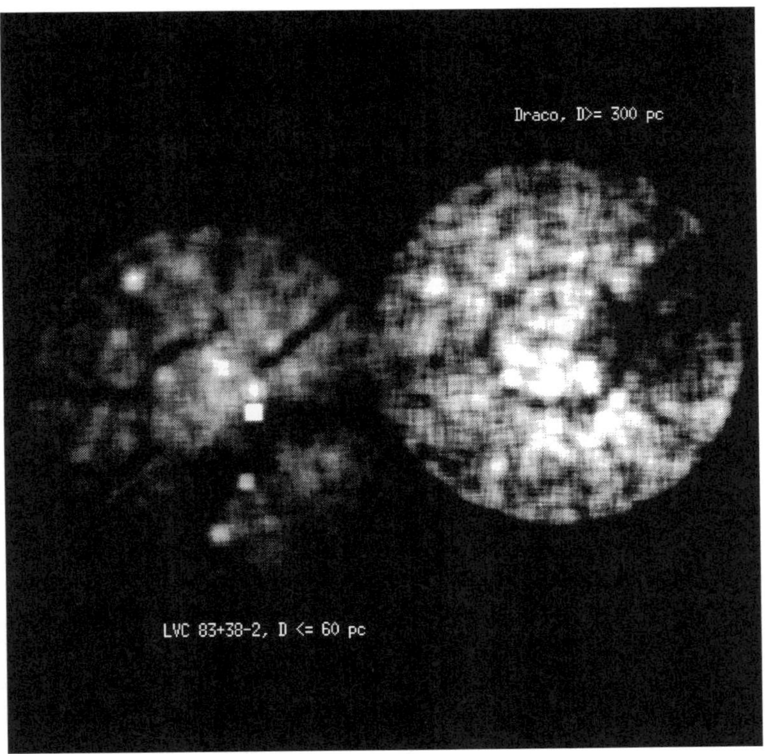

Figure 1: The SXRB towards the "nose" of the Draco cloud (right-hand side). Two pointed *ROSAT* PSPC observations are displayed. The effective field of view is indicated by the circular shape of both maps. On the left-hand side the X-ray shadow of LVC 86+38−2 is shown. This cloud is located within the LHB and closer than 60 ± 20 pc to the sun. On the right-hand side the X-ray absorption feature of the Draco nebula is presented. The Draco cloud is located at a distance greater than 300 pc from the sun. By analysing these combined PSPC pointings Kerp (1994) presented evidence for a two-temperature plasma model and showed that the soft X-ray radiation transport can be solved accurately.

to be <u>harder</u> than at the "OFF" shadow positions. The contrary behaviour is observed by Burrows & Mendenhall (1993)! The disagreement between expectation and observation disappears if we take into account that the entire SXRB radiation is a superposition of two plasmas at different temperatures and the softer (cooler) one is associated with the LHB. Then, at the deepest parts of the X-ray shadow the harder (hotter), distant plasma is heavily blocked. The contribution of the softer plasma dominates the observed X-ray radiation. At the "OFF" shadow positions we observe a mixture of both plasmas which appears to be harder.

Moreover, this simple two-temperature plasma model of the SXRB can easily explain the constant B-band and Be-band intensity ratio. The band averaged cross sections are $\sigma_{\mathrm{Be}} = 10 \cdot 10^{-20}\,\mathrm{cm}^2\,\mathrm{HI}^{-1}$ and $\sigma_{\mathrm{B}} = 1.7 \cdot 10^{-20}\,\mathrm{cm}^2\,\mathrm{HI}^{-1}$. A typical HI 21-cm column density at the northern high-galactic-latitude sky is about $1\text{--}2 \cdot 10^{20}\,\mathrm{cm}^{-2}$. This amount of neutral hydrogen attenuates the distant (hotter) X-ray plasma radiation within the Be-band $\simeq 100\%$ (exactly -65 dB) and within the B-band by about 92%. For this reason, within the statistical uncertainties of the observations of Bloch et al. (1986) and Juda et al. (1991), the contribution of the distant X-ray source is undetectable. In conclusion, a simple two-component plasma model, consisting of a cooler LHB and a hotter distant X-ray plasma, is in quantitative agreement with the observations.

3   Evidence for a galactic X-ray halo

Up to now we restricted the discussion towards the northern galactic sky close to the Draco cloud. Evidence for a galactic X-ray corona, enclosing the entire galactic disk, is not presented. Recently Sidher et al. (1996) presented the analysis of several pointed *ROSAT* PSPC observations along a slice centred at $l \sim 90°$ galactic longitude, while the galactic latitude is scanned from the galactic north to south pole. They derive exactly the same temperatures for the LHB and the distant X-ray source along this slice as Kerp (1994). In addition, the detailed analyses of *ROSAT* all-sky survey data by Snowden et al. (1994), Herbstmeier et al. (1995), and Kerp et al. (1996) suggest the existence of a galactic X-ray corona.

As suggested by Snowden et al. (1994) the intensity distribution of the galactic X-ray corona is probably a superposition of scattered hot-spots and reveals a patchy intensity pattern. This scenario also implies that the soft X-ray radiation transfer can only be accurately solved within an individual X-ray patch.

A unique solution across the entire galactic sky should be impossible.

3.1   The X-ray absorbing ISM

At this point we leave for a moment the topic of X-ray emission and focus on the neutral or the dominant X-ray absorbing phase of the galactic interstellar medium. The patchiness of the attenuated soft X-ray emission is strongly related to the distribution and the quantity of the absorbing matter. The analysis of pre-*ROSAT* X-ray observations suggested that the measured HI 21-cm column density was not in quantitative agreement with the corresponding photoelectric absorption cross section. In order to claim that the soft X-ray radiation transport equation cannot be solved quantitatively one has to take care of two important points: (i) the angular resolution of the X-ray detectors was low, and so far it was impossible to exclude unresolved X-ray sources from the SXRB data, (ii) the HI data basis used does not sufficiently sample the sky and has not been corrected for the contribution of stray-radiation (Kalberla et al. 1980, Hartmann et al. 1996). *IRAS* 100-μm data offer a high angular resolution and a complete coverage of the sky but are not adequate for the quantitative evaluation of the X-ray radiation transport. This is caused by the unknown absolute infrared background level of the *IRAS* 100-μm maps. Accordingly, the diffusely distributed galactic neutral hydrogen layer (Dickey & Lockman 1990, and Lockman, these proceedings) is not correctly represented by the 100-μm maps. The attenuation of soft X-rays can be mainly attributed to this neutral hydrogen layer. Figure 2 shows the attenuation of the soft X-ray emission by the galactic neutral hydrogen. The figure consists of four panels where in each panel the count rate, measured at a specific sky position, is drawn at the $N_{\rm H}$-value representing different velocity intervals relative to the local-standard-of-rest. The solid line represents the two-component SXRB plasma model of Kerp (1994). The X-ray data points are extracted from a pointed *ROSAT* PSPC observation located roughly 5° apart from Draco's nose. In panel A, the column density of the diffuse galactic hydrogen layer determines the position of the X-ray count rate markers in the scatter diagram. Panel B shows the arrangement of the count rate markers if we include the intermediate-velocity clouds (IVCs) as soft X-ray absorbers. The total amount of absorbing matter increases with each panel and can be traced by the motion of an individual count rate marker towards the right-hand side of the scatter diagram. The third panel (C) includes as X-ray absorber the column density of the high-velocity clouds (HVCs). Accounting for a reduced metallicity of

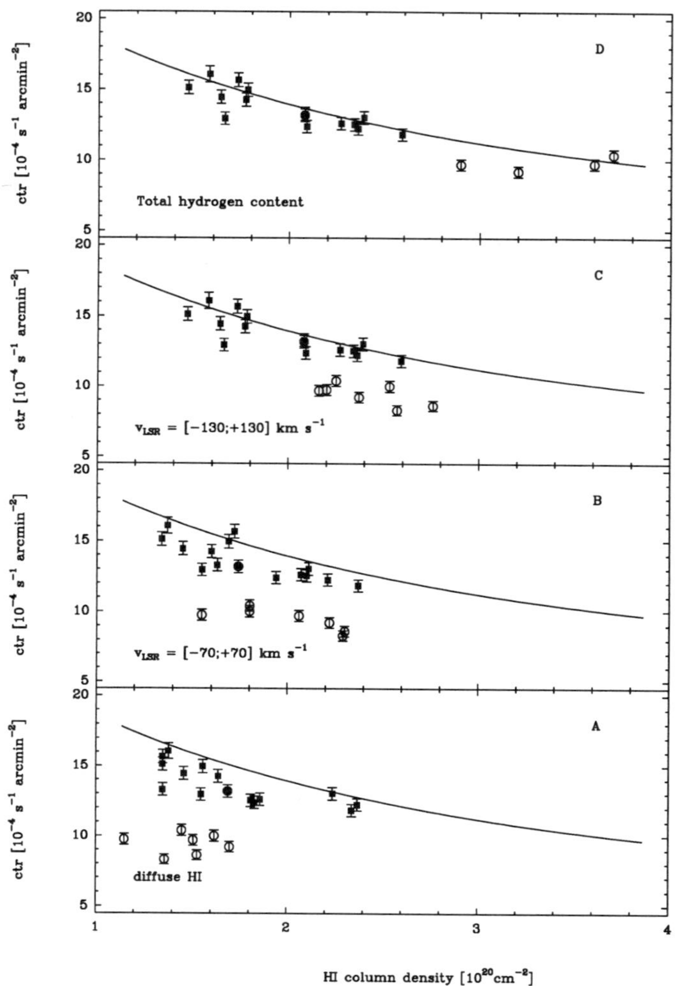

Figure 2: Scatter diagrams showing the SXRB count rate versus absorbing HI column density are presented. The velocity interval included in the determination of the amount of absorbing matter is noted in each panel. Depending on the amount of absorbing column density, measured towards the individual sky positions, the count rate is constant while the count rate markers are moving to the right-hand side within the different panels. The errorbars indicate the 1-σ uncertainty level of the X-ray data points. The diagram will be discussed in the text below.

about 30% (de Boer et al. 1987) most of the count rate markers fit the curve quite well. Finally, the open circles indicate an area where a deep X-ray shadow is associated with strong *IRAS* 100-μm emission, suggesting the presence of molecular hydrogen. In panel D the amount of molecular hydrogen is evaluated by the *IRAS* data and added to the amount of atomic hydrogen derived from H I 21-cm line observations. This leads to a good alignment of the residual data points close to the solid line. This result indicates the importance of H I 21-cm line data when analysing SXRB observations. The attenuation of soft X-rays by the galactic neutral hydrogen layer determines the absolute count rate level (panel A). Individual clouds, like IVCs and HVCs, only modulate the general X-ray absorption caused by the galactic neutral hydrogen layer. *IRAS* 100-μm data can only serve as a supplement to the H I 21-cm line observations. If we then take into consideration that stray-radiation contributes up to 50% of the detected H I 21-cm line emission it becomes obvious that one has to correct for its contribution by evaluating the SXRB radiation transport.
The patchiness of the SXRB intensity distribution can be a result caused artificially by an insufficient quality of the H I 21-cm line data. *IRAS* 100-μm data alone is not sufficient to evaluate the soft X-ray radiation transport because of the unknown absolute 100-μm background intensity level.
We finally need to address the contribution of the ionized hydrogen layer (Reynolds 1991) to the absorption of the SXRB radiation. If we take into account that one half of the helium is in a single-ionization state (Domgörgen & Mathis 1994) we can derive an effective absorbing column density of $N_{\mathrm{H}+} \sim 5 \cdot 10^{19}$ cm^{-2} at northern high-galactic-latitudes. Its contribution lies between one half and one third of that of the atomic neutral hydrogen layer. Neglecting its absorption leads to a lower brightness estimate of the galactic halo X-ray plasma. Assuming the positional dependence of H$^+$ on a *cosec(b)*-law (Reynolds 1991), the difference in attenuation at high-galactic-latitudes is at most $\Delta N_{\mathrm{H}+} \sim 4 \cdot 10^{19}$ cm^{-2} across $\Delta b = 50°$. For instance, analysing a deep pointed *ROSAT* PSPC observation of $T = 10$ ksec integration time and an observed SXRB count rate of about $I = 10 \cdot 10^{-4}$ cts s^{-1} arcmin^{-2}, yields a 1-σ uncertainty of about $\Delta I = 4 \cdot 10^{-5}$ cts s^{-1} arcmin^{-2}. This can be transformed into an uncertainty of the amount of absorbing matter of $\Delta N_{\mathrm{HI}}(1\sigma) = 1.8 \cdot 10^{19}$ cm^{-2}. Only deep PSPC observations towards very different lines of sight can clearly reveal the absorption of the H$^+$ layer. Therefore, we have to keep in mind that the galactic halo plasma is most probably brighter than calculated on the basis of H I data.

3.2   Validity of the plasma temperatures and emissivities

Finally, we have to discuss one of the most important questions: Do we really observe a galactic X-ray emitting plasma in collision ionized equilibrium? This question cannot correctly be answered yet. Rochia et al. (1984) found evidence for thermal emission lines of OVII–OVIII from spectral investigations of the high-latitude sky. Also Hasinger (1991) reported the detection of a line emission bump at $0.6\,\mathrm{keV}$ in deep pointed *ROSAT* data. The diffuse X-ray spectrometer (DXS) reveals emission lines within the LHB spectrum (Sanders et al. 1996). Hence, we most probably observe thermal plasma radiation. Moreover the DXS data suggest (Sanders et al. 1996) that the LHB plasma is <u>not</u> in collision ionized equilibrium. According to these results we have to keep in mind that the derived temperatures and emission measures of the galactic X-ray plasmas are not precisely fixed values. Other plasma models have to be taken into account like the adiabatic cooling plasma model of Breitschwerdt & Schmutzler (1994, these proceedings). Spectral observations of the SXRB at high latitudes are strongly required.

4   SXRB emission and high-velocity clouds

At a moderate angular resolution level of $\sim 30'$ the dense molecular cores of the high-galactic-latitude clouds have only a low area filling factor. As a first approximation we can assume that the soft X-ray radiation is dominantly attenuated by the neutral matter traced by the neutral atomic hydrogen. We can further assume that the chemical composition of the absorbing matter is close to solar abundances. These assumptions, however, are no oversimplifications because of the above mentioned statistical limitations of the X-ray data.
By combining two unique data sets, the *ROSAT* all-sky survey (Snowden et al. 1995) and the Leiden/Dwingeloo HI 21-cm line survey (Hartmann & Burton 1996) we obtain a first view of the soft X-ray sky. In its present form the *ROSAT* all-sky survey offers an angular resolution of $12'$; it has been cleaned for the contribution of non-cosmic backgrounds and of unresolved sources (Snowden et al. 1995). The Leiden/Dwingeloo HI 21-cm sky-survey covers the entire sky north of $\mathrm{DEC} = -30°$. It is the first HI 21-cm line all-sky survey which is corrected for stray-radiation (Hartmann et al. 1996).

First results of the correlation analysis of the *ROSAT* all-sky survey and the Leiden/Dwingeloo data are presented by Herbstmeier et al. (1995) and Kerp et al. (1996). Both papers are focussed on high-velocity cloud complexes at the northern galactic sky. They showed that the soft X-ray radiation transfer can be evaluated accurately. Evidence for a smooth SXRB intensity distribution is also presented. A smooth intensity distribution denotes that at $l \sim 90°$ I_{Halo} is about $18 \cdot 10^{-4}\,\mathrm{cts\,s^{-1}\,arcmin^{-2}}$, and at $l \sim 160°$ I_{Halo} is about $11 \cdot 10^{-4}\,\mathrm{cts\,s^{-1}\,arcmin^{-2}}$, while $b \sim 40°\!-\!50°$. A variation of the galactic halo plasma emissivity by about a factor of two is obvious.

In addition, Herbstmeier et al. (1995) demonstrated that (i) large areas of HVC complex M are associated with soft X-ray shadowing, and (ii) close to very steep intensity gradients in the HI 21-cm column density distribution, enhanced soft X-ray emission is detected (see also Herbstmeier, these proceedings). The detection of soft X-ray shadows caused by HVCs is of special interest because HVCs are objects of the distant galactic halo. Fortunately, Danly et al. (1993) could determine an upper distant limit of HVC complex M of $|z| \leq 4.4\,\mathrm{kpc}$. So, we obtain a lower distance limit of the galactic X-ray halo plasma of about $4\,\mathrm{kpc}$.

Herbstmeier et al. (1995) found indications that the HVCs M I and M III are closely located to soft X-ray enhancements. This suggests that also the HVCs next to the X-ray emission of the LHB and the galactic corona are probably soft X-ray emitters. Kerp et al. (1996) verified that HVCs indeed are soft X-ray emitters. They show the existence of soft X-ray enhancements across large areas of the sky (several degrees) close to the boundaries of the HVCs of complex C (first detection see Hirth et al. 1985). Moreover, Kerp et al. (1996) found a positional correlation between the soft X-ray enhancements and the so-called "velocity bridges" in the HI 21-cm line radiation. Examples of velocity bridges are published by Pietz et al. (1996) and by Kalberla et al. (these proceedings). This positional correlation between soft X-ray enhancements and indicators of cloud deceleration is up to now best interpreted by an impact of HVCs onto the galactic gas layers. It is important to note that this interaction scenario suggests that the infalling HVCs being objects of the nearby galactic halo. In addition, the velocity bridges indicate that the HVCs are located within a z-height range of 1–2 kpc. de Boer (these proceedings) presents a complete list of optical studies of absorption line measurements towards HVCs. Close to the location of the lower galactic longitude and latitude ends of the HVC complex C, de Boer (these proceedings) reports the detection of an upper distance limit of HVC gas within this distance boundary.

As suggested by Kerp et al. (1994, 1995) magnetic reconnection seems to be the most likely mechanism to transform kinetic energy into thermal radiation. As shown by Raymond (1992), Kahn & Brett (1993), and Kahn (these proceedings) the reconnection of magnetic field lines can induce thermal energy into the magnetized ISM close to the HVC. Since the temperature of the lower galactic halo is at least $T \sim 10^5$ K (Sembach & Savage 1992) the sound speed is in the order of the radial velocity of the HVCs. Consequently, it is very difficult to generate a shock. Moreover, if shock heating occurs, the temperatures obtained by transforming the entire HVC motion into heat is $T_{\text{heat}} \leq 10^6$ K. Especially the detection of enhanced $ROSAT$ $\frac{3}{4}$ keV radiation (Kerp et al. 1994) suggests the existence of a several million degree plasma. We need a very hard shock with relative velocity differences of about 300–400 km s^{-1}. This does not exclude the probability of shock heating by the collision of HVCs with a fast galactic wind (Völk 1991, Breitschwerdt, priv. com.). The detection of a strong magnetic field of $B_{||} = 25 \pm 6 \,\mu$G in HVC 90.5+42.5–130 (Kerp & Güsten, in prep.) biases the theoretical interpretation towards the magnetic reconnection scenario.

Further observational investigations of the HI 21-cm line data will yield constraints to the theories. If the impact scenario proves right we should detect a distortion in the velocity field of the galactic interstellar medium. Then we know whether the collision scenario of a fast galactic wind with the HVCs or the impact hypothesis is valid. Moreover, the galactic wind scenario can be proved by the search for dust emission close to the rims of HVCs, because heavy elements have to be blown into the galactic halo by the wind emerging from the galactic disk. In both interaction scenarios we anticipate radiation of ionized hydrogen close to the interaction areas.

5   Summary

The SXRB radiation $E \leq 0.28$ keV is subdivided into two major sources, the LHB and a thermal plasma localized within the galactic halo. The galactic halo plasma does not have a constant emissivity across the entire sky. Moreover, the analysis of the SXRB radiation transport suggests that the galactic halo emissivity changes by a factor of roughly five across several degrees in galactic coordinates. We can exclude the hypothesis that the X-ray halo intensity distribution is caused by the superposition of individual hot spots. The evaluation of the soft X-ray radiation-transport equation is feasible by using the correlation analysis of *ROSAT* PSPC data plus stray-radiation corrected HI 21-cm line data. *IRAS* 100-μm maps are by now only supplementary data to the HI 21-cm line measurements. HVCs seem to be sources of soft X-ray radiation and, in addition, modulate the SXRB intensity distribution.

References

Bloch J.J., Jahoda K., Juda M. et al., 1986, ApJ 308, L59

Bowyer C.S., Field G.B., Mack J.E., 1968, Nat, 217, 32

Breitschwerdt D., Schmutzler T., 1994, Nat 371, 774

Burrows D.N., Mendenhall J.A., 1991, Nat 351, 629

Burrows D.N., Mendenhall J.A., 1993, Adv. in Space Res. Vol. 13, No. 12, 83

Danly L., Albert C.E., Kuntz K.D., 1993, ApJ 416, L29

Dickey J.M., Lockman F.J., 1990, ARA&A 28, 215

de Boer K.S., Jura M.A., Shull S.M., 1987, in "Exploring the Universe with the IUE Satellite", Ed. Kondo Y., Reidel, p. 485

Domgörgen H., Mathis J., 1994, ApJ 428, 647

Hartmann D., Burton W.B., 1996, An Atlas of Galactic Neutral Hydrogen Emission, Cambridge University Press, under contract

Hartmann D., Kalberla P.M.W., Burton W.B., Mebold U., 1996 A&AS, in press

Hasinger G., 1991, in "X-ray emission from galactic nuclei and the cosmic X-ray background" Eds. Brinkmann W., Trümper J.E., p. 321

Herbstmeier U., Mebold U., Snowden S.L. et al., 1995 A&A, 298, 606

Hirth W., Mebold U., Müller P., 1985, A&A 153, 249

Juda M., Bloch J.J., Edwards B.C. et al., 1991, ApJ 367, 182

Kahn F.D., Brett L., 1993, MNRAS 263, 37

Kalberla P.M.W., Mebold U., Reich W., 1980, A&A 82, 275

Kerp J., Herbstmeier U., Mebold U., 1993, A&A 268, L21

Kerp J., 1994, A&A 289, 597

Kerp J., Lesch H., Mack K.-H., 1994, A&A 286, L13

Kerp J., Lesch H., Mack K.-H., Pietz J., 1995, Adv. in Space Res. Vol. 16, No. 3, 119

Kerp J., Mack K.-H., Egger R. et al., 1996, A&A, in press

Lilienthal D., Wennmacher A., Herbstmeier U., Mebold U., 1991, A&A 250, 150

Morrison R., McCammon D., 1983, ApJ 270, 119

Pietz J., Kerp J., Kalberla P.M.W. et al., 1996, A&A 308, L37

Raymond J., 1992, ApJ 384, 502

Reynolds R.J., 1991, IAU Symp. 144, Ed. Bloemen H., 67

Rochia R., Arnaud M., Blondel C. et al., 1984, A&A 130, 53

Sembach K.R., Savage B.D., 1992, ApJS 83, 147

Sanders W.T., Edgar R.J., Liedahl D.A., 1996, "Röntgenstrahlung from the Universe", Eds.: Zimmermann H.U., Trümper J.E., Yorke H., MPE–Report 263, p. 329

Sidher S.D., Sumner T.J., Quenby J.J., Gambhir M., 1996, A&A 305, 308

Snowden S.L., Cox D.P., McCammon D., Sanders W.T., 1990, ApJ 354, 131

Snowden S.L., Mebold U., Hirth W., Herbstmeier U., Schmitt J.H.M.M., 1991, Sci. 252, 1529

Snowden S.L., Hasinger G., Jahoda K. et al., 1994, ApJ 430, 601

Snowden S.L., Freyberg M.J., Schmitt J.H.M.M. et al. 1995, ApJ 454, 643

Snowden S.L., 1996, in "Röntgenstrahlung from the Universe", Eds.: Zimmermann H.U., Trümper J.E., Yorke H., MPE–Report 263, p. 299

Spitzer L., 1956, ApJ 124, 20

Völk H.J., 1991, IAU Symp. 144, Ed. Bloemen H., p. 345

Wennmacher A., Lilienthal D., Herbstmeier U., 1992, A&A 261, L9

Acknowledgements

I thank K.-H. Mack, H. Rottmann, and U. Runkel for carefully reading the manuscript. This work was supported by the Deutsche Forschungsgemeinschaft under grant-No. ME 745/17- 2.

Address of the author:

JÜRGEN KERP Radioastronomisches Institut der Universität Bonn, Auf dem Hügel 71, 53121 Bonn, Germany
and
Max-Planck-Institut für extraterrestrische Physik, Postfach 1603, 85740 Garching, Germany

The large-scale structure of the X-ray halo

M.J. Freyberg

1 Introduction

Recent progress in the understanding of the local interstellar medium has been made using data of the ROSAT PSPC All-Sky Survey (Snowden *et al.* 1995). In the soft X-ray band $(0.1 - 2.4\,\mathrm{keV})$ for the first time X-ray shadows had been found. This had established the existence of soft X-ray emission from beyond the bulk of the absorbing neutral gas of our Galaxy. Additionally, around external spiral galaxies (e.g. NGC 4631, NGC 891 – often referred to as a twin of our Galaxy) extended soft X-ray emission has been detected with temperatures $T \sim 2.8 - 3.6 \times 10^6$ K (Wang *et al.* 1995, Bregman & Pildis 1994). Does our Galaxy have a large X-ray halo as well? If it does exist, what are its physical parameters (spatial distribution of hot gas, extent, temperature, etc.)? In the following we present the first all-sky analysis of the galactic soft X-ray halo.

2 Soft X-ray background with the ROSAT PSPC

The nature of soft X-ray background (SXRB) is still unsolved (cf. Snowden (1996) and references therein): how much of the SXRB is of local origin, what is its spectrum, what fraction is made up of a galactic halo and what does the (often neglected) extragalactic background contribute. Here we adopt the following global model: the local environment consists of a hot bubble ($T \sim 10^6$ K) of unknown shape, the Galaxy is embedded in an extended X-ray halo made up of hot gas, and overlaid is a (practically) isotropic extragalactic X-ray background (eg. AGN, clusters of galaxies). Both extragalactic and halo component are absorbed by the full hydrogen column density $N_{\mathrm{H}}(l, b)$ while the emission from the local hot bubble (LHB) is absorbed by less than $N_{\mathrm{H}} \sim 10^{19}$ cm^{-2}. This model can be written as

$$I(E; l, b) = I_{0,\mathrm{LHB}}(E; l, b) \; + \; I_{0,\mathrm{halo}}(E; l, b) \times \exp\left[-\sigma_T(E, N_{\mathrm{H}})N_{\mathrm{H}}(l, b)\right]$$
$$+ \; I_{0,\mathrm{extragal}}(E) \times \exp\left[-\sigma_\alpha(E, N_{\mathrm{H}})N_{\mathrm{H}}(l, b)\right] \quad (1)$$

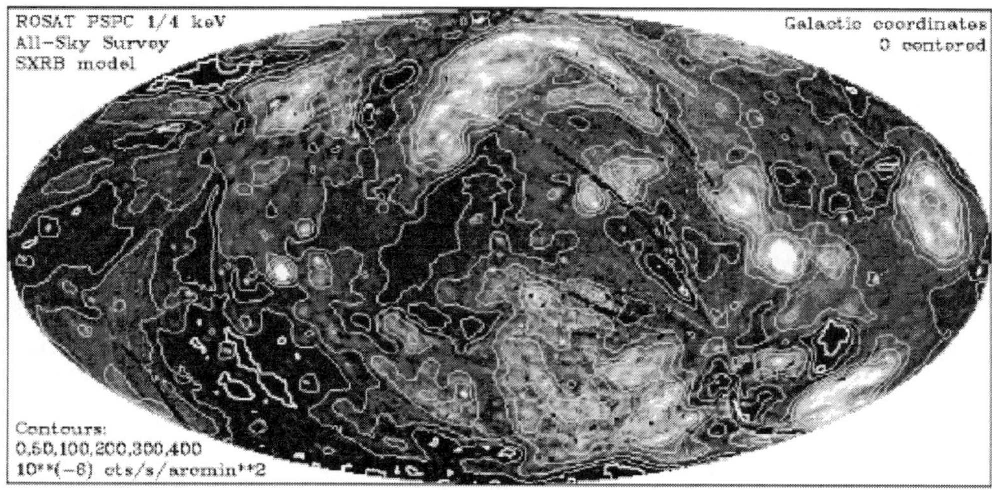

Figure 1: Difference between observation [ROSAT PSPC All-Sky Survey (C band)] and model according to Eq.(1); difference contours range from $0 - 400 \times 10^{-6}$ cts/s/arcmin^2, the zero level is indicated by thick lines. The best-fit halo has an extent of $R \sim 25$ kpc and $T \sim 3 \times 10^6$ K.

The intensities $I_0(E)$ denote the *unabsorbed* values while $I(E)$ is the spectrum incident on the X-ray telescope. Since the ROSAT PSPC has only a moderate energy resolution the data have been binned into several energy bands ("C band" with $E \sim 1/4$ keV, "M band" with $E \sim 3/4$ keV, "IJ band" with $E \sim 1.5$ keV, cf. Snowden *et al.* 1995). Band-averaged absorption cross sections $\sigma(E)$ depend on the incident spectrum (temperature T, power-law index α) as well as on N_{H}. This becomes very important if one considers regions with a large variety of column densities (especially for full-sky analysis, cf. Freyberg 1994). X-ray colours (*hardness ratios*) such as HR1 $=$ (M+IJ$-$C)/(M+IJ+C) or HR2 $=$ (M $-$ IJ)/(M $+$ IJ) are a useful probe to study spectral features of the SXRB (eg. Breitschwerdt *et al.* 1996). These hardness ratios cover a wide energy range and any modeling has to take into account the different energy dependence of the band-averaged absorption cross sections for halo and extragalactic component, $\sigma_T(E)$ and $\sigma_\alpha(E)$, respectively.

To disentangle the halo from the extragalactic background one has to search for spatial variations of the absorbed component. These spatial variations can be assigned to the halo while the constant component is assumed to be due to AGN etc. In a recent analysis (Freyberg 1994) certain symmetry assumptions for the

LHB and the halo have been used and a best-fit model for the distribution of the galactic halo has been determined.

3   Results and Discussion

Figure 1 contains a map of residuals between observations [ROSAT PSPC All-Sky Survey (C band)] and model according to Eq.(1). Best-fit model parameters were found for a spherical halo with extent $R \sim 25\,\mathrm{kpc}$, temperature $T \sim 3 \times 10^6\,\mathrm{K}$, density $n \sim 10^{-4}\,\mathrm{cm}^{-3}$ – the statistical significance for the parameters is quite poor. Prominent features that have not been included in the modeling are the *North Polar Spur*, *Vela* SNR, *Monogem Ring*, *Eridanus enhancement*. A significantly too bright halo is found only at a few places, at (l, b) $\sim (105°, -40°)$, $(l, b) \sim (160°, +45°)$ (*Lockman Hole*), $(l, b) \sim (210°, +22°)$, and $(l, b) \sim (235°, -25°)$. These regions lie generally in directions of very low N_H which indicates that the halo deviates from a pure sphere but may be flattened (however, a disk-shaped halo gave worse results). Additionally, the assumed extragalactic power law may steepen at softer energies which would lead to a weaker halo. The derived large extent is due to the weak large-scale variations of the absorbed X-ray background components.

Besides this approach also a multipole analysis using spherical harmonics has been performed to determine spatial background variations (Freyberg 1994). All those regions of the sky had been excluded that appeared to be non-typical for the soft X-ray background such as the aforementioned features as well as the galactic plane, the galactic center and anti-center regions. In the M band a significant dipole has been found in the general direction of the galactic center (note, that this region had been excluded). The dipole was very stable against further omissions of regions in the fitting procedure. This may be interpreted in terms of a large-scale temperature gradient in the *local* component like in Snowden, Schmitt & Edwards (1990). However, we prefer a *distant* origin and assign the dipole to a galactic X-ray halo. Higher multipole moments could in principle constrain the shape of such a halo. Unfortunately, quadrupole and even higher moments seem to have reached the noise level which means that these contributions are affected by small-scale fluctuations like unresolved sources, dim supernova remnants etc. The IJ band data did not show any significant dipole, all moments but the monopole were of approximately equal strength. This may be caused by the insufficient source detection algorithm that had been used to clean the background data.

We conclude, that the various X-ray observations (e.g. shadows) and studies (e.g. large-scale spatial variations of absorbed component) require the existence of a galactic soft X-ray halo. This halo consists of hot gas with $T \sim 3$ million degrees. The weak large-scale variations point towards an extended halo of $R \sim 25\,\mathrm{kpc}$. Evidence for such an extended halo ($R \sim 50\,\mathrm{kpc}$) is given by the detection of Hα emission from the Magellanic Stream (Weiner & Williams 1996). A more elaborate analysis using ROSAT PSPC All-Sky Survey data with higher spatial and spectral resolution is in preparation.

References

Bregman, J.N., Pildis, R.A., 1994, ApJ 420, 570

Breitschwerdt, D., Egger, R., Freyberg, M.J., Frisch, P.C., Vallerga, J.V., 1996, rapporteur paper of WG2 at ISSI workshop *The Heliosphere in the Local Interstellar Medium* in Bern

Freyberg, M.J., 1994, Dissertation, LMU München

Snowden, S.L., Schmitt, J.H.M.M., Edwards, B.C., 1990, ApJ 364, 118

Snowden, S.L., Freyberg, M.J., Plucinsky, P.P., Schmitt, J.H.M.M., Trümper, J., Voges, W., Edgar, R.J., McCammon, D., Sanders, W.T., 1995, ApJ 454, 643

Snowden, S.L., 1996, in Proc. of the International Conference on *Röntgenstrahlung from the Universe* in Würzburg, eds. H.U. Zimmermann, J.E. Trümper & H. Yorke, MPE Report 263

Wang, Q.D., Walterbos, R.A.M., Steakley, M.F., Norman, C.A., Braun, R., 1994, ApJ 420, 570

Weiner, B.J., Williams, T.B., 1996, AJ 111, 1156

Address of the author:

M.J. FREYBERG, Max-Planck-Institut für extraterrestrische Physik, Postfach 1603, D-85740 Garching, Germany. Email: mjf@mpe-garching.mpg.de

The Rôle
of the Magnetic Field

Professor Dr. Richard Wielebinski, Direktor of the Max Planck Institut für Radioastronomie in Bonn, whose 60th birthday was celebrated during the workshop.

Magnetic Field Effects in Galactic Fountains

F.D. Kahn

1 Introduction

The Galactic fountain carries interstellar material from the disk to a height
of some 3 kpc above the plane. The gas involved is originally shocked in the
inner parts of supernova remnants (SNRs) and heated so strongly that it still
has not cooled at late times, when the remnant has expanded enough to collide
with neighbouring SNRs. The interstellar magnetic field is carried along with
the gas and eventually the lines of force from the different remnants disconnect
from the field in the disk, and link up to form a large scale field which is
carried upwards in the fountain. The upward flow has insufficient energy to
escape from the Galaxy and material returns to the disk, most probably in the
form of filaments or compact clouds. The downflow therefore does not return
all the magnetic flux that was carried up in the fountain, and the net effect is
to produce a magnetic pump which removes flux from the Galactic disk. The
compact clouds themselves carry an internal field but no net flux. There must
be field reversals within such clouds, and it seems probable that significant
amounts of energy are released there. It is argued that some of the plasma can
then be heated to a high enough temperature to produce the X-ray emission
that has been reported in some HVC's.

2 The Energy Supply for the Upflow

About 10^6 years after a supernova explosion the remnant has grown to a radius
$r_s = 70$ pc and contains some $4 \times 10^4 M_\odot$ of gas, of which about 0.5 per cent
by mass is still hot. The bulk of the mass has cooled off and is confined to a
small fraction of the volume, where the density is high and the temperature
low. The major part of the interior volume is filled with tenuous gas at low
density (about 10^{-26} gm cm^{-3}) and high temperature (about 10^6K). The cold
and dense gas stays behind in the disk; the surviving hot gas merges with the

hot gas from other comtemporary remnants nearby and then flows off into the Galactic halo, being too hot to be gravitationally confined to the disk. On the other hand, this gas does not in general have thermal energy enough to escape from the Galaxy as a whole, and rises to a maximum height of some 3 kpc and then falls back.

In the upward flow the thermal energy of the gas is partly converted to kinetic energy and partly lost by radiation. Beyond a height of about 1 kpc the gas has cooled off completely, and the flow is ballistic. According to the simplest model the gas returns to the disk in a dense and mainly cool layer, bounded at its base by a shock which sweeps up more fountain gas as time goes on. Eventually all the material returns to the disk, after an excursion lasting on average some 10^8 years. This model has been discussed before several times (Kahn, 1981, 1994, Kahn and Brett, 1993) and describes the flow pattern most appropriate to our region of the Galaxy.

There are other possibilities, but they probably apply elsewhere in this Galaxy, or in another part of the Universe. If there are a large enough number of supernova explosions and if they occur close enough together in space and time then a chimney can be opened up through the overlying matter in a galactic disk, and a more concentrated upflow is created (Heiles 1989, Norman 1991, Normandeau, Taylor and Dewdney 1996). Alternatively, it has been suggested that the cosmic ray gas provides a substantial fraction of the energy content of the hot interstellar medium (Zirakashvili et al 1996). The cosmic rays cannot lose energy by radiation and the flowing ISM consequently stays buoyant longer and may be expelled from the Galaxy altogether. The flow then becomes a wind rather than a fountain.

To return to the present model, only some $200 M_\odot$ of the remnant is still hot at the end of Phase III, after time $t = 3 \times 10^{13} \mathrm{s} = 10^6$ year. This gas must originally have occupied a volume of $2 \times 10^{59} \mathrm{cm}^3$, given that the average interstellar density is $\rho_o = 2 \times 10^{-24}$ gm cm^{-3}. The hot gas that fills the interior of the remnant at the end of Phase III must have initially been contained in a sphere of radius $R_o = 3.7 \times 10^{19} \mathrm{cm} = 12$ pc. An important step is now to discover what happened to the magnetic field during the expansion.

It is simplest to assume that the supernova explosion occurred in a region where the magnetic field was uniform, with strength $B_o = 3 \times 10^{-6}$ gauss. By the end of Phase III, the gas of the hot bubble has expanded by a factor 6 in linear measure. The field strength B has therefore dropped by a factor $6^2 = 36$, and the gas density ρ by a factor $6^3 = 216$.

The ratio

$$\lambda = B/\rho \tag{1}$$

is important in the physics of the subsequent evolution. With the values that have been assumed here, λ equals 10^{19} (gauss cm^3 gm^{-1}). At the end of Phase III the thermal energy density is 2×10^{-10} erg cm^{-3} and the magnetic energy density

$$\frac{B^2}{8\pi} = 2.8 \times 10^{-16} \text{erg cm}^{-3} \tag{2}$$

with $B = B_o/36 = 8.3 \times 10^{-8}$ gauss. The comparison of thermal and magnetic energy densities shows that the magnetic field is dynamically quite insignificant in the evolution of the hot part of a SNR, and in the later upflow from the Galactic plane.

3 The Return Flow

The crudest model treats the fountain flow as one-dimensional. While this procedure is a vast oversimplification, it does bring out some important features. The gas returning from the halo forms a flat sheet, bounded below by a shock. The gas that is still coming up passes through the shock and becomes part of the descending sheet. The model shows that the downward velocity of the sheet, at any height, is just half the upward velocity of the fountain at that height. So let Φ be the upward mass flux (gm cm^{-2}s^{-1}) in the fountain and v the local speed, then the descending sheet has a speed

$$V = \frac{3}{2}v \tag{3}$$

relative to the upward flow, the rate of addition of mass to the sheet, per unit area, is

$$\dot{\Sigma} = \frac{3}{2}\Phi \tag{4}$$

and pressure behind the shock is

$$P_s = \frac{9}{8}\dot{\Sigma}v = \frac{27}{16}\Phi v \tag{5}$$

The post-shock temperature and density are, respectively, given by

$$\frac{kT_s}{\bar{m}} = \frac{3}{16}\left(\frac{3}{2}\right)^2 v^2 = \frac{27}{32}v^2 \tag{6}$$

and

$$\rho_s = 4\Phi/v \tag{7}$$

With the standard values $\Phi = 10^{-19}$ gm cm^{-2}s^{-1} and $v = 100$ km s^{-1}, the post shock pressure, density and temperature are $P_s = 1.68 \times 10^{-12}$ dyne cm^{-2}, $\rho_s = 4 \times 10^{-26}$ gm cm^{-3} and $T_s = 6 \times 10^5$K. A simple calculation shows that the newly-shocked gas cools in about half a million years, or 1.7×10^{13}s. Consequently the hot layer just above the shock has surface density $\Sigma_h = 2.5 \times 10^{-6}$ gm cm^{-3}, and a thickness of about 6×10^{19} cm or 20 pc. In the one-dimensional model the gas that has cooled lies above the layer that is still hot. But this gas will be compressed to higher density, and the stress taken up by magnetic pressure. With the parameters assumed here, the magnetic field in the cooled gas is given by

$$\frac{B_c^2}{8\pi} \equiv P_s = 1.68 \times 10^{-12} \text{ dyne cm}^{-2} \tag{8}$$

so that B_c equals 6.5×10^{-6} gauss, and relation (1) then shows that the gas density is 6.5×10^{-25} gm cm^{-3}. Provisionally this gas may be identified with the descending HI clouds in the halo. Its pressure is

$$P_g = \frac{k}{m_a}\rho_g T_g = 4.6 \times 10^{-15} \text{dyne cm}^{-2} \tag{9}$$

with $T_g = 100$K; the magnetic pressure clearly dominates in the cool layer, but not in the newly shocked layer, which is much hotter. The cool layer would accumulate a surface density of some 2×10^{-4} gm cm^{-2} over a period of 50 million years and grow to a thickness of some 100 pc.

The temperature stratification makes the descending layer subject to the Rayleigh-Taylor instability, since the hot and tenuous gas, lying below, supports the cool and dense gas further up. The instability produces motions in at least two dimensions, and so the one-dimensional treatment must be abandoned. The presence of the magnetic field tends to favour perturbations in which the layer fragments into filaments parallel to the lines of force. The filaments continue to fall towards the Galactic disk, but now at higher speed because the upflow can move around a filament, and is not necessarily swept up by it. Observation shows that clouds of high and intermediate velocity gas do indeed have a filamentary structure (Wakker and Schwarz 1991). The present description seems therefore to be appropriate.

4   The Fountain as a Magnetic Pump

In a recent paper Brandenburg, Moss and Shukurov (1995) make the interesting point that the Galactic fountain can act as a magnetic pump, in the following sense. The upward flow leaves the disk and reconnection takes place between the lines of force trailing from the bubbles that have originated in the individual SNR's (Kahn and Brett 1993). The field in the flow thus detaches itself from that left in the cooled parts of remnants which remain behind in the disk. The fountain therefore transports with it a reconnected set of lines of force, largely parallel to the Galactic plane. In other words, the upward flow carries with it not only material but also magnetic flux, on a length scale comparable with that of the Galaxy. The gas later returns to the disk in a fountain flow. Most probably it does so not in an organized sheet but in small fragments, because the sheet has an unstable structure as has been explained. If these fragments have filamentary form and are aligned with the magnetic field, and if the filaments stretch from one side of the flow to the other, then as much flux is returned to the disk as went up with the fountain, in the first place. The only proviso here is that none of the gas in the fountain must escape from the Galaxy.

The situation is entirely different if some portion of the gas returns to the disk in compact clouds, rather than in the filaments described above. The observational evidence on HVC's and IVC's, which one can identify with returning material, certainly indicates that this is the case. There is, as a result, an important distinction between the MHD properties of the upflow and the return flow. The upflow carries magnetic flux linked to the gas, but the downflow carries some of its mass, at least, in compact clouds, and therefore returns less magnetic flux than was previously removed by the upward flow. The combined flow thus pumps magnetic flux from the disk to the halo.

The compact clouds still have an interior magnetic field which takes up the external pressure, the gas in the cloud being cool. Typically, let an HVC descend at speed $v_d = 100$ km s^{-1} in a fountain flow rising at speed $v_u = 100$ km s^{-1}. Let the mass flux in the fountain be $\Phi = 10^{-19}$ gm cm^{-2} s^{-1} once more, then the ram pressure on the cloud is

$$P_{ext} = \left(\frac{\Phi}{v_u}\right)(v_u + v_d)^2 = 4 \times 10^{-12} \text{ dyne cm}^{-2} \tag{10}$$

The magnetic field strength within the cloud is given by

$$\frac{B_*^2}{8\pi} = P_{ext} \tag{11}$$

and so B_* equals 10^{-5} gauss. Setting λ equal to 10^{19}, in CGS units, as before, the density ρ_g is found to be 10^{-24} gm cm^{-3}. Most of this gas is in atomic form, with only a small fraction of it being ionized. On a large enough scale the gas is still tied to the lines of force, through the coupling provided by the ionized part. The Alfvén speed for the material as a whole, that is neutral as well as ionized, is

$$v_{A,n} = \left(\frac{B_o^2}{4\pi\rho_g} \right)^{1/2} = 28 \text{ km s}^{-1} \tag{12}$$

with the present values. This calculation applies for disturbances on a large enough linear scale, so that there is good coupling between the plasma.

Coupling occurs because atoms are polarized by the charge on passing ions or electrons. The induced dipole deflects the passing charged particle; the effective cross-section is such that the collision rate is $\tau_{coll,i}^{-1} = 10^{-9}n_n$, for ions, and $\tau_{coll,e}^{-1} = 2.5 \times 10^{-7}n_n$ for electrons. Here n_n is the density of neutrals, in cm^{-3}, and τ is measured in seconds. Conversely the collision frequency for a neutral with an ion is $\tau_{coll,n}^{-1} = 10^{-9}n_i$, and here n_i equals n_e, the electron density.

Suppose that a MHD disturbance, with wavenumber k, is passing through the combined neutral/ionized medium. Its (angular) frequency will be $kv_{A,n}$ and the plasma-neutral coupling will be good enough for this description to apply only if

$$kv_{A,n} \ll \tau_{coll,n}^{-1} = 10^{-9}n_i \tag{13}$$

or

$$k \ll 10^{-9}n_i/v_{A,n} \tag{14}$$

The MHD treatment can be used for the medium as a whole only for disturbances with a length scale in excess of $5 \times 10^{15}f^{-1}$ cm or $1.7 \times 10^{-3}f$ pc. For a typical value f equal to 0.01, the shortest allowable length scale is 0.2 pc, in round numbers.

5   Reconnection of the Magnetic Field, and Heating of the Plasma

The important question in this context is whether reconnection can take place within an HVC or IVC. Certainly reconnection takes place on length scales much smaller than 0.2 pc, and consequently under conditions where the neutrals and the plasma are (almost) uncoupled. The disturbances which affect the plasma alone are propagated with a larger Alfvén speed

$$v_{A,i} = v_{A,n} f^{-1/2} = 28 f^{-1/2} \text{ km s}^{-1} \tag{15}$$

A disturbance with wavenumber k now corresponds to angular frequency $k v_{A,i}$. Such waves will propagate without experiencing serious friction against the neutrals if

$$k v_{A,i} \gg \tau_{coll,i}^{-1} = 10^{-9} n_n \tag{16}$$

or

$$k \gg 10^{-9} \frac{n_n}{v_{A,i}} = 10^{-9} \frac{n_n f^{-1/2}}{v_{A,n}} = 2 \times 10^{-16} f^{-1/2} = 2 \times 10^{-15} \text{ cm}^{-1} \tag{17}$$

for $f = 0.01$. So the plasma is uncoupled from the neutrals on length scales much below 5×10^{14} cm, or 0.0002 pc, in round numbers.

Let the reconnection occur in a zone of width L, parallel to the field lines, and depth D, perpendicular to the field. Let the plasma approach the reconnection zone from either side with speed v. The classical result is that the characteristic timescale τ_* on which the plasma can cross the field lines is

$$\tau_* = \frac{\sigma D^2}{c^2} \, , \tag{18}$$

where σ is the conductivity, and so

$$D = v\tau_* = \frac{\sigma v D^2}{c^2} \tag{19}$$

or

$$D = \frac{c^2}{\sigma v} \tag{20}$$

is the characteristic depth. The conductivity σ_* in a plasma at, say, 10^4K is about 10^{13}s^{-1}. To get an estimate, take the speed v to be 10^4 cm s^{-1} and then D equals about 10^4 cm, which is absurdly small.

But there is a flaw in this calculation. The conductivity is anisotropic in a low density plasma, being orders of magnitude larger along the field lines than across them. So let the mean field in the reconnection zone

$$\mathbf{B} = (B, 0, 0) \tag{21}$$

be stratified in the x_3 – direction and aligned in the x_1 direction, with B positive (negative) far away in the positive (negative) x_3 – direction. Then the current required by the field gradient is

$$\mathbf{j} \equiv (0, j, 0) = \frac{c}{4\pi} \nabla \wedge \mathbf{B} = \frac{c}{4\pi} \frac{\partial B}{\partial x_3} (0, 1, 0) \tag{22}$$

and runs in the x_2 direction. So the depth D has to be found using the right value for $\sigma_\perp$, the conductivity across the field lines. The transverse conductivity is much enhanced if there are Alfvén waves present; here it will be assumed that the waves set up fluctuating transverse components B_2 and B_3 of the field, that

$$\langle B_2 \rangle = \langle B_3 \rangle = 0 \tag{23}$$

and that

$$\langle B_2^2 \rangle = \langle B_3^2 \rangle = B_\perp^2 \tag{24}$$

The conductivity of the plasma is then specified by the tensor

$$\sigma_* \begin{vmatrix} 1 & 0 & 0 \\ 0 & B_\perp^2/B_\parallel^2 & 0 \\ 0 & 0 & B_\perp^2/B_\parallel^2 \end{vmatrix} \tag{25}$$

Accordingly the estimate for D should be made with the conductivity set equal to $\sigma_* B_\perp^2/B_\parallel^2$, with the result that

$$D = \frac{c^2 B_\parallel^2}{\sigma_* v B_\perp^2} \tag{26}$$

Reconnection changes the topology of the magnetic field. Before being reconnected the field lines run parallel to the x_1-axis and cross the reconnection zone from left (right) to right (left) above (below) the plane $x_3 = 0$. After reconnection the lines of force are still largely parallel to the x_3-axis but now they enter the region from left (right) and double back within it to exit right (left). The magnetic stresses are consequently not balanced, and the plasma is squeezed sideways out of the zone, and moves away with velocity $v_{A,i}$. Equating the rates of inflow and outflow of energy for the zone leads to

$$\frac{B_\parallel^2}{8\pi} v L = \frac{1}{2} \rho_i v_{A,i}^3 D = \frac{B_\parallel^2}{8\pi} v_{A,i} D \tag{27}$$

or

$$v L = v_{A,i} D \tag{28}$$

and now set

$$\delta = \frac{D}{L} \tag{29}$$

The term ρ_i above stands for the density of the ionized component.

One more condition can be imposed on the reconnection zone: the random component B_3 of the field must be large enough that field lines from one side of the reconnection zone are taken across to the other side by the random motions, so that further reconnection can take place on smaller length scales. The formal statement of this requirement is

$$\frac{\langle B_\perp^2 \rangle}{B_\parallel^2} \approx \frac{D^2}{L^2} = \delta^2 \tag{30}$$

It follows from (26), (28), (29) and (30) that

$$L = \frac{c^2 v_{A,i}^3}{\sigma_* v^4} \text{ and } D = \frac{c^2 v_{A,i}^2}{\sigma_* v^3} \tag{31}$$

Typical values are then

v	10^2	10^3	10^4	10^5	cm s^{-1}
D	7×10^{16}	7×10^{13}	7×10^{10}	7×10^7	cm
L	2×10^{22}	2×10^{18}	2×10^{14}	2×10^{10}	cm

The values given in the third and fourth columns seem reasonable; in particular 10^5 cm s^{-1} is roughly the speed that one would expect from an argument based on the Petschek (1964) mechanism applied on length scales where the neutrals and ions are well coupled.

But the coupling is poor on the shorter length scales, typically D, where the magnetic energy is converted to thermal energy. So the plasma component picks up enough energy there to be raised to temperature T_i where

$$\frac{3}{2} \frac{kT_i}{m} = \frac{B_\parallel^2}{8\pi \rho_i} = \frac{B_\parallel^2}{8\pi \rho f} \tag{32}$$

and with the present parameters $T_i = 2 \times 10^6$K so the plasma can be raised to a high enough temperature to explain the emission of X-rays within an HVC.

6   Possible Heating of Cosmic Ray Electrons

The initial energy release during reconnection results in the heating of the plasma which supports the magnetic field, because the charged particles are not coupled to the neutrals on the short length scales involved, less than 10^{14} cm. Suppose now that the individual patches where reconnection occurs are separated by much greater distances, typically $\mathcal{L} \sim 0.3$ pc. On this larger scale the plasma is well coupled to the neutrals. The field now relaxes with typical speed $v_{A,n}$ rather than $v_{A,i}$, that is at about 30 rather than 300 km s^{-1}. The time-scale becomes

$$\tau_n \sim \mathcal{L}/v_{A,n} \sim 3 \times 10^{11} \text{ s}$$

or 10^4 years rather than

$$\tau_i \sim L/v_{A,i} \sim 3 \times 10^6 \text{ s}$$

or about a month.

Much more energy is available for release on the larger scale, but the process takes much longer. Some of the energy goes to heat the cosmic ray electrons in the cloud. An electron trapped in a closed loop of the field will increase its energy in inverse proportion to the length of the loop. Since the loop can contract by a large factor, there is scope for a large increase in the energy of the electrons. It is feasible that the electrons are raised to such a high enegy that synchrotron radiation will occur, even at X-ray frequencies. To produce photons with energy 0.25 keV in a magnetic field of 10^{-5} gauss requires that the emitting electron have an energy of order 10^4 GeV. Such an electron has a radiative lifetime of order 10^4 years so the process is just about feasible. It would be most interesting to look more closely at the X-ray spectrum of the HVC's. In particular, a line spectrum would show that the radiation is emitted by a hot plasma, and a continuum spectrum that it comes from c. r. electrons.

References

Brandenburg A., Moss D., Shukurov A., 1995, MNRAS, 276, 651
Heiles C., 1989, in IAU Coll. 120, eds. G. Tenorio-Tagle et al., p.484
Kahn F.D., 1981, in Investigating the Universe, D. Reidel Publishers, p.1
Kahn F.D., 1994, in Kinematics and Dynamics of Diffuse Astrophysical Media,
    eds. J.E. Dyson & E.B. Carling, Kluwer Academic Publishers, p.325

Kahn F.D., Brett L., 1993, MNRAS, 263, 37

Kerp J., Mack K.-H., Egger R., Pietz J., Zimmer F., Mebold U., Burton B.W., Hartmann D., 1996, A&A, in press

Norman C.A., 1991, in IAU Symp. 144, p.337

Normandeau M., Taylor A.R., Dewdney P.E., 1996, Nature, 380, 687

Petschek H.E., 1964, NASA SP-50, US Govt. Printing Office, p.425

Wakker B.P., Schwarz U.J., 1991, A&A, 250, 499

Zirakashvili V.N., Breitschwerdt D., Ptuskin V.S., Völk H.J., 1996, A&A, in press

Address of the author:

F.D. KAHN, Department of Physics & Astronomy, University of Manchester, Manchester, M13 9PL, UK.

Radio Studies of Cosmic Rays and Magnetic Fields in Galactic Halos

Rainer Beck

1   Introduction

Some spiral galaxies are surrounded by diffuse radio continuum emission, as was first discovered in NGC4631 (Ekers & Sancisi 1977) and NGC891 (Allen, Baldwin, & Sancisi 1978). Our own Galaxy has a "thick radio disk" (Webster 1978; Beuermann, Kanbach, & Berkhuijsen 1985). Systematic studies showed that the distribution can generally be described by *thick disks* (Hummel 1990; Harnett & Reynolds 1991; Dumke et al. 1995) while almost spherical *radio halos* are rare.

Radio emission from thick disks is mostly of nonthermal (synchrotron) origin and thus indicates that cosmic-ray electrons and magnetic fields exist at considerable heights above the plane. Radio observations are essential to study the disk-halo interaction.

2   Radio Morphology

Until today there is no statistical investigation about the thickness of radio disks in relation to other galactic properties. Observations are biased towards nearby and bright edge-on galaxies. According to the compilation of 13 galaxies by Hummel (1990), the full width at half maximum of the radio disks at 1.5 GHz varies between 0.5 and 3.7 kpc. The distributions of radio surface brightness perpendicular to the plane can be best described by exponential functions. Scale heights z_0 have been determined for several galaxies (Table 1). Thin disks have $z_0 \leq 0.6$ kpc (2 galaxies), thick disks $z_0 \simeq 1$ kpc (3 galaxies), and halos $z_0 \geq 1.5$ kpc (4 galaxies).

Table 1 includes only the best-observed objects and blurs the fact that bright, extended radio halos are rare. Hummel, Beck & Dettmar (1991a) observed 181 edge-on galaxies from the Uppsala catalogue smaller than $4'$, but found no

Table 1: Exponential radio scale heights of edge-on galaxies

Galaxy	Wavelength (cm)	Scale height (kpc)	Reference
NGC 253	90	2.3 (NE and SW)	Carilli et al. (1992)
	90	3.3 (NW and SE)	Carilli et al. (1992)
	2.8	3.0 (NW and SE)	Beck et al. (1994)
NGC 891	49, 20	0.8 (E)	Hummel et al. (1991c)
	49, 20	1.0 (W)	Hummel et al. (1991c)
	2.8	0.6	Dumke et al. (1995)
NGC 3628	2.8	0.4	Dumke et al. (1995)
NGC 4565	20	0.6	Sukumar & Allen (1991)
	2.8	$\simeq 0.7$	Dumke et al. (1995)
NGC 4631	90	2.0 (N), 3.2 (S)	Hummel & Dettmar (1990)
	20	2.3 (N), 2.0 (S)	Hummel & Dettmar (1990)
	2.8	1.3	Dumke et al. (1995)
NGC 4666	20	1.7	Dahlem et al. (1996b)
	6	1.8	Dahlem et al. (1996b)
	2.8	$\simeq 1.1$	Dumke et al. (in prep.)
NGC 5775	20	1.8	this paper
	6	1.7	this paper
NGC 5907	2.8	1.0	Dumke et al. (1995)
NGC 7331	2.8	1.4	Dumke et al. (1995)

other case similar to NGC4631. Hence prominent radio halos are exceptional rather than normal.

NGC253 is the edge-on galaxy with the brightest and largest radio halo observed so far (Carilli et al. 1992), extending to more than 10 kpc above the plane at 330 MHz (λ90 cm). The flux density at 5 kpc above the plane is $\simeq 45\mu$Jy per 70″beam, $\simeq 1.5$ larger than that in the halo of NGC4631 at the same height and at the same frequency (Hummel & Dettmar 1990). The irregular appearance of the NGC253 halo is mainly due to the lower sensitivity compared with the map of NGC4631. A better map is urgently required. At 1.5 GHz (λ20 cm) NGC253 is much larger than the VLA primary beam so that any diffuse emission with more than $\simeq 10$ kpc extent is missing in Fig.1. NGC253 also has a bright X-ray halo (Pietsch 1994) so that a strong outflow from the disk or the nucleus driven by the high star-formation rate is probable.

The radio halo of NGC4631 is due to interaction with NGC4656 causing a galactic wind, either by gravitational forces or by enhancement of star forma-

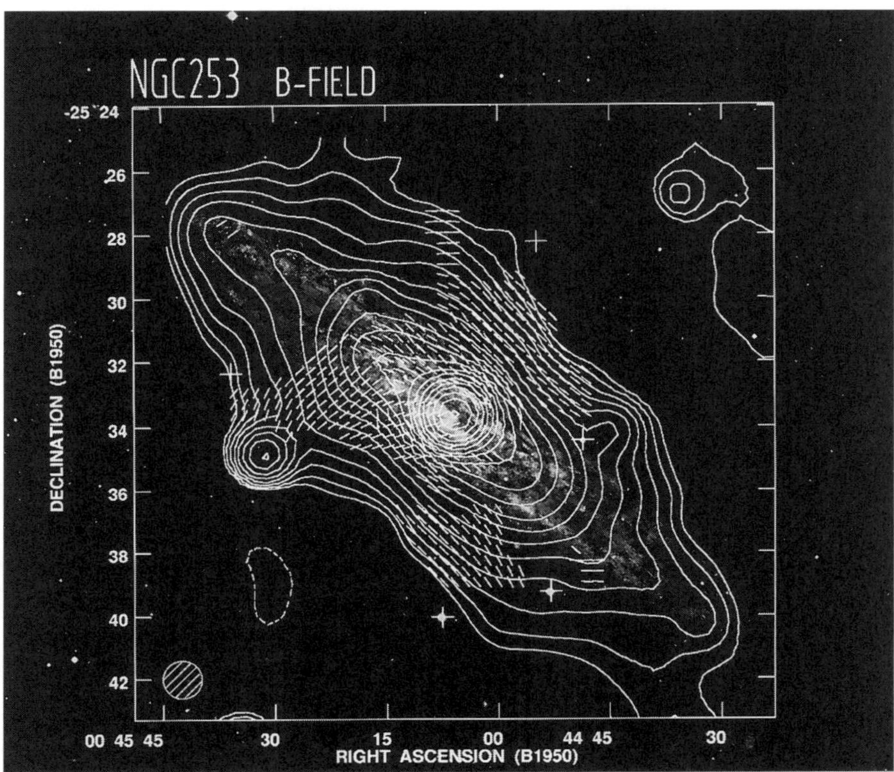

Figure 1: Total radio continuum emission and magnetic field structure of NGC253, observed at $\lambda 20\,$cm with the VLA and smoothed to $70''$resolution. The field structure in the disk was derived from $\lambda 2.8\,$cm ($10.5\,$GHz) observations with the Effelsberg telescope (from Beck et al. 1994).

tion in the disk (Hummel & Dettmar 1990; Hummel, Beck & Dahlem 1991b). The vertical magnetic field lines (see Section 4) may further support the outflow. Diffuse X-ray emission has been detected in the halo of NGC4631 (Wang et al. 1995; Vogler & Pietsch 1996) as well as extraplanar $H\alpha$ emission (Rand, Kulkarni & Hester 1992). Radio and diffuse $H\alpha$ emission from the halo of NGC4666 has been studied by Dahlem et al. (1996b).

The other extremes are NGC4565 (Fig.2) and M31. The radio emission from any thick disk of M31 is not detectable and is ≥ 200 times weaker than from NGC891 (Berkhuijsen et al. 1991). Either the low star-formation rates are below the threshold for "chimney" outflows (Dahlem, Lisenfeld & Golla 1995) or the plane-parallel magnetic-field structure suppresses the propagation of the cosmic rays.

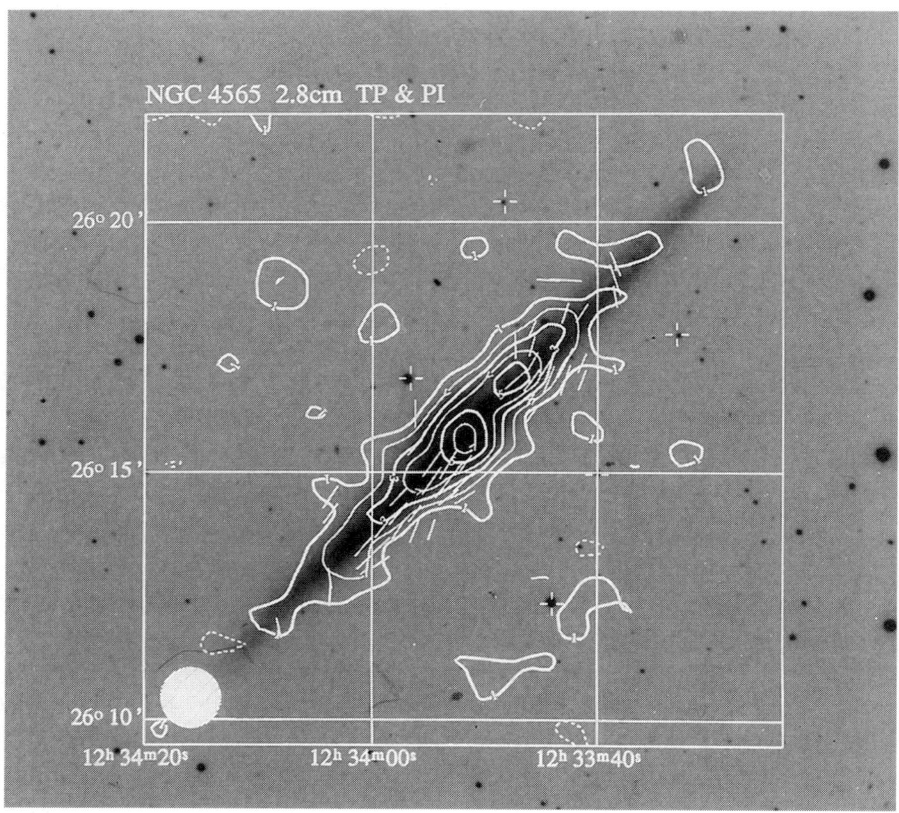

Figure 2: Total radio emission and magnetic field structure of the galaxy NGC 4565, observed at a $\lambda 2.8$ cm (10.5 GHz) with the Effelsberg 100-m telescope (69″ resolution) (from Dumke et al. 1995).

The occurrence of halos in radio continuum and in X-rays is possibly correlated, e.g. due to the outflow of hot gas, cosmic rays and magnetic fields. However, the detailed structures of the radio and X-ray halos of NGC253 and NGC4631 show little resemblance. NGC4565 and NGC3628 are galaxies with X-ray halo emission (Vogler, Pietsch & Kahabka 1996; Dahlem et al. 1996a), but no detected radio halos. Magnetic fields affect the formation of radio halos (see Section 4), but the regions of densest hot gas do not coincide with the regions of strongest fields.

Dahlem et al. (1995) claimed that the galaxies with the most extended radio halos are those with the highest energy input by supernovae per surface area of the underlying star-forming disk. However, in their sample of 9 galaxies NGC5775 (Fig.3) and NGC891 are those with the highest energy input rates, but not those with the largest scale heights (see Table 1). The halo galaxies NGC4631 and NGC4666 have significantly lower energy input rates than NGC5775.

According to Dumke at al. (1995) the halo extent is correlated with the "star-formation efficiency" (SFE) defined as the infrared luminosity per unit molecular mass. The thin-disk galaxy NGC4565 has the lowest SFE ($\simeq 1.5 L_o M_o^{-1}$), while the SFE of the radio-halo galaxy NGC4631 is about 10times larger. The halo galaxy NGC253 also reveals an exceptionally large SFE value ($\simeq 11 L_o M_o^{-1}$; Young et al. 1989) compared with the average value of $\simeq 5 L_o M_o^{-1}$ for non-active spiral galaxies (Chini et al. 1995).

3 Cosmic Rays in Thick Disks and Halos

Cosmic rays are believed to be injected by supernova remnants and to diffuse into interstellar space. Their propagation speed along interstellar field lines is limited by the Alfvén velocity which is $\simeq 100\,$km/s in the hot interstellar medium so that the cosmic rays propagate $\simeq 1\,$kpc within $10^7\,$y. The propagation speed perpendicular to the field lines is smaller.

Cosmic rays lose their energy due to nonthermal bremsstrahlung, ionization and Coulomb interactions with the neutral and ionized interstellar gas, adiabatic cooling in an expanding flow, synchrotron radiation, and inverse Compton scattering. Some particles leak out of the galaxy before they lose their energy. The observed radio continuum emission in the GHz range is radiated by cosmic-ray electrons with energies of a few GeV where the synchrotron and the inverse Compton (IC) losses dominate.

The inner disk of a galaxy has initially the same radio spectral index as the supernova remnants, but energy losses may change the spectral slope. In the presence of a thick disk or halo most of the integrated radio emission emerges from this extended component. Hence the integrated radio spectrum of a galaxy depends not only on the energy losses of the cosmic-ray electrons, but also on their diffusion characteristics and convective transport ("chimneys" or winds).

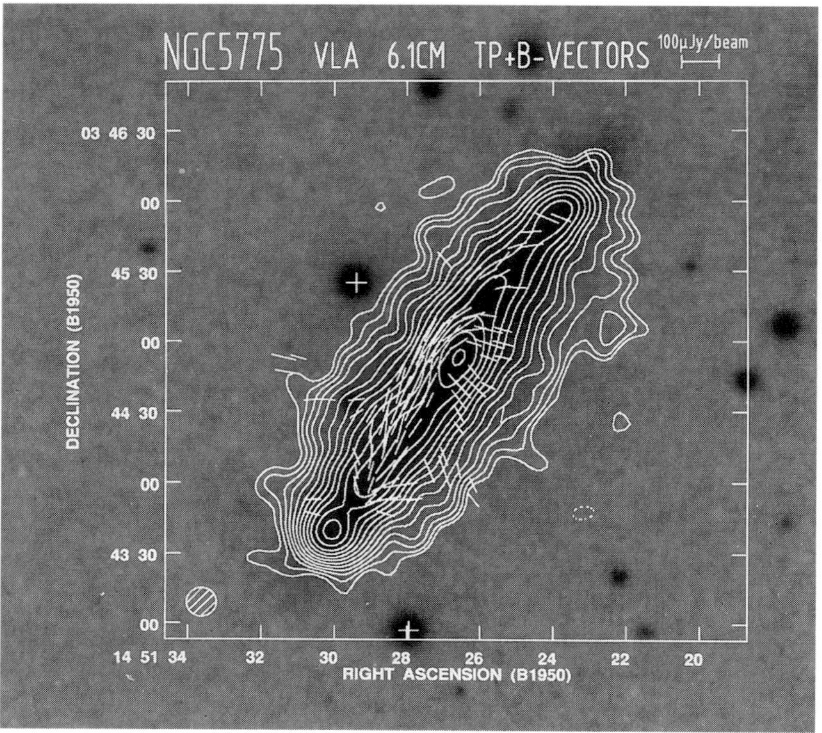

Figure 3: Total radio emission and magnetic field structure of the galaxy NGC5775, observed at $\lambda 6.1\,\mathrm{cm}$ (4.9 GHz) with the VLA (12.5″ resolution) (Golla, unpublished).

Any variation in the energy spectrum of the cosmic-ray electrons is reflected in the radio synchrotron spectrum ($S_\nu \propto \nu^{-\alpha}$). However, due to the broad profile of the synchrotron spectral emissivity in frequency for an individual electron and to the nonlinear relation between the observation frequency ν and the electron energy E ($\nu \propto E^2 B_\perp$), even an abrupt change in the electron spectrum causes a smooth turnover in radio frequency. Spatial variations of the total magnetic field strength $B_\perp$ within the integration volume further smear out spectral breaks.

3.1 Halo models

In the *diffusion-halo model* the diffusion coefficient of cosmic rays out into the halo is energy-dependent ($D_\perp \propto E^\delta$), so that more energetic particles have a higher escape probability (leakage time $\propto E^{-\delta}$). From Galactic cosmic rays near the Earth, $\delta \simeq 0.6$ (Engelmann et al. 1990) or $\delta \simeq 0.33$ (Biermann 1995) have been inferred. Two spectral breaks are expected:

a) $E < E_i$: Ionization losses dominate, and the spectrum of the integrated radio emission is flatter ($\alpha = \alpha_0 - 0.5$) than the injection spectrum (α_0). The local spectral index does not vary from the plane to the halo, and the halo size does not vary with frequency.

b) $E_i < E < E_D$: Diffusive losses dominate, and the integrated radio spectrum is steeper than the injection spectrum ($\alpha = \alpha_0 + \delta/2$). The local spectrum steepens from the plane (α_0) to the halo ($\alpha_0 + \delta/2$). The halo size does not vary with frequency.

c) $E_D < E$: Synchrotron radiation and inverse Compton losses dominate, and the integrated radio spectral index is $\alpha = \alpha_0 + 0.5$. The local radio spectrum steepens continuously with increasing distance from the plane, and the halo size decreases with $\nu^{-(1-\delta)/4} \simeq \nu^{-0.1}$.

In the *dynamical halo model* the cosmic rays are transported by a galactic wind with a velocity increasing linearly with distance from the plane (Lerche & Schlickeiser 1982; Pohl & Schlickeiser 1990a,b). Two spectral breaks are expected:

a) $E < E_i$: Ionization losses dominate, and the radio spectral index behaviour is the same as in the diffusion-halo model.

b) $E_i < E < E_D$: Convective transport dominates, and energy losses are mainly due to adiabatic cooling. The radio spectrum is the injection spectrum with $\alpha = \alpha_0$. A *flattening* of the local radio spectrum from disk to halo is predicted, and the halo size *increases* with $\nu^{\delta/4}$.

c) $E_D < E$: Diffusive transport, synchrotron and inverse Compton losses dominate. The radio spectral index behaviour is the same as in the diffusion-halo model.

3.2 Test: Integrated synchrotron spectra

The average spectral index of the integrated synchrotron emission from spiral galaxies ($\geq 400\,\mathrm{MHz}$) is $\alpha = 0.85 \pm 0.02$ (Niklas, Klein, & Wielebinski 1996), compared with an injection spectral index of $\alpha_0 \simeq 0.6$ (Engelmann et al. 1990) or $\alpha_0 \simeq 0.7$ (Biermann & Strom 1993). Spectral breaks are smooth in frequency and thus difficult to observe. Hence it cannot be excluded that the average synchrotron spectral indices reflect the transition between the injection spectrum and the high-frequency spectrum dominated by synchrotron and IC losses ($\alpha \simeq 1.1...1.2$), indicative of dynamical halos.

However, Niklas et al. (1996) found little evidence for spectral steepening towards higher frequencies so that synchrotron and IC losses seem to be small, except for some interacting galaxies in the Virgo cluster, the interacting galaxy NGC2276 with its exceptionally strong magnetic field (Hummel & Beck 1995) and the thin-disk galaxies (see below). Furthermore, in case of strong synchrotron losses the synchrotron spectral index in a fixed frequency range should become steeper with increasing magnetic field strength, which is *not* observed (Niklas & Beck 1996).

Hence the results by Niklas et al. (1996) are in favour of the diffusion-halo model. The difference between the injection and observed spectra indicates an exponent of the energy dependence of diffusion of $\delta = 2\Delta\alpha \simeq 0.3...0.5$, close to the Galactic value.

The observed integrated synchrotron spectra of the thin-disk galaxies NGC4565 and NGC7331 are significantly steeper than for the sample average ($\alpha \simeq 1.1$). In thin-disk galaxies the time scale for synchrotron and IC losses is smaller than that for escape, because the pressure from low star formation rate cannot drive outflows and/or the plane-parallel magnetic field suppresses the vertical diffusion of cosmic rays.

On the other hand, the synchrotron spectra of the halo galaxies NGC253, NGC891, NGC4631 and NGC4666 are much flatter ($\alpha \simeq 0.78$). A significant spectral flattening towards low frequencies is seen only for NGC4631 (Hummel & Dettmar 1990; Pohl, Schlickeiser & Hummel 1991). None of the halo galaxies shows indications for a steepening at high frequencies, probably because the wind flow makes the time scale for escape losses shorter than that for synchrotron and IC losses.

The statistical significance of spectral variations strongly depends on the accuracy of the flux density measurements. Most published values at low frequencies are based on radio synthesis telescopes of the early days. The new GMRT array in India will improve the accuracy of the data. Data at 32 GHz are expected soon with the new multi-beam receiver in the Effelsberg telescope.

3.3   Test: Halos sizes

The observed halo sizes remain roughly constant between 330 MHz and 1.5 GHz (Table 1), consistent with both halo models. Between 1.5 GHz and 10.5 GHz the sizes decrease faster than predicted, probably due to a decrease in magnetic field strength. The halo size of NGC253 remains constant between 330 MHz and 10.5 GHz, another indication of its exceptional properties.

3.4   Test: Spectral profiles

A third test of halo models is the variation of local radio spectra from the plane to the halo. All models predict a spectral steepening at high frequencies in the outer halo where cosmic-ray electrons lose their energy by the synchrotron and IC processes. This is in agreement with observations, e.g. in NGC891 (Hummel et al. 1991c) and in NGC4631 (Hummel & Dettmar 1990). However, Bloemen, Duric & Irwin (1993) observed NGC3556 with high resolution and conclude from their radio spectra that the electron energy spectrum remains unmodified from the plane to the halo.

The observations show that the radio spectral index remains roughly constant ($\alpha \simeq 0.9$) in regions where the radio emission is most extended (the northern halo of NGC4631 and the NE side of NGC891). This fact can be understood in galactic-wind models with a velocity field of non-constant acceleration (Breitschwerdt 1994).

The inner disk of NGC253 has an exceptionally high star-formation rate and a strong magnetic field ($\simeq 20\mu$G, Beck et al. 1994) so that cosmic-ray electrons radiating at 10GHz have a lifetime against synchrotron losses of only $4 \cdot 10^6$y. However, the observed spectrum between 1.5 GHz and 10.5 GHz is flat ($\alpha \simeq 0.6$) due to the fast outflow.

The strongest galactic winds emerge from active galactic nuclei, e.g. in M82 with a flow velocity in excess of 1000 km/s and a field strength of $\simeq 50\mu$G (Klein et al. 1988), and in NGC3079 with $v \simeq 5000$ km/s (Duric & Seaquist 1988). The radio spectrum steepens outwards in M82 (Reuter et al. 1992), but flattens in NGC3079.

At low frequencies, the dynamical-halo model predicts a spectral flattening from the plane to the halo (Pohl & Schlickeiser 1990a). Reliable high-resolution data at low frequencies are still lacking in external galaxies. The spectral flattening observed in the local halo of our Galaxy (Reich & Reich 1988) can be due to the dynamical halo. Note that this effect occurs below some critical electron energy. As this energy depends on the velocity gradient of the wind flow and on the magnetic field strength, and as the relation between electron energy and observation frequency also depends on field strength, the interpretation of radio spectra is not at all straight forward.

High-resolution observations of edge-on galaxies show that the structure of radio halos is not uniform, but consists of several spurs with aligned magnetic fields, e.g. in M82 (Reuter et al. 1992) and in NGC4631 (Golla & Hummel 1994). This calls for better halo models with realistic magnetic field structures and velocity fields.

4 Magnetic Field Structure in Disks and Halos

Plane-parallel magnetic fields in disks are probably generated by the dynamo process (Section 5). Parker instabilities and "chimney" outflows produce field loops with alternating field directions. Such loops lead to significant radio polarization with vectors oriented either parallel or perpendicular to the plane, depending on their elongation. In diffusion halos no polarization is expected as the field is mostly irregular. A galactic wind V_z perpendicular to the plane drags the field out, but an azimuthal $(\phi-)$ gradient of V_z is required to produce B_z from B_ϕ, or a radial $(r-)$ gradient of V_z to obtain B_z from B_r. A global wind accelerated only in z-direction lifts up the plane-parallel disk field into the halo.

NGC891, NGC5907 and NGC7331 and many other edge-on galaxies possess *thick disks*. In these galaxies the observed field orientations are mostly parallel to the disk (Table 2). NGC4565 (Fig.2) has the most regular plane-parallel field (Sukumar & Allen 1991), probably suppressing the diffusion of cosmic rays out of the disk· and thus reducing the synchrotron scale height.

The regular magnetic field in the disk of NGC253 (Fig.1) is also predominantly parallel to the plane (Beck et al. 1994), due to strong differential rotation and strong dynamo action even close to the center. In contrast to the other radio-halo galaxies, the regular field is also parallel to the plane in the lower halo (where data are still available), although the radio spectrum and the huge X-ray halo are clear indicators of a strong outflow. In NGC253 the halo has a cylindrical shape so that the relevant gradients in wind velocity are probably not strong enough to generate a significant component of B_z. The field structure in the outer halo is not yet known.

Table 2: Magnetic field structures of (almost) edge-on galaxies

Galaxy	Wavelength (cm)	Field structure	Ref.
M82	6, 3.6	Radially away from the nucleus	Reuter et al. (1994)
NGC 253	20, 6 2.8	$\parallel$ plane	Carilli et al. (1992), Beck et al. (1994)
NGC 891	20, 6 2.8	$\parallel$ & inclined to plane	Sukumar & Allen (1991), Dumke et al. (1995)
NGC 3628	20, 6 2.8	$\parallel$ & inclined to plane	Reuter et al. (1991), Dumke et al. (1995)
NGC 4565	20, 6 2.8	$\parallel$ plane	Sukumar & Allen (1991), Dumke et al. (1995)
NGC 4631	6, 2.8 20, 6, 3.5	$\perp$ plane (inner region), $\parallel$ & inclined to plane (outer regions)	Hummel et al. (1991b), Golla &Hummel (1994)
NGC 4666	20, 6 2.8	inclined to plane inclines to plane	Dahlem et al. (1996b), Dumke et al. (in prep.)
NGC 5775	6	$\parallel$ and inclined to plane	this paper
NGC 5907	6	$\parallel$ plane	Dumke & Krause (in prep.)
NGC 7331	2.8 20, 6	Almost $\parallel$ plane	Dumke et al. (1995), Hummel (unpubl.)

NGC4631, NGC4666 and M82 (Reuter et al. 1994) are halo galaxies with dominating vertical field components. Magnetic spurs in these halos are connected to star-forming regions in the disk. The field is probably dragged out by the strong, inhomogenous galactic wind. The magnetic field lines in the NGC4631 halo are roughly perpendicular to the inner disk, which is almost rigidly rotating so that dynamo action is weak (Hummel et al. 1991b; Golla & Hummel 1994). A few regions with field orientations parallel to the disk are visible in the (differentially rotating) outer disk.

In the disks of spiral galaxies, the observed polarized emission is weak due to depolarization effects. Faraday depolarization alone is insufficient to explain the low degrees of polarization near the plane of NGC4631 (Golla & Hummel 1994). In their sample of edge-on galaxies, Dumke et al. (1995) found $p \simeq 5\%$ in the plane at 10.5 GHz where Faraday effects are negligible. The field structure in the plane is mostly turbulent due to star-forming processes and cannot be resolved by the telescope beam (beam depolarization). The degree of polarization at high frequencies increases with increasing distance from the plane because the star-forming activity and thus the field turbulence decrease. At 5 kpc height the correlation length of the turbulent field increases to typically 2 kpc, compared with 50-100 pc in the plane. A large correlation length of the turbulent magnetic field in the halo also solves the problem how to explain the strong Faraday depolarization observed at 1.5 GHz in the radio halos of NGC253 and M83 (Beck et al. 1994; Ehle et al. 1996).

The increase of the degree of polarization at 1.5 GHz with height above the plane of NGC891 has been modeled by Hummel et al. (1991b) by Faraday depolarization. Thermal gas of (observed) scale height $\simeq 1$ kpc is required together with a turbulent magnetic field of scale height $\simeq 4$ kpc. This is consistent with equipartition between the field and cosmic-ray energy densities, where $z_{\mathrm{b}} = 2z_{\mathrm{CR}} = (3 + \alpha)z_0 \simeq 3.6$ kpc for a synchrotron scale height of $z_0 \simeq 0.9$ kpc (Table 1) and $\alpha \simeq 1.0$.

5 Halo Dynamos

Turbulent motions and differential rotation in the halo drive a galactic dynamo which may generate a magnetic field with a symmetry different from that in the disk. A dynamo in a differentially rotating thin disk generates preferably quadrupolar fields (even symmetry with respect to the plane), while dipolar fields (odd symmetry) can be generated in quasi-spherical halos (Sokoloff & Shukurov 1990). Superposition of disk and halo fields may explain why the observed field structure is complicated in many face-on galaxies. In M51, for example, the directions of the disk and halo fields are opposite to each other (Berkhuijsen et al. 1996).

Polarization vectors only give the field orientation, not its *direction*. Rotation measures (RM) in galactic halos are thus important to distinguish between halo fields consisting of loops with alternating field directions and large-scale field patterns generated by a dynamo. The distribution of RMs yields the field parity with respect to the midplane. Beck et al. (1994) determined rotation measures at a few positions in the lower halo of NGC253 and found weak evidence for RMs of the same sign at $\simeq 5$ kpc above and below the plane, as expected for an even-symmetry mode. Golla & Hummel (1994) could not find a clear RM pattern from their data of NGC4631.

All dynamos in differentially rotating bodies generate magnetic fields which are plane-parallel, irrespective of their symmetry. In order to explain the observed vertical field components, radial or azimuthal gradients of the vertical velocity component are required, i.e. an expanding wind (Brandenburg et al. 1993). Dominating vertical fields can only be generated by a highly non-uniform ("spiky") wind (Elstner et al. 1995) where any regular RM pattern is destroyed, as observed in NGC4631.

6 Magnetic Field Strengths in Halos

There is no direct way to determine field strengths in galactic halos. Faraday rotation measures yield the strength of the regular field component along the line of sight if the thermal electron densities are known, e.g. from X-ray data. However, reliable rotation measures in halos are not yet available.

Synchrotron intensities can be used to estimate the strength of the total field if equipartition between the energy densities of cosmic rays and magnetic field holds, which is not obvious for galactic wind flows. Faraday depolarization data in NGC891 gave some support for the equipartition assumption (see Section 4), but this may not be the case in other galaxies.

Ehle (1995) determined field strengths in the halos of several galaxies and compared the magnetic energy densities with that of the hot thermal gas. The magnetic energy density was found to be always larger than that of the hot gas so that magnetic fields cannot be neglected in studies of halo dynamics. One may assume that the magnetic energy is balanced by the sum of the thermal energy of the hot gas plus its kinetic energy, whereas the warm ($\simeq 10^4$ K) halo gas is dynamically unimportant.

7 Summary

Edge-on spiral galaxies reveal thin disks, thick disks or halos which indicate that different physical processes dominate. An undisturbed plane-parallel magnetic field, generated by a thin-disk dynamo, suppresses cosmic-ray diffusion out of the plane and leads to steep radio spectra due to strong synchrotron and inverse Compton losses. A higher star-formation efficiency per unit gas mass drives "chimney" outflows which form field loops and allow cosmic rays to diffuse into a thick disk within their synchrotron lifetime. Beyond some critical value of the star formation efficiency a global galactic wind sets in and builds up an X-ray and radio halo. The magnetic field structure depends on the gradients of the wind flow.

References

Allen R.J., Baldwin J.E., Sancisi R., 1978, A&A 62, 397

Beck R., Carilli C.L., Holdaway M.A., Klein U., 1994, A&A 292, 409

Berkhuijsen E.M., Golla G., Beck R., 1991, in *The Interstellar Disk-Halo Connection in Galaxies*, ed. H. Bloemen, Kluwer, Dordrecht, p. 233

Berkhuijsen E.M., Horellou C., Krause M., et al., 1996, A&A, in press

Beuermann K., Kanbach G., Berkhuijsen E.M., 1985, A&A 153, 17

Biermann P.L., 1995, Sp.Sci.Rev. 74, 385

Biermann P.L., Strom R.G., 1993, A&A 275, 659

Bloemen H., Duric N., Irwin J., 1993, Proc. 23rd ICRC, Vol. 2, p. 279

Brandenburg A., Donner K.J., Moss D., et al., 1993, A&A 271, 36

Breitschwerdt D., 1994, *Dynamik des Interstellaren Mediums in Halos von Galaxien*, Habilitation Thesis, University of Heidelberg

Carilli C.L., Holdaway M.A., Ho P.T.P., de Pree C.G., 1992, ApJ 399, L59

Chini R., Krügel E., Lemke R., Ward-Thompson D., 1995, A&A 295, 317

Dahlem M., Lisenfeld U., Golla G., 1995, ApJ 444, 119

Dahlem M., Heckman T.M., Fabbiano G., et al., 1996a, ApJ 461, 724

Dahlem M., Petr M., Lehnert M.D., et al., 1996b, A&A, in press

Dumke M., Krause M., Wielebinski R., Klein U., 1995, A&A 302, 691

Duric N., Seaquist E.R., 1988, ApJ 326, 574

Ehle M., 1995, PhD Thesis, University of Bonn

Ehle M., Klein U., Pietsch W., Beck R., 1996, A&A, submitted

Ekers R.D., Sancisi R., 1977, A&A 54, 973

Elstner D., Golla G., Rüdiger G., Wielebinski R., 1995, A&A 297, 77

Engelmann J.J., Ferrando P., Soutoul A., et al., 1990, A&A 233, 96

Golla G., Hummel E., 1994, A&A 284, 777

Harnett J.I., Reynolds J.E., 1991, A&A Suppl. Ser. 88, 73

Hummel E., 1990, in *Windows on Galaxies*, eds. G. Fabbiano et al., Kluwer, Dordrecht, p. 141

Hummel E., Beck R., 1995, A&A 303, 691

Hummel E., Dettmar R.-J., 1990, A&A 236, 33

Hummel E., Beck R., Dettmar R.-J., 1991a, A&A Suppl. Ser. 87, 309

Hummel E., Beck R., Dahlem M., 1991b, A&A 248, 23

Hummel E., Dahlem M., van der Hulst J.M., Sukumar S., 1991c, A&A 246, 10

Klein U., Wielebinski R., Morsi H.W., 1988, A&A 190, 41

Lerche I., Schlickeiser R., 1982, A&A 107, 148

Niklas S., Beck R., 1996, A&A, in press

Niklas S., Klein U., Wielebinski R., 1996, A&A, in press

Pietsch W., 1994, in *Panchromatic View of Galaxies*, eds. G. Hensler et al.,
    Editions Frontières, Gif-sur-Yvette, p. 137

Pohl M., Schlickeiser R., 1990a, A&A 234, 147

Pohl M., Schlickeiser R., 1990b, A&A 239, 424

Pohl M., Schlickeiser R., Hummel E., 1991, A&A 250, 302

Rand R.J., Kulkarni S.R., Hester J.J., 1992, ApJ 396, 97

Reich P., Reich W., 1988, A&A 196, 211

Reuter H.-P., Krause M., Wielebinski R., Lesch H., 1991, A&A 248, 12

Reuter H.-P., Klein U., Lesch H., et al., 1992, A&A 256, 10

Reuter H.-P., Klein U., Lesch H., et al., 1994, A&A 282, 724

Sokoloff D., Shukurov A., 1990, Nature 347, 51

Sukumar S., Allen R.J., 1991, ApJ 382, 100

Vogler A., Pietsch W., 1996, A&A 311, 35

Vogler A., Pietsch W., Kahabka P., 1996, A&A 305, 74

Wang Q.D., Walterbos R.A.M., Steakley M.F., et al., 1995, ApJ 439, 176

Webster A., 1978, MNRAS 185, 507

Young J.S., Xie S., Kenney J.D.P., Rice W.L., 1989, ApJ Suppl. Ser. 70, 699

Address of the author:

RAINER BECK, Max-Planck-Institut für Radioastronomie, Auf dem Hügel 69,
D-53121 Bonn, Germany.

Magnetic fields in the disk and halo of M51

E.M. Berkhuijsen, C. Horellou, <u>M. Krause</u>, N. Neininger, A.D.Poezd,
A. Shukurov and D.D.Sokoloff

1 Introduction

The large-design spiral galaxy M51 (NGC 5194) is the first external spiral
galaxy from which linearly polarized radio emission was detected (Mathewson
et al. 1972; Segalovitz et al. 1976), and for which the global structure of
the regular magnetic field highlightend by these observations was investigated
(Tosa & Fujimoto 1978). However, the global magnetic field structure remained
unclear up to now since the standard analysis of the rotation measures between
polarization observations at different wavelengths led to inconclusive results
(Horellou et al. 1990; Horellou et al. 1992; Neininger 1992).

We now developed a method to analyze all polarization observations of this
galaxy (i.e. at the wavelengths $\lambda\lambda 2.8$, 6.2, 18.0, 20.5 cm) *simultaneously* with
allowance for a multi-layer distribution of the magnetic field and thermal electron density which is extensively described by Berkhuijsen et al. (1996).

2 The method

The magnetic field is parametrized in terms of coefficients of Fourier series in
the azimuthal angle measured in the plane of M51. Our method allows for both,
horizontal and vertical components of the *regular* magnetic field. We attempted
to build a coherent, self-consistent picture of the global magnetic structure
including other available information coming from, e.g., intrinsic polarization
angles, depolarization data, total synchrotron emission, thermal radio emission,
the morphology of the galaxy,etc.

M51 is not completely transparent for polarized emission at $\lambda\lambda 18$ and 20 cm.
This allows us to analyze the line-of-sight structure of the magneto-ionic medium by taking into account the depth in the disk from which the polarized
emission is observed. At each wavelength the observed polarization angles were
averaged over sectors of $20°$ width in rings at radial distances 3–6, 6–9, 9–12

and 12–15 kpc. We fitted these observed azimuthal distributions at all four wavelengths simulaneously.

In the two outer rings the data are consistent with one magnetic field configuration along the line of sight. However, in the two inner rings we found two completely different magnetic field configurations in the two regions along the line of sight sampled at the small wavelengths ($\lambda\lambda 2.8$ and 6.2 cm) and long wavelengths ($\lambda\lambda 18$ and 20 cm) respectively. Since it is unlikely that the regular magnetic field may have such a complicated vertical structure within the upper part of the disk, we concluded that the change of the direction of $B_\parallel$ takes place in the halo. Then the observed Faraday rotation at the short wavelengths is dominated by the direction of the magnetic field in the disk whereas that at the long wavelengths is mainly determined by the halo. Interestingly, recent observations of M51 in X-rays (Ehle et al. 1995) provide an independent confirmation of the presence of a hot thermal halo of about 10 kpc in radius.

We introduced a formalism accounting for the different contributions of the halo and the disk to the total rotation measure and estimated the line of sight into M51 at the long wavelengths from depolarization arguments. We also derived estimates of the volume density of thermal electrons, their scale height and filling factor for the disk based on the thermal radio emission from M51 (Klein et al. 1984) and an analogy with the Milky Way. Using these, we estimated the strength of the regular magnetic field in the disk and halo from our fits.

3   Results

Our analysis indicates the existence of a *magneto-ionic halo* in M51 with a radial extent of about 10 kpc. The regular magnetic fields in the disk and the halo have different structures (see Fig. 1). In the halo the regular magnetic field is axisymmetric and horizontal. Its field lines are spirals pointing *inwards* and generally opposite to those of the disk. In the *disk* the azimuthal structure of the magnetic filed is neither axisymmetric nor bisymmetric, but it can be satisfactorily represented by a superposition of both modes with about equal weights. Magnetic lines of the disk are spirals generally directed *outwards*.

With the available resolution of about 3.5 kpc, the vertical component of the magnetic field is negligible ($|B_z| \lesssim 0.3\,\mu\text{G}$) inside the galactocentric radius of 12 kpc. In the outermost ring ($r = 12$–15 kpc) a small vertical component exists which may be a result of strong three-dimensional distortions in the regular magnetic field in this ring caused by the companion.

The magnetic field pattern in the disk of M51 shows a discontinuity at $r \simeq$ 9 kpc, the position at which the inner and outer spiral structure join. The relatively strong, coherent magnetic field in the inner part occurs in the system of spiral arms excited by density waves, whereas the weaker and partly distorted field in the outer rings exists in the area of the material spiral arms produced by the encounter with the companion.

The azimuthally averaged strength of the regular magnetic field obtained for the disk decreases from about $7\,\mu$G at radius 3–6 kpc to about $4\,\mu$G at 12–15 kpc. These averages are in good agreement with the independant estimates of the regular magnetic field strength from the total synchrotron emission and the degree of polarization.The strength of the regular field in the halo decreases from $3\,\mu$G at 3–6 kpc to zero beyond 9 kpc.

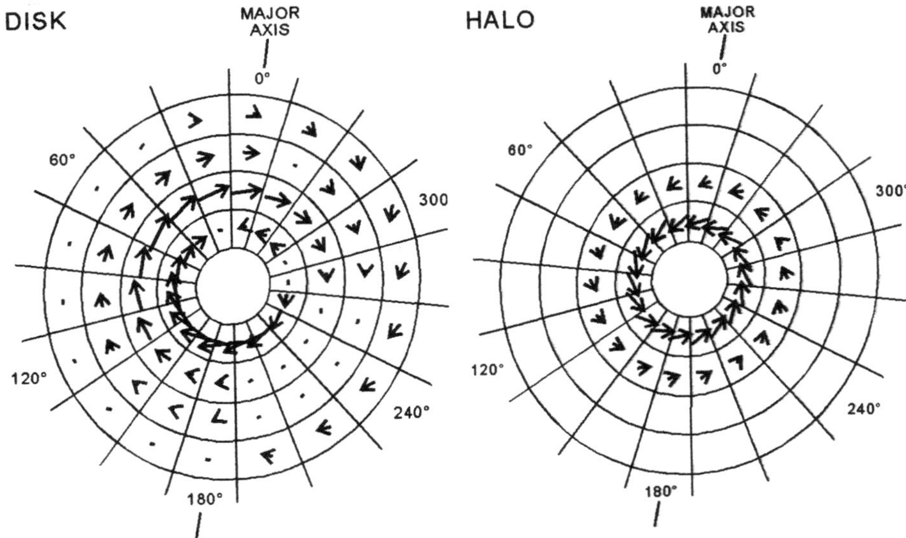

Figure 1: Directions of the horizontal regular magnetic field in the disk (a) and halo (b) of M51 according to our fits. The lengths of the vectors are proportional to B in the inner two rings in the disk, to $3B$ in the two outer rings in the disk, to $2.5B$ for 3–6 kpc in the halo, and to $5B$ for 6–9 kpc in the halo. The vertical component at $r = 12$–15 kpc is not included.

References

Berkhuijsen E.M., Horellou C., Krause M., Neininger N., Poezd A.D., Shukurov A., Sokoloff D.D., 1996, A&A (in press) (and electronic publication)
Ehle M., Pietsch W., Beck R., 1995, A&A 295, 289
Horellou C., Beck R., Klein U., 1990, in IAU Symp. 140 "Galactic and Intergalactic Magnetic Fields", eds. R. Beck, P.P. Kronberg, R. Wielebinski, Kluwer, Dordrecht, p. 211
Horellou C., Beck, R., Berkhuijsen E.M., Krause M., Klein U., 1992, A&A 265, 417
Klein U., Wielebinski R., Beck R., 1984, A&A 135, 213
Mathewson D.S., van der Kruit P.C., Brouw W.N., 1972, A&A 17, 408
Neininger N., 1992, A&A 263, 30
Segalovitz A., Shane W.W., de Bruyn A.G., 1976, Nature 264, 222
Tosa M., Fujimoto M., 1978, PASP 30, 315

Addresses of the authors:

E.M. BERKHUIJSEN, M. KRAUSE, N. NEININGER, Max-Planck-Institut für Radioastronomie, Auf dem Hügel 69, 53121 Bonn, Germany

C. HORELLOU, Onsala Space Observatory, 43992 Onsala, Sweden

A.D. POEZD, D.D. SOKOLOFF, Physics Department, Moscow University, Moscow 119899, Russia

A. SHUKUROV, Computing Center, Moscow University, Moscow 119899, Russia

The relationship between magnetic field strength and gas density from observations

Elly M. Berkhuijsen

1 Introduction

Although magnetic fields play an important role in structuring the ISM of a galaxy, the relationship between the total magnetic field strength, B, and the total gas volume density, ρ_g, is poorly known. Theoretical considerations like flux freezing or a balance between the various pressures in the ISM suggest $B \propto \rho_\mathrm{g}^\kappa$ with $\kappa \simeq 0.5$, but values between 0 and 1 may occur depending on the local circumstances (Troland & Heiles 1986). To test this relationship observationally measurements of B as well as of ρ_g are required. So far two methods have been used:

a) The Zeeman effect yields $B_\parallel$, the line-of-sight component of B, from which B can be computed, and excitation models give ρ_g. In the Milky Way these measurements probe the ISM on scales < 100 pc with densities $\gtrsim 100\,\mathrm{cm}^{-3}$. They indeed yield $\kappa \simeq 0.5$ (Fiebig & Güsten 1989).

b) The nonthermal radio continuum emission yields an average of $B_\perp$ along the line of sight in the source from which the volume-averaged estimate of B, $\bar{B}$, is obtained. As the nonthermal intensity is $I_\mathrm{N} \propto N_\mathrm{r} B_\perp^{1+\alpha} L$ (N_r = density of relativistic electrons, α = nonthermal spectral index, L = line of sight) both α and L must be known. Furthermore, energy equipartition between cosmic-ray particles and magnetic fields must be assumed, $N_\mathrm{cr} \propto N_\mathrm{r} \propto B^2$. Values of $\bar{\rho}_\mathrm{g}$, the mean volume density of the total gas, are derived from H I and CO data with a conversion factor $X = N_\mathrm{H2}/I(\mathrm{CO})$.

In this paper some new results for the Milky Way and M31 are presented based on nonthermal radio continuum data. In each case significant correlations (correlation coefficients > 0.7) were found between the volume emissivity, ε_N, and $\bar{\rho}_\mathrm{HI}$, $\bar{\rho}_\mathrm{H2}$ and $\bar{\rho}_\mathrm{g} = \bar{\rho}_\mathrm{HI} + 2\bar{\rho}_\mathrm{H2}$. These densities are equivalent volume densities rather than true volume densities and they are mean values along the line of sight. As we are dealing with observed quantities on both axes bisectors were fitted through the points. Details of the calculations will be given elsewhere.

2   The Milky Way

For the longitude range $10° < \ell < 60°$, $-0\overset{\circ}{.}5 \leq b \leq 0\overset{\circ}{.}5$ the nonthermal radio continuum emission at 1420 MHz and α were taken from the survey of Reich & Reich (1986, 1988). L is the line of sight through the disk of the MW extending to a radius of 15 kpc.

Figure 1 presents the correlation of $\varepsilon_N(1420)$ with $\bar{\rho}_g$ and the derived values of κ are given in Table 1. The value for H I is significantly larger than for H_2. As H_2 is the dominant neutral gas component in the Galactic plane, the value of κ for the total gas is close to that of H_2 and much smaller than 0.5.

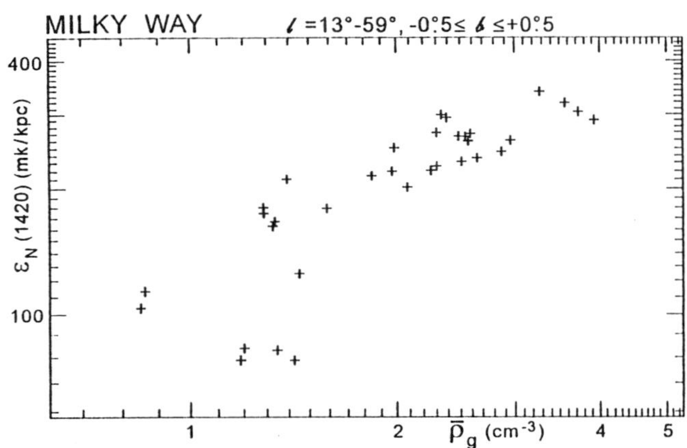

Figure 1: Dependence of the volume emissivity at 1420 MHz on the mean volume density of the total gas in the Galactic plane.

The value of κ could also be determined on a larger linear scale by using data averaged in 1 kpc wide rings around the centre. In this case $\bar{B}$ was computed from the radial variation of the surface brightness at 408 MHz and $L = h_N$, the exponential scale height (Beuermann et al. 1985). The values of κ in Table 1 again show that the dependence of $\bar{B}$ on $\bar{\rho}_{HI}$ is stronger than on $\bar{\rho}_{H2}$.

3   M31

In an area in the SW spiral arms of M31 Berkhuijsen et al. (1993) found good correlations between $\bar{B}$ and gas volume densities on the scale of 250×1200 pc. The values of κ given in Table 1 were redetermined using the bisector instead of the mean of the regression lines. Again the value of κ for H I is larger than for H_2, but the values for H_2 and the total gas are significantly larger than those found in the MW.

Table 1: Values of κ in $\bar{B} \propto \bar{\rho}^{\kappa}$

Area	$\bar{\rho}_{\mathrm{HI}}$	$\bar{\rho}_{\mathrm{H2}}$	$\bar{\rho}_{\mathrm{g}}$	$2\bar{\rho}_{\mathrm{H2}}/\bar{\rho}_{\mathrm{g}}$	Linear scale (pc)
MW					
$\ell = 13°$–$59°$, $b \simeq 0°$	0.89±0.11	0.23±0.03	0.26±0.03	0.89	< 300
Rings 3–17 kpc	0.47±0.06	0.15±0.01	0.27±0.02	0.58	1000
M31					
SW arms	1.04±0.13	0.44±0.03	0.68±0.03	0.57	250×1200
Rings 2–20 kpc	0.59±0.05	0.25±0.01	0.39±0.02	0.46	2000

To investigate the value of k on the scale of 2 kpc wide rings the distribution of the nonthermal continuum emission at 2702 MHz and the spectral index of Beck & Gräve (1982) were used with scale heights equal to those of H I derived by Braun (1991). H I volume densities were computed from the H I map of Cram et al. (1980) and those of H_2 from the CO map of Koper et al. (1991) using scale heights half of those of H I and $X = 2.5 \cdot 10^{-20}$ cm^{-2} (K km s^{-1})$^{-1}$. As for the other samples, the value of κ for H_2 is smaller than for H I (Table 1).

4   Discussion

The correlation between $\bar{B}$ and $\bar{\rho}$ holds on both small and large scales which indicates a close coupling between magnetic field lines and gas clouds.

Two different trends are visible in Table 1:
1. For each galaxy the values of κ decrease with increasing linear scale. This may be due to the inclusion of non-correlating gas in the averages in rings. A plot of the nonthermal emission against N_{HI} for M31 shows that the correlation breaks down below $N_{\mathrm{HI}} \simeq 10^{20}$ cm^{-2}. This is close to the critical value below which star formation does not occur (Kennicutt 1989).
2. Both in the MW and in M31 $\bar{B}$ depends more strongly on $\bar{\rho}_{\mathrm{HI}}$ than on $\bar{\rho}_{\mathrm{H2}}$. Hence the value of κ for the total gas depends on the fraction of H_2 of the neutral gas, which is also given in Table 1. A large fraction of H_2 leads to a small value of κ as is most clearly seen when comparing the results for the MW at $b \simeq 0°$ and the SW arms in M31, which are for similar linear scales.

Presently it is not clear why $\bar{B}$ seems to be more strongly coupled to $\bar{\rho}_{\mathrm{HI}}$ than to $\bar{\rho}_{\mathrm{H2}}$. Several effects could play a role:
a)   An underestimate of $\bar{\rho}_{\mathrm{HI}}$ and/or an overestimate of $\bar{\rho}_{\mathrm{H2}}$ increasing with density. The value of κ for H_2 would increase if the conversion factor X would

decrease with H_2 density instead of being (kept) constant.

b) A volume filling factor of H_2 decreasing faster with ρ than that of $H\,\textsc{i}$.

c) The degree of ionization is smaller in H_2 clouds than in $H\,\textsc{i}$ clouds (Myers & Khersonsky 1995). Therefore the magnetic field may be less well coupled to H_2 than to $H\,\textsc{i}$.

The latter point seems most attractive. Points a) and b), however, cannot be ruled out as data to test them are lacking.

The value of κ for the total gas in the MW is about half that derived from observations of the Zeeman effect (Fiebig & Güsten 1989). This difference may be related to the difference in linear scales sampled. In M31 the mean value of κ is ≈ 0.5. It is interesting to note that Niklas & Beck (1996) derived $\kappa = 0.48 \pm 0.05$ for the global average of the total gas of a sample of 43 nearby galaxies. This is consistent with a mean fraction of H_2 of 0.5 as in M31.

References

Beck R., Gräve R., 1982, A&A 105, 192

Berkhuijsen E.M., Bajaja E., Beck R., 1993, A&A 279, 359

Beuermann K., Kanbach G., Berkhuijsen E.M., 1985, A&A 153, 17

Braun R., 1991, ApJ 372, 54

Cram T.R., Roberts M.S., Whitehurst R.N., 1980, A&A Suppl. 40, 215

Fiebig D., Güsten R., 1989, A&A 214, 333

Heiles C., 1995, in "The Physics of the ISM and IGM", eds. A. Ferrara et al., ASP Conf. Ser. 80, p. 507

Kennicutt Jr. R.C., 1989, ApJ 344, 685

Koper E., Dame T.M., Israel F.P., Thaddeus P., 1991, ApJ 383, L11

Myers P.C., Khersonsky V.K., 1995, ApJ 442, 186

Niklas S., Beck R., 1996, A&A (submitted)

Reich P., Reich W., 1986, A&A Suppl. 63, 205

Reich P., Reich W., 1988, A&A 196, 211

Troland T.H., Heiles, C., 1986, ApJ 301, 339

Address of the author:

E.M. BERKHUIJSEN, Max-Planck-Institut für Radioastronomie, Auf dem Hügel 69, 53121 Bonn, Germany.

The radio – far infrared correlation of spiral galaxies

S. Niklas

1 Introduction

The tight and universal correlation between the far infrared (FIR) and the radio continuum emission of galaxies is one of the most important results of the IRAS mission. The correlation does not just reflect a 'richness effect' (Xu et al. 1994), but can be attributed to a coupling between the dust heating photons, emitted by young, massive stars, and the synchrotron-emitting relativitic electrons of the cosmic rays. Some models try to explain the existence and slope of the correlation for galaxies which are assumed to be optically thick for cosmic ray electrons and UV photons (Völk 1989; Lisenfeld et al. 1996). An attempt to model the correlation for the transition from optically thick to thin was made by Helou & Bicay (1993). A large set of radio continuum data of Shapley-Ames galaxies shows that the thermal radio emission is linearly correlated to the FIR emission. The synchrotron radio-FIR correlation is given by $P_{sync} \propto L_{FIR}^{1.25 \pm 0.08}$ (Niklas 1995; Niklas et al. 1995a) The same data are now used for a test of existing models and for an attempt to explain the correlation even for galaxies with large radio halos and flat synchroton spectral indices indicating a high escape probability of the cosmic ray electrons.

2 Test of models

2.1 The synchrotron spectral indices

A critical test for the calorimeter model by Völk (1989) is the mean value of the non-thermal spectral indices of galaxies. The model assumes that the electrons completely lose their energy due to synchrotron and/or inverse Compton losses in their host galaxies. In this case one expects a synchrotron spectral index distribution $\alpha_{nth} \geq 1.0$ ($S_{nth} \propto \nu^{-\alpha_{nth}}$) for an injection spectra with $\alpha_0 \geq 0.5$. The latter is predicted by diffusive shock acceleration of the cosmic-ray electrons in supernova remnants (Bogdan & Völk 1983). However, the distribution of non-thermal spectral indices of 74 galaxies peaks sharply at $\langle \alpha_{\mathrm{nth}} \rangle = 0.85$ with $\sigma_{\alpha_{\mathrm{nth}}} = 0.13$ (Niklas et al. 1995b). The model by Lisenfeld et al. (1996) which allows finite escape probabilities predicts in the case of diffusive electron transport non-thermal spectral indices between 100 MHz and 10 GHz in the range of $\alpha_{\mathrm{nth}} \simeq 0.9 \cdots 1.0$. According to Niklas et al. (1995b) two thirds of the galaxies exhibits flatter synchrotron spectra.

2.2 Radio halos and electron transport

The absence of large radio halos indicates large energy losses of the relativistic particles within the disk and thus efficient confinement of the electrons in the galaxy. Additionally, if the galactic magnetic field is oriented mainly parallel to the plane and the star formation activity is low so that no strong outflows occur, the convective escape probability of the electrons is also low. A prominent example for a calorimeter galaxy is NGC4565 with its thin radio disk and its steep non-thermal spectrum ($\alpha_{\mathrm{nth}} = 1.18 \pm 0.11$). On the other hand the non-thermal spectral indices of galaxies with large radio halos (e.g. NGC253, NGC891 and NGC4631) are clearly in disagreement with the calorimeter model. Especially NGC4631 ($\alpha_{\mathrm{nth}} = 0.78 \pm 0.04$) shows strong signs of star formation and its magnetic field is oriented perpendicular to the plane. The detection of bright X-ray halos of NGC253 and NGC891 (Bregman & Pildis 1994; Pietsch 1994) indicates gas outflows from these galaxies. Other galaxies of the sample of Niklas et al. (1995b) with a star formation efficiency SFE $\geq 10 \, \mathrm{L_\odot/M_\odot}$ exhibit flat synchrotron spectral indices ($\alpha_{\mathrm{nth}} \leq 0.75$). In these active galaxies convective outflow by a wind or the occurence of galactic fountains are expected. The influence of convective electron transport may flatten a spectrum significantly.

2.3 Transition from optically thick to thin

In the model of Helou & Bicay (1993) the transition from optically thin (low FIR luminosity) to optically thick galaxies (high FIR luminosity) should be accompanied by an increase in spectral index by 0.5 because bright galaxies should be those with strong energy losses and therefore low escape probability. If synchrotron losses affect the non-thermal spectra, a correlation between α_{nth} and the average magnetic field strength is expected. The latter quantity can be determined by assuming equipartition between the energy density of the cosmic rays and the magnetic field. Fig. 1 shows that neither a correlation between

α_{nth} and L_{FIR} nor between α_{nth} and B_{eq} exist. Therefore, one can conclude that the variations in synchroton spectral index are due to both energy-loss processes and propagation properties of the cosmic ray electrons.

3   A new approach

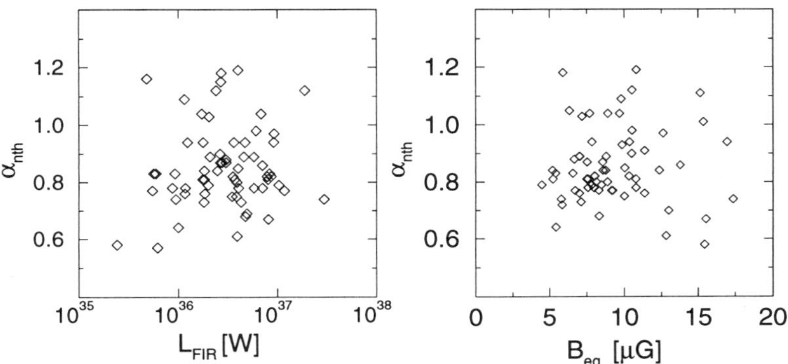

Figure 1: Non-thermal spectra indices α_{nth} of 74 Shapley-Ames galaxies plotted versus the FIR luminosity L_{FIR} (left) and versus the equipartition magnetic field strength B_{eq}(right)

The conclusion from the analysis of the galaxy survey data by Niklas et al. (1995a) is that a significant number of galaxies do not behave like calorimeters and the data are also in conflict with the optically-thin model by Helou & Bicay (1993). One clue to the understanding of the correlation is the relation between the average magnetic field strength and the average volume density of neutral gas ρ in a galaxy. Such a relation has been found from observational data with various spatial resolutions. Additionally, relations between ρ, the star formation rate SFR and L_{FIR} are required and equipartition between the energy density of the cosmic-ray electrons and B is assumed (Fig. 2). Combination of the assumed and measured proportionalities leads to a slope of the non-thermal radio-FIR correlation of 1.32 ± 0.29, which is in good agreement with the observed value. A detailed discussion of these relations and their implications can be found in Niklas & Beck (1996). This approach proposes that the gas content is the primary factor controlling the physics of the interstellar medium. The tight radio-FIR correlation is a byproduct.

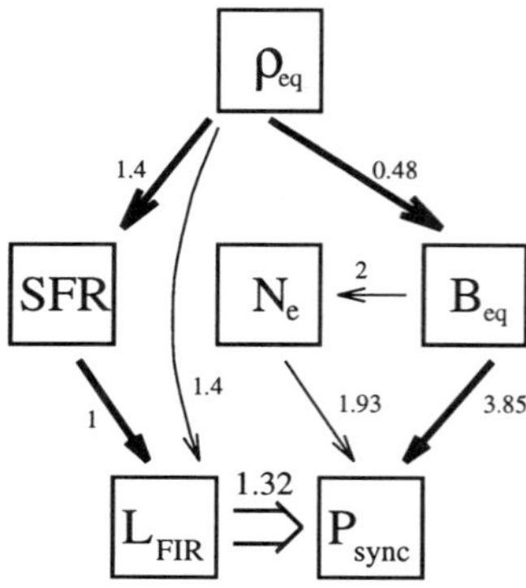

Figure 2: Diagram of the relations between the different parameters controlling the radio-FIR correlation. The *observed* correlation (thick arrows) are between the volume density of the neutral gas ρ_{eq}, the SFR and the strength of the equipartition magnetic field B_{eq}, and betweem SFR and L_{FIR}. Equipartition between the energy densities of the cosmic-electrons N_e and of B is assumed. The correlation follows from these relations.

References

Bogdan T.J., Völk, H.J., 1983, A&A 122, 129

Bregman J.N., Pildis R.A., 1994, ApJ 420, 570

Helou G., Bicay M.D., 1993, ApJ 415, 93

Lisenfeld U., Völk H.J., Xu C., 1996, A&A 306, 677

Niklas S. 1995, PhD Thesis, University of Bonn

Niklas S., Beck R. 1996, A&A, subm.

Niklas S., Klein U., Braine J., Wielebinski R., 1995a, A&AS 114, 21

Niklas S., Klein U. ,Braine J., Wielebinski R., 1995b, in *New Lights on Galaxy Evolution*, Bender, R., Davies, R. (eds.), IAU Symp. No. 171, Kluwer Academic Publishers, in press

Völk H.J., 1989, A&A 218, 67

Xu C., Lisenfeld U., Völk H.J., 1994, A&A 285, 19

Address of the author:

S. Niklas, Max – Planck – Institut für Kernphysik, Postfach 10 39 80, 69029 D-Heidelberg, Germany

The vertical structure of the galactic gaseous disk and its relation to the dynamo problem

H.-E. Fröhlich, M. Schultz, D. Elstner

1 Introduction

Following the standard approach (cf. Krause & Rädler 1980; Krause 1994) for the numerical simulation of a galactic dynamo, one has to specify the input physics: the source parameter $\alpha(r, z)$ and the destruction parameter, the magnetic diffusivity $\eta(r, z)$. These two functions, containing the effects of turbulent motions on the large-scale magnetic field, essentially determine, for example, the pitch angle and the strength of the galactic dynamo. They are intrinsically averages over scales larger than the largest eddies, but, of course, smaller than the system considered.

To calculate these material quantities one needs besides the turnover-time τ_{corr} and angular velocity Ω both, the smoothed density, $\rho(r, z)$, of the gas and the turbulent velocities, $u'(r, z)$, as functions of the place (r, z) within the Galaxy. Especially the thickness of the gas layer and the behaviour of turbulent velocity with height above the mid-plane are important.

The following expressions for α and η are from Rüdiger & Kitchatinov (1993):

$$\alpha = -\frac{8}{15}\,\tau_{\mathrm{corr}}^2\,\Omega\,\langle u'^2\rangle\left(\frac{3}{2}\frac{d}{dz}\log\rho + \frac{1}{2}\frac{d}{dz}\log\langle u'^2\rangle\right), \tag{1}$$

$$\eta = \frac{1}{3}\,\tau_{\mathrm{corr}}\,\langle u'^2\rangle, \quad \text{where} \tag{2}$$

$$\langle u'^2\rangle = 3\,\langle u_z'^2\rangle. \tag{3}$$

The density distribution taken is the observed one. To get the turbulent velocities we need the turbulent gas pressure. The total pressure follows from the assumption of hydrostatic equilibrium. It can be computed from the observed density distribution and the z-component of the gravitational acceleration. Because we are lacking detailed information concerning the non-kinetic pressure distributions, magnetic pressure and cosmic ray pressure, we have taken that *minimal* turbulent pressure, which is necessary to prevent onset of Parker instability.

The criterion used is that delineated by Lachièze-Rey et al. (1980). Hence we have obtained with that assumption only *lower* limits for the α- and η-coefficients. Upper limits for α and η result from identifying the turbulent pressure with the total one, thus neglecting magnetic as well cosmic ray pressure gradients.

2   Observed properties of the galactic gas disk

For the density stratification the empirical H I distribution from Dickey & Lockman (1990) has been taken: a combination of two Gaussians, and an exponential. This profile is valid at least for $3\,\mathrm{kpc} \le r \le r_\odot$ (where $r_\odot = 8\,\mathrm{kpc}$.) Then we have added a further Gaussian to include the molecular gas layer. The extended ionised gas, which is the source of the diffuse Hα emission at high latitudes, has been described by a further exponential with a scale-height 1500 pc (Reynolds 1989).
For the sake of simplicity, this normalised gas density profile, which is appropriate to describe the gas in the solar neighbourhood, is assumed to be valid in other parts of the Galaxy too.

For smaller radial distances, i.e. $r \le 3\,\mathrm{kpc}$, the same shape of the density profile has been taken, but the vertical scale-height is scaled down linearly in such a manner that it diminishes at $r = 0$.

In order to account for the observed flaring-up at $r > 8\,\mathrm{kpc}$, we have let the vertical scale-height in the outer part of the Galaxy to be proportional to r.

It should be noted that as long as self-gravitation of the gas can be neglected only the shape of the density distribution is of relevance for our task.

The gravitational acceleration towards the plane k_z,(Oort 1960) is approximated by the expression:

$$k_z(r,z) = \frac{2\sigma(r)^2}{z_0} \tanh\left(\frac{z}{z_0}\right) + \left(\frac{v_{\mathrm{rot}}}{r}\right)^2 z \,. \tag{4}$$

σ^2 is the vertical velocity dispersion of old disk stars.

The first contribution is from a self-gravitating isothermal sheet of old stars with constant half-thickness ($z_0 = 600\,\text{pc}$). The second term comes mainly from the dark matter halo of the Galaxy. To model the radial dependence of the force k_z an exponential Freeman disk is assumed:

$$\sigma(r)^2 = (20\,\text{km s}^{-1})^2 \, \exp(-(r - r_\odot)/0.44\,r_\odot). \tag{5}$$

The velocity dispersion, somewhat rounded, is that for the population of normal-metallicity disk K giants (Bahcall 1984), which seems to be well described by an isothermal distribution function.

The second term is computed from the Bahcall-Soneira model (cf. Bahcall 1986). Of course, $v_{\text{rot}}(r)$ is the rotational velocity the galaxy would have if there were *no* stellar disk.

With the assumption of hydrostatic equilibrium, at least in the time average, and ignoring any external pressure, the total pressure is computed and that fraction of it which seems to be necessary for hydro-magnetic stability. From that minimal pressure and the given density distribution the turbulent velocities follow:

$$\frac{d}{dz}\,p_{\text{tot}} = -\rho\,k_z, \tag{6}$$

$$p_{\text{min}}/p_{\text{tot}} = \frac{1}{\gamma_c}\,\frac{d\,\ln p_{\text{tot}}}{d\,\ln \rho}. \tag{7}$$

As usual $\gamma_c = \delta \ln p_{\text{gas}}/\delta \ln \rho$ describes how density and pressure perturbations are related. Following Bloemen (1987) we take $\gamma_c = 1$.

In order to calculate the unquenched α- and η-coefficients, we have to specify the characteristic time-scale for the turbulent eddies — the turnover time. The following parameterisation has been adopted:

$$\tau_{\text{corr}}(r) = c_\tau / \sqrt{\left(\frac{d\,k_z}{d\,z}\right)_{z=0}}. \tag{8}$$

Except the dimension-less c_τ, which remains the only adjustable turbulence parameter, $\tau_{\text{corr}}(r)$ corresponds to the local hydrostatic time-scale.

The angular velocity $\Omega(r)$ is according to the Bahcall-Soneira model (Bahcall 1986).

3   Resulting α- and η-distributions

The resulting distributions (minimal values and maximal ones) of $\alpha(r, z)$ and $\eta(r, z)$ are given in the Figures.

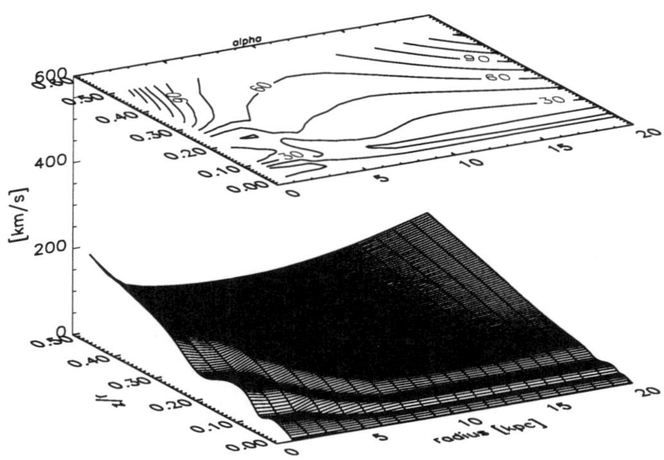

Figure 1: The minimal α-effect for $\gamma_{\mathrm{c}} = 1$ and $c_\tau = 1$

(The coefficients are plotted over a r-z/r-plane.) A lot of the fine-structure in z-direction is apparently artificial and unintended. It reflects merely the very complicated gas density profile taken — a superposition of diverse Gaussian and exponential contributions.

The turbulent velocity u' rises with vertical distance z. This is due to the exponential tail in the density profile. To get such an exponential decline within a harmonic potential the velocity dispersion u'^2 must rise linearly with height z.

Because we have taken the scale-height of the gas constant, this means independent of radial distance, at half the sun's distance much higher turbulent velocities are necessary to support the gas against the much deeper potential wall there. Such higher velocities are not excluded by observations, as Dickey & Lockman (1990) stated.

The quantity $D(r, z) = \int_0^z \alpha\left(-\Omega' r\right) \tau_{\mathrm{corr}} \eta^{-1}\, dz$ can be regarded (in vertical

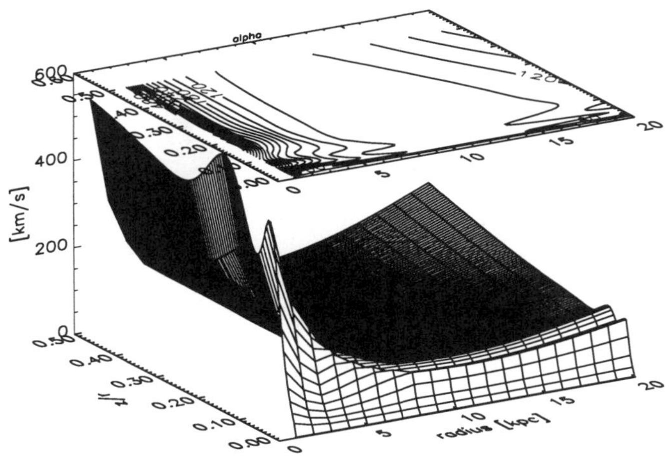

Figure 2: The maximal α-effect for $c_\tau = 1$

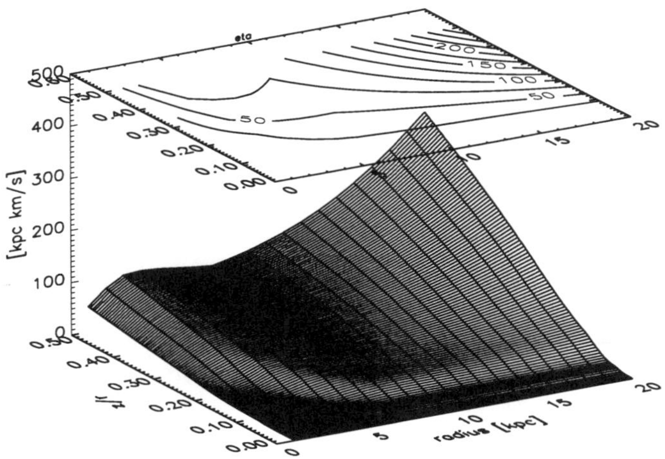

Figure 3: The minimal eddy diffusivity for $\gamma_c = 1$ and $c_\tau = 1$

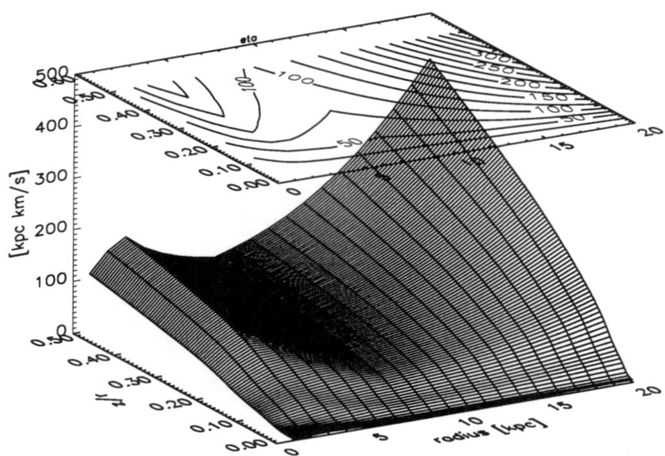

Figure 4: The maximal eddy diffusivity for $c_\tau = 1$

direction) as a cumulative dynamo number. As Fig. 5 reveals, the galactic dynamo is expected to work preferentially in heights some 100 pc away from the mid-plane, but obviously the higher regions contribute too. There is no saturation discernible.

The strong increase towards the centre is mainly due to the fact, that the rotational speed remains high except very near the centre. With a much more extended region of rigid rotation this feature would vanish. By the way, this dynamo number $D(r, z)$ does not depend strongly on whether the minimal gas pressure has been taken or the total one.

There is, of course, a thin, but nevertheless important halo component (cf. Spitzer 1990). Incorporation of an exponential halo, which is assumed to reach down to the bottom of the gaseous disk ($z = 0$), would alter the α-coefficients only marginally as compared with the no-halo case. Fig 6. shows for the solar neighbourhood the run of α (filled region) for a range of halo models (Bloemen's model 4 with halo scale-heights: $2\,\mathrm{kpc} \leq h_{\mathrm{halo}} \leq 10\,\mathrm{kpc}$). The standard case is indicated, too.

But a thicker and more pronounced H_2 disk (still well within the HI layer), as claimed for by Dame & Thaddeus (1994) and Malhotra (1994), who have detected a population of smaller clouds having a larger scale-height, would raise the α-values somewhat. To show this raising in α (dashed line) observed mid-plane H_2 densities and FWHM-values from Dame & Thaddeus, which

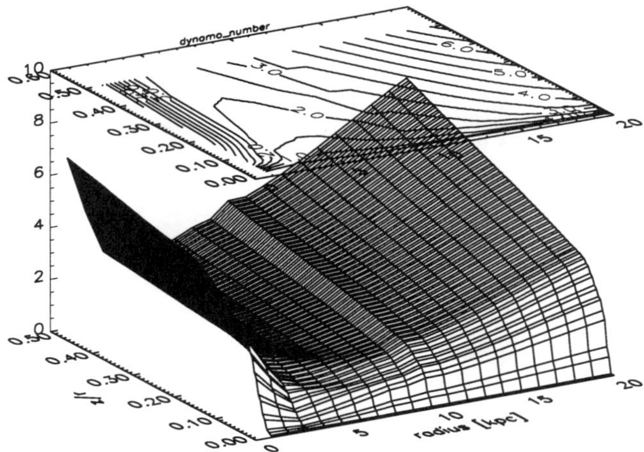

Figure 5: The cumulative dynamo number, $D(r, z)$, for minimal gas pressure and $\gamma_c = 1$, $c_\tau = 1$

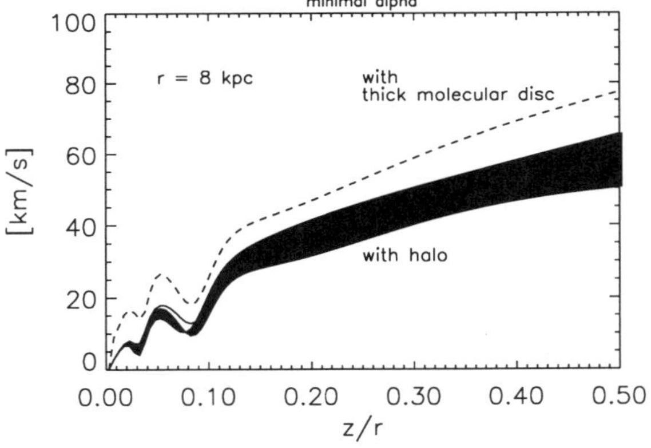

Figure 6: An additional halo component would not alter α, but a thick H_2 disk would raise it somewhat. The standard case (full line) runs generally within the filled region. $r = 8\,\mathrm{kpc}$

have been derived for the Sagittarius spiral arm, have been taken instead of the standard values.

Self-gravity of the gas is then not longer negligible and, therefore, has been taken into account. Of course this situation is representative for a spiral-arm region only.

4   The dynamo

Within this concept the incoming parameters for the turbulent dynamo are given. We consider an axisymmetric differentially rotating disk. For the observed density distribution and rotation law the correlation time is the free parameter in our model, which is varied from $2 \cdot 10^7$ to $8 \cdot 10^7$ years at 8kpc ($c_\tau = 1 - 4$ cf. Eq. 8). The full expressions for the α-tensor including the off diagonal elements magnetic buoyancy and diamagnetism are given in Elstner et al. (1996). The dependence of the turbulence parameters from the mean magnetic field goes mainly over the field strength normalised with the equipartition value. Because of the exponential decay of the equipartition field with distance from the galaxy, the linear increasing α doesn't support dynamo action far outside the galaxy in the nonlinear case. In contrary to the common practice, to consider only the influence of the mean magnetic field to α, the action on to the turbulent diffusivity has to be considered as well.

Following Kitchatinov et al.(1994) the magnetic diffusivity decreases with increasing mean field but new terms arise in the EMF

$$\mathcal{E} = ... + \mathrm{rot}(-\eta_\mathrm{T}(\bar{\mathrm{B}})\mathrm{rot}\bar{\mathrm{B}} + \mathbf{U}^\mathrm{mag} \times \bar{\mathrm{B}}) \tag{9}$$

with

$$\mathbf{U}^\mathrm{mag} = \hat{\eta}(\bar{\mathbf{B}}) \ \nabla \log \bar{\mathbf{B}}^2 + \eta_\mathbf{z}(\bar{\mathbf{B}}) \ \frac{\mathbf{j} \times \bar{\mathbf{B}}}{\bar{\mathbf{B}}^2}. \tag{10}$$

The terms with η_T and η_z are acting as magnetic field killers and the term with $\hat{\eta}$ concentrates magnetic field. The effective diffusivity is reduced with increasing magnetic energy density. Now it depends if the α effect is quenched strong enough to get a finite magnetic energy. The calculations show that we find finite energies of the magnetic field, but orders of magnitude different for stationary and oscillating solutions.

To illustrate the effects of magnetic diffusivity quenching we compare the results with calculations, where the effect is switched off.

Under the chosen parameters the models without η-quenching give stationary quadrupole solutions for short correlation times and oscillating dipoles for longer correlation times. The magnetic field strength for these models agrees well with the equipartition value, which is in the order of several μG. The solutions for longer correlation time show dipole symmetry, which seems to be dependent on the assumed vertical density stratification. Other models with slightly changed density stratification can result in oscillating quadrupoles.

The main point are the reversals of the magnetic field due to the oscillations, which are present for large correlation times.

The inclusion of η-quenching results in unacceptable large magnetic fields for small correlation times. The stationary quadrupole solution does not stop at equipartition field strength with turbulence quenching only.

Also the new terms arising in the electromotive force for non vanishing magnetic fields are not able to stop the dynamo action for reasonable field strength. The only term which could help in this case is (rot $\mathbf{B} \times \mathbf{B}$) $\times \mathbf{B}$. It acts strong in regions, where the magnetic field changes sign. That seems to be the reason for much weaker fields for the oscillating modes (large correlation times).

The nonlinearity only acting on to the turbulence shows rather good results for the case without η-quenching. Taking into account the full nonlinearity to the turbulence, only the oscillating solutions remain as a possible explanation of galactic magnetic fields within the axisymmetric models.

The results of the calculations are summarised in Table 1.

Table 1: Characteristics for dynamo models without η-quenching and with η-quenching (in brackets). The only free parameter c_τ is varied

c_τ	B_{max} [μG]	β_{max}	parity	pitch angle	period [Gyr]
1	14 ($>$600)	1.1	quad. (quad.)	-5	∞ (∞)
2	13 ($>$600)	1.1	quad. (quad.)	-10	∞ (∞)
3	11 (65)	0.9 (5.4)	dip. (dip.)	-25 (-10)	0.50 (0.75)
4	14 (90)	1.1 (7.3)	dip. (dip.)	-30 (-5)	0.40 (1.40)

The magnetic field strength for the stationary quadrupole without η-quenching and the dipole with η-quenching are plotted in Figure 8. The maximum field is located at about 500pc above and below the galactic plane. Because of the dipole character the gape in the plane is more pronounced in the solution for $c_\tau = 3$ than in the models with smaller c_τ. The radial extension of the field strength is varying with time for the oscillating solutions.

Figure 7 shows the magnetic field at six times in the half cycle of the model with η-quenching and $c_\tau = 3$. The toroidal field component has different signs for northern and southern half planes (dipole) and there are reversals along the radius, which move during the oscillation radially outwards. The poloidal field lines close within a radius of about 10kpc most of the time, but there are

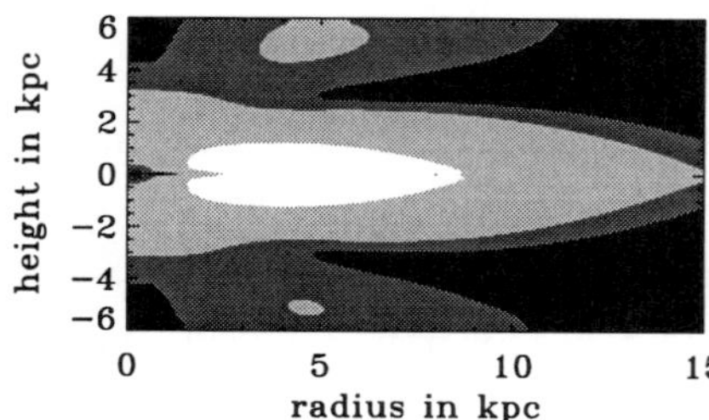
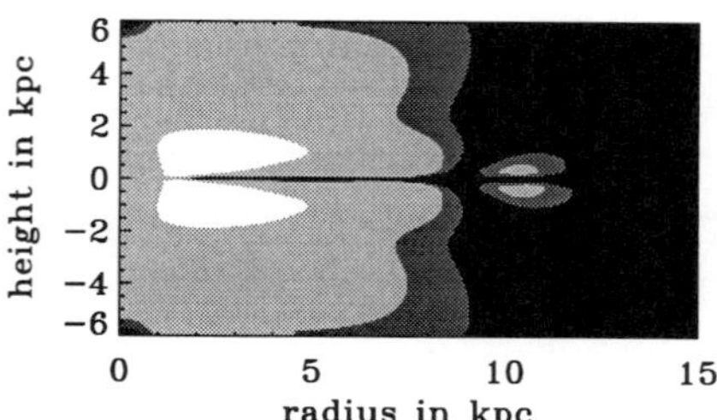

Figure 7: Magnetic field strength distribution for the model without η-quenching (left panel) and with η-quenching (right panel). The scale is $(0.001, 0.05, 0.1, 0.5)B_{\mathrm{max}}$ from dark to light

periods, where the field lines close radially far outside the galaxy.

The models without η-quenching show more or less good agreement with observed magnetic fields of galaxies. The α deduced from observed density-profile is large enough to reproduce observed pitch angles of about $30°$ (cf. Elstner et al. 1996). The vertical extension of the magnetic field in these models have scale heights of the order of 1kpc. The field would appear mostly parallel to the disk in an edge-on polarisation view.

Inclusion of η-quenching rules out the stationary quadrupolar solutions at least in the axisymmetric case, if no further nonlinear back reactions of the magnetic field are considered. One could imagine that magnetic driven winds can damp the dynamo action in these cases. This idea is promising, because these models have strongest field induction above the disk, where the wind should be already effective enough to carry away the magnetic field. A stronger α-quenching alone could also stop the dynamo, but then the pitch angles would be very small or the dynamo wouldn't work at all. Other diffusive processes like reconnection are thinkable to solve the problem.

Another conclusion of this calculations is, that quadrupolar symmetry of the magnetic field is not the only possibility for todays galactic dynamo models.

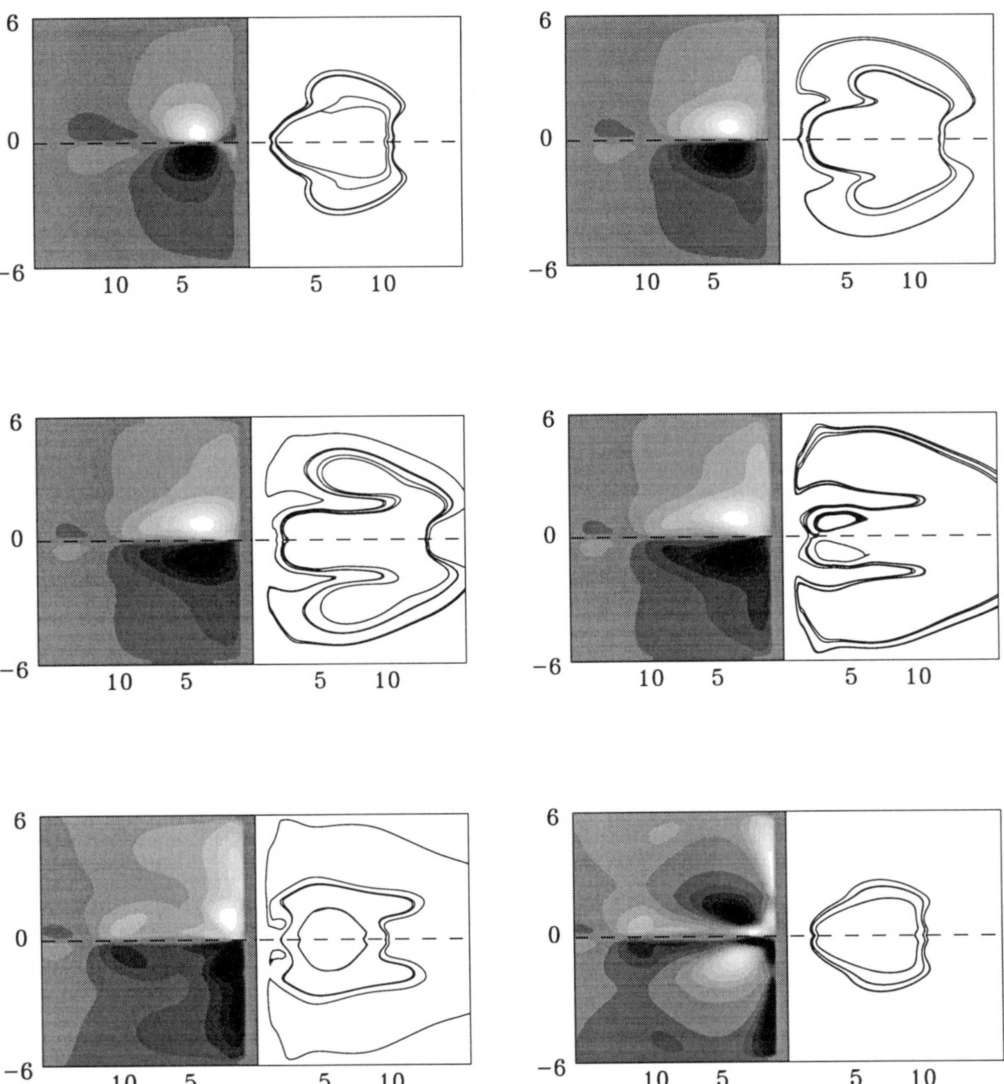

Figure 8: Snapshots of the oscillating magnetic field from the model with $c_\tau = 3$ and η-quenching. The greyscale plot shows the field strength of the toroidal component, with white for positive and black for negative direction. The field lines of the poloidal field are plotted in the right panel. The units are kpc

References

Bahcall J.N., 1984, ApJ 287, 926

Bahcall J.N., 1986, ARA&A 24, 577

Bloemen J.B.G.M., 1987, ApJ 322, 694

Dame T.M., Thaddeus P., 1994, ApJ 436, L173

Dickey J.M., Lockman F.J., 1990, ARA&A 28, 215

Elstner D., Rüdiger G., Schultz M., 1996, A&A 306, 740

Kitchatinov L.L., Pipin V.V., Rüdiger G.,1994, Astron. Nachr. 2, 157

Krause F., 1994, in: Cosmical Magnetism, ed. D. Lynden-Bell, Kluwer Academic Publishers, p. 161

Krause F., Rädler K.-H., 1980, Mean-Field Magnetohydrodynamics and Dynamo Theory, Akademie-Verlag, Berlin, and Pergamon Press, Oxford

Lachièze-Rey M., Asséo E., Cesarsky C.J., Pellat R., 1980, ApJ 238, 175

Malhotra S., 1994, ApJ 433. 687

Oort J.H., 1960, Bull. Astr. Inst. Netherlands 15, 45

Reynolds R.J., 1989, ApJ 339, L29

Rüdiger G., Kitchatinov L.L., 1993, A&A 269, 581

Spitzer L., 1990, ARA&A 28, 71

Address of the authors:

H.-E. FRÖHLICH; M. SCHULTZ; D. ELSTNER, Astrophysikalisches Institut Potsdam, An der Sternwarte 16 D-14482 Potsdam, Fed. Rep. Germany

Magnetization of the Intergalactic Medium: by Primeval Galaxies

Philipp P. Kronberg & Harald Lesch

1 Introduction

Recent radio, X-ray and optical observations of galactic scale winds have revealed a number of surprises. Nearby-universe starburst galaxies have (i) extensive radio and X-ray halos, resulting from vigorous outflow, and (ii) in cases such as M82 where faint, polarized radio emission and co-extensive X-ray emission have been imaged in great detail, we find that strong and organized magnetic field structures exist throughout the radio halo. The field strengths, up to $10\mu G$ or more are as strong as in the much denser disk of our Milky Way. The halos of nearby starburst galaxies such as M82 are constrained systems, in the sense that we can put firm limits on their age, total number of ionizing photons, time taken to evacuate the inner starburst region, and hence snuff out the starburst. These constraints, based on stellar evolution, limit the timescale of the outflow phenomenon.

A relatively large volume of halo was very "inflated up" in a short period of time – probably less than 10^8 years. Since we observe the halo, *e.g.* of M82 to be highly magnetized, some very rapid field regeneration process must have been associated with the outflow (*cf.* Lesch et al. (1989)) to fill it with sufficient magnetic flux. It seems reasonable on several arguments to expect that M82-like starbursters have, over their lifetime, experienced several such starburst events – perhaps every few x 10^8 years, each episode lasting a few x 10^7 years. This suggests that each event filled the surrounding i.g.m. with ejected (and metal-enriched) gas, cosmic rays and magnetic fields. In addition to a likely past history of repeated starbursts, it is reasonable to expect that most galaxies in the nearby universe have undergone one or more periods of vigorous ejection of magnetized cosmic ray gas in an episode of intense star formation and consequent bursts of supernovae. Further, if the first generations of galaxies were starbursting dwarfs, these more numerous, smaller galaxies also presumably had vigorous outflow halo phases, aided by their lower escape velocities. Many will have later merged to form larger galaxies.

What we want to ask here is whether this sort of galactic "volcanic" activity in the early, more compact universe had a significant impact in magnetizing the intergalactic medium. If so, to what extent, and over what cosmic epochs? Obviously the answers to such questions have potentially far-reaching implications for galaxy, and group/cluster evolution. A key interesting question is whether a sizable number of present-universe galaxies formed from a pre-magnetized i.g.m.

In this presentation, we briefly explore the question of starburst seeding of the i.g.m., and attempt to arrive at some first-order conclusions based on plausible scenarios of early galaxy formation, and outflow models. We extrapolate what we have recently learned from detailed studies of *nearby* galaxy-scale outflows back into earlier epochs of galaxy formation, and we examine some magnetic field regeneration mechanisms in galaxies that have a vigorous outflow.

2   The "Volcanic Early Universe" of Starburst Galaxy Halos

The question of how much of the "available i.g.m." is filled with a magnetic field can be posed in the form of the following model: The outflow zones of starbursting galaxies in the more compact earlier universe inflated magnetized halos by analogy with what we observe in M82. We can use the outflow timescales of M82 and other, analagous "closeby" systems as a benchmark to infer how much halo volume is inflated over the lifetime of a single starburst. The outflow velocity is of interest, along with the question of whether it is greater that the escape velocity. Another parameter of interest is the repeat interval of a starburst for a given galaxy –*i.e.* due to subsequent galaxy-galaxy interactions, and other dynamical effects which can serve to feed fresh gas into the nuclear region, and hence trigger the next starburst.

For a detailed model we need to know the redshift range over which the first galaxies formed, and the cosmological model, i.e. R(t) where R is the cosmological scale factor as a function of proper time. In addition, we want to know whether all comoving volume elements had an equal probability of galaxy formation, hence outflow activity, and also the hierarchical evolutionary scenario, in the course of which some (cosmic time-dependent) fraction of dwarf galaxies subsequently merged into larger galaxies.

Since we do not yet have good answers to the specific questions above, we shall attempt some crude estimates of the fraction of the i.g.m. which could be "magnetized" by starburst-driven outflow halos, based on some simple assumptions and recent observations at the present cosmological epoch. We can imagine a universe model in which a typical co-moving volume has a radius, $r(z)$, defined as one half the distance to the nearest neighbouring galaxy. Thus at a given epoch, $t(z)$, a co-moving volume of extragalactic space can be modelled as a combination of contiguous spheres, each having one galaxy at its centre. (If two galaxies were to subsequently merge, then their individual, "proprietry" spheres also merge to form a new combined spherical volume.) Thus, the volume of each of these spheres represents the volume of intergalactic space, V_A, which is *available* to be filled with outflow gas. The i.g. volume which gets filled with magnetized outflow halo gas (V_F) constitutes some fraction ($f(z)$) of V_A at a given z. V_F will depend on (i) the elapsed $t(z)$, and (ii) the parameters that specify the outflow.

We choose parameters which can be related to observed outflows in nearby galaxies having enhanced star formation: Our model calculates V_F by assuming the outflow "pipe" has a fixed average cross section, σ, effective outflow velocity, v_o, of magnetized CR + thermal gas, and a total duration of the outflow (T). The volume of the resulting halo after time τ is defined by the radius, $r(0)$, of an equivalent sphere whose volume is V_F. Thus $V_F(\tau) = v_0.\sigma.\tau$.

We adopted cosmological parameters that are consistent with a "realistic" universe age (t_0, at $z = 0$) that is, consistent with minimum globular cluster ages of not less than ~ 12 Gyr.

The question of whether the i.g.m. has been substantially magnetized by early starburst galaxies amounts to asking whether the ratio, $f(z) = V_F(z)/V_A(z)$ can be greater than a negligibly small number for a reasonable range of the parameters just discussed. Many parameters are not observationally well constrained, such as those describing the detailed scenario of hierarchical galaxy formation and its distribution over proper time, and how much of the ejected halo gas participates in the Hubble flow. However, recent improvements in observations of outflow halos in the nearby universe provide us with some of the first "handles" on the relevant parameters for a simple model. These include the frequency, likely duration, the rate of outflow, and the degree of magnetic flux increase (see below) in strong starburst outflows.

In Figure 1 we show the interrelation between redshift, proper time, and $f(z)$ for $H_0 = 60$, $\Omega = 0.3$, $\Lambda = 0.7$. The calculations show the fraction of the i.g.m. that is *filled* with magnetized halo gas by extrapolating backward from the present epoch (at which the relevant parameters were estimated). It makes the conservative assumption that *all* the outflow halo material is detached from the Hubble flow.

In the illustrative calculation shown in Figure 1, we have assumed a present-epoch average inter-galaxy separation ($=2\,r_0$) of 0.2 Mpc, an average outflow velocity of 800 km/s, and a total outflow time of 500 Myr $--$ *i.e.* the total time over what may have been repeated outflow episodes. Since we don't know the merger scenario since the time of formation of the first galaxies, our chosen value of r_0 refers to a higher local density of galaxies than we actually see at $z = 0$. It should also be noted that r_0 refers to the typical (pre-mergers) galaxy separation within those zones of the Universe where galaxies are found, *i.e.* not within the large intergalactic voids, whose volume we also don't "count" in computing $V_A(z)$ and $f(z)$.

Assume, for example that most of the galaxies formed, and generated vigorous starburst outflows over a period from $z = 8$ to $z \sim 3.0$, (a range in t of 1.73 Gyr for $H_0 = 60$, $\Omega = 0.3$, $\Lambda = 0.7$). Then projecting backwards to redshift 3.0 (at which the Universe is ~ 2.5 Gyr old), roughly 10% of the i.g.m. will be filled with starburst halo gas and magnetic field. Fig. 1 also shows that the elapsed cosmic time at $z = 3.0$ is ample to have allowed sufficient outflow, provided most galaxies are at least a few hundred Myrs old by this time. In Figure 1 we have stopped the backwards (in time) calculation of $f(z)$ at $z = 4$, since at higher z's the elapsed cosmic time becomes too short to justify a prediction of $f(z)$ without a more elaborate early universe galaxy formation model. (Refer to the right hand scale and the $t(z)$ curve.)

This calculation ignores the likely possibility that at least some of the ejecta will have co-expanded. In reality, a higher $f(z)$ would result at the lower redshifts (albeit with lower $B_{i.g.}$ values). That is because a high $f(z)$ at large redshifts, say $z \geq 4$, will make it more likely that the outflow material, extending well beyond the parent galaxy's gravitational influence, will subsequently co-expand with the Universe. This would tend to make $f(z)$ significantly larger at low redshifts than Figure 1 shows.

We conclude from the sample calculation above that a substantial fraction of intergalactic space could plausibly have been permeated by magnetized outflow material from the stellar energy driven in galaxies. The extent to which this is the case will be greater, the earlier the first generations of galaxies formed, and the older the universe is. If true, and given that diffuse magnetic fields are not easily destroyed once generated, the i.g.m. in those regions of the universe containing galaxies will be permeated by a magnetic field at some level. A further consequence of this outcome is that intergalactic magnetic fields could originate in stars, and not require a primordial, pre-formation seed field.

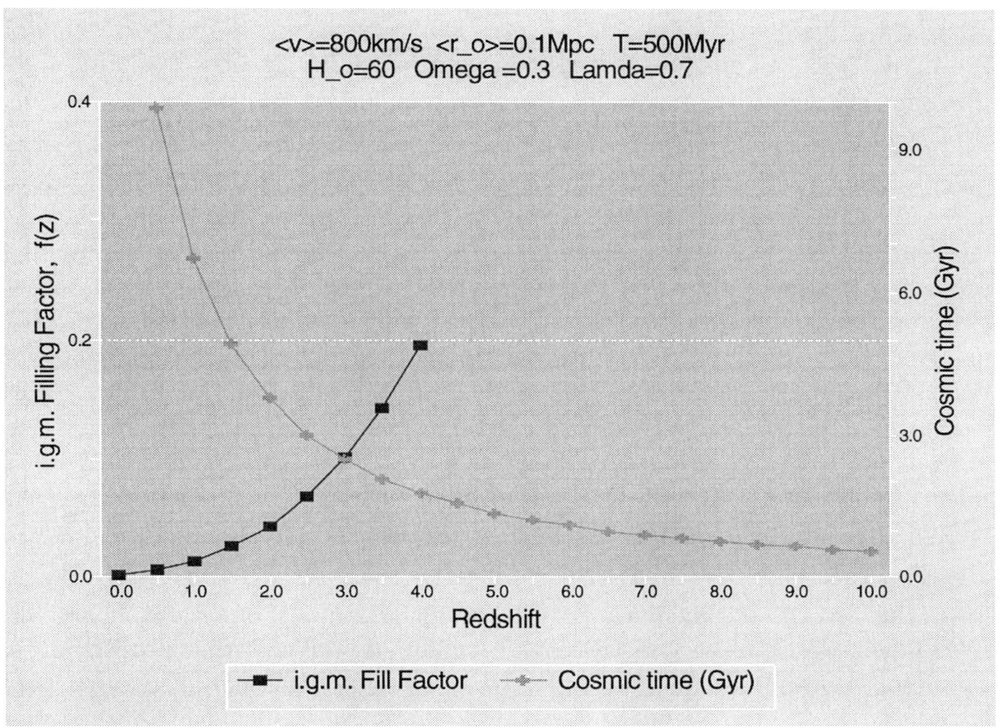

Figure 1: Plot showing the interrelation between cosmic time, redshift, and $f(z)$ for a universe with H_0=60, Ω=0.3, and Λ=0.7. The i.g.m. filling factor is extrapolated backwards unde r the assumption of an effective outflow "chimney" area of 2 kpc^2, an average outflow velocity of $800km/s$, and an aggregate outflow time of $500Myrs$. The average galaxy-galaxy separation at $z = 0$ is $0.2Mpc$, (*i.e.* twice r_0.)

3   Early Universe Analogues of Jet-Lobe Radio Sources as an Additional Source of I.G.M. Fields?

The second type of astrophysical "system" which could seed a large part of the i.g.m. with widespread magnetic field are extended, jet-lobe radio galaxies (*cf.* Rees (1987)). We shall not consider these *additional* candidates for i.g.m. magnetization at present. They are more difficult to model than starbursters, since we have even less evidence as yet for their occurence rate, and for the aggregate lobe volume (of magnetic fields and cosmic rays) that have been inflated by jets at earlier epochs. Jet-CR lobe sources could seed a fraction of the i.g.m. with a substantial magnetic field in addition to the starburst-halo phenomenon discussed above.

At redshifts below ~ 1 we have evidence that extended radio lobes can very quickly (in $\leq 10^8$ years) fill a large volume of the i.g.m. with cosmic rays, and magnetic fields of ~ 10 to $100\mu G$. By contrast, extended lobes largely "disappear" at $z \geq 2$. Two possible reasons why we haven't observed extended radio lobes at large z are: (i) that they are more tightly confined by a denser i.g.m. environment and never develop, or (ii) that extended, magnetized lobes are generated at the earlier cosmological epochs, but that the much higher rate of inverse Compton cooling of the CR electrons ($\propto E^2$), due to the higher background photon density strongly attenuates their synchrotron emissivity. For our purposes possibility (ii) is very relevant, since the magnetic fields, and most of the CR nuclei will still exist; but are just not visible because of the lack of cm-wave synchrotron radiation. Once created, the magnetic fields will remain in the i.g.m. for a very long time, since they are not easily destroyed.

Since the true co-moving density of FRI and FRII radio sources at $2 \leq z \leq 8$ is not known, we have no observational "handle" in modelling their contribution to a more widespread i.g.m. field. An interesting comment in passing here is that, if possibility (ii) is the case, we could detect high redshift radio lobes at very low radio frequencies, which are relevant to the lowest energy cosmic rays. The fact that their emission will be redshifted to even lower frequencies emphazises the significance of searching for radio images at arcsecond-level resolution *at the very lowest radio frequencies*. This is at present a largely unexplored observational window.

4   Magnetic Field Amplification in Gas Rich Dwarf Galaxies

In this section we discuss and briefly review some mechanisms for field amplification in galaxies as it relates to the outflow process. Magnetic fields are the result of electrodynamic processes acting on the highly conducting cosmic plasmas. One distinguishes between two qualitatively different process classes. The first class is related to the so called "seed-fields". These are produced by electric currents, which originate from forces acting differentially on protons and electrons: the interaction of a plasma with a neutral gas, or the reaction of a plasma to an intense radiation source, or a rotating plasma in a gravitational field. The seed fields are typically of the order of 10^{-18} G (Rees 1987). In order to reach the observed field strengths of several μ G in galaxies, a very efficient amplification is necessary.

This amplification is provided by the second class of electrodynamic mechanism, magnetohydrodynamic (MHD) processes, which amplify a given magnetic flux via induction, i.e. $\mathbf{v} \times \mathbf{B}$. The concept of a dynamo is related to these MHD mechanisms. A galactic dynamo consists of a chain of processes which finally results in an exponential growth of the magnetic field (e.g. Kronberg 1994; Beck et al. 1996 for recent reviews). A galactic disk dynamo acts in two ways: first poloidal magnetic field lines (perpendicular to the disk) are stretched into the disk plane via axisymmetric differential rotation, transforming into toroidal magnetic fields. The second step is the action of turbulent motions perpendicular to the disk, which transforms the toroidal fields into poloidal fields. Such axisymmetric dynamos have growth times of the order of several Gigayears (Camenzind and Lesch 1994). In the light of the recent observations of μG fields in high-redshift galaxies, which are supposed to be progenitors of galaxies, the axisymmetric dynamos are too slow to amplify the tiny seed fields onto the μG-level (Kronberg et al. 1992).

Chiba and Lesch (1994) considered an alternative approach by adding the non-axisymmetric velocity components into the dynamo chain and neglecting the turbulent part. The nonaxisymmetric motions are driven by spiral arms and bars, which are the result of dynamical disk instabilities driven by self-gravity and differential rotation. The incorporation of non-axial motions increases the efficiency of the dynamo action significantly. Nonaxisymmetric perturbations lead to angular momentum transport outward and thereby induce radial gas motions inward. The time scale for the angular momentum transport is given by (Lynden-Bell and Kalnajs 1972; Larson 1984)

$$\tau \simeq 2 \left(\frac{M}{M_{disk}} \right) \left(\frac{\Sigma}{\delta\Sigma} \right)^2 \frac{2\pi}{\Omega}. \tag{1}$$

$M \simeq v_{rot}^2 R/G$ is the total mass interior to radius R; $M_{disk} \simeq 2\pi\Sigma$ is the disk mass interior to R and $v_{rot} = R\Omega$ is the rotation velocity. The non-axisymmetric disturbance of the gravitational potential produces a disturbance in the surface density $\delta\Sigma$.

Eq. (1) shows that if most of the mass is in the disc, $(M/M_{disk} \sim 1)$ and if a non-axisymmetric disturbance of large amplitude and wavelength is present

$\Sigma/\delta\Sigma \sim 1$, the time scale for redistribution of angular momentum is of the same order as the orbital period.

The condition $\delta\Sigma \sim \Sigma$ shows the importance of the gaseous component. Since protogalaxies can be expected to be very gas rich, the influence of non-axisymmetric perturbations is essential for the gas motion in such objects. The numerical simulations of Katz and Gunn (1991) clearly reveal that nonaxisymmetric structures appear very rapidly in the early formation of disk galaxies. It has been shown by Lubow et al. (1986) that for a gas to star mass ratio of only 15%, the gas contributes seven times more to self-gravity than the stars. The main reason for the importance of gas gravity is that a stellar disturbance is a fractionally small perturbation on an otherwise axisymmetric stellar disc, whereas the gas is fully participating. In that case only a small fraction of the stellar density generates a bar or a spiral wave, whereas a spiral field or a bar are influenced by the gaseous component. So that the assumption $\delta\Sigma \sim \Sigma$ is easily fulfilled for the gas, for even weak disturbances in the stars.

It is possible to give an approximate estimate for the reaction of the magnetic field onto a velocity field that is driven by non-axisymmetric flows (Chiba and Lesch 1994). Any density disturbance in a magnetized gas, which moves with high velocity, intensifies the component of magnetic field which is tangential to the disturbance. Since any non-axisymmetric perturbation leads to angular momentum transfer and a radial plasma inflow, it seems reasonable to approximate the dynamics of a proto-disk galaxy which contains spiral arms or bars or is interacting, by a two-dimensional velocity field $(v_r, , v_{rot}, 0)$. We use the induction equation, with diffusion neglected (Ruzmaikin et al. 1988, Sect. VII),

$$\frac{\partial \mathbf{B}}{\partial t} = \nabla \times (\mathbf{V} \times \mathbf{B}). \tag{2}$$

For the azimuthal field component B_φ one gets

$$\frac{\partial B_\varphi}{\partial t} = -\frac{\partial}{\partial r}(v_r B_\varphi) + B_r r \frac{d\Omega}{dr}. \tag{3}$$

with $B_r \sim const.$

The solution to this equation grows exponentially as

$$B_\varphi = \left[B_{\varphi 0}(r_0) + \tau B_r r \frac{d\Omega}{dr} \right] exp\left(\frac{t}{\tau}\right) - \tau B_r r \frac{d\Omega}{dr}, \tag{4}$$

where the amplification time is

$$\tau = -\left(\frac{\partial v_r}{\partial r}\right)^{-1}. \tag{5}$$

Eq. (5) describes the same time scale as Eq. (1), i.e. $\tau \propto t_{\rm rot}$. For a dwarfish galaxy with a radius of some kpc and a rotation speed of $100 \, km \, s^{-1}$ we have a rotation time of about $5 \cdot 10^7$ years for a 1 kpc radius. Having executed about 10 revolutions and taking into account compressional effects during the collapse phase which increase a seed field of 10^{-18} G to 10^{-13}G (Lesch and Chiba 1995), we easily reach a disk field strength of about 10μG after 0.5 Gigayears.

The second amplification step is the increase of the magnetic field by a galactic wind (Lesch et al. 1989, Reuter et al. 1992). Galactic winds are driven by intense star formation in the galactic disk. Since galaxies are well known to experience a phase of very efficient star formation during their early evolution, we can expect that protogalaxies are characterized by strong galactic winds (e.g. Lanzetta et al. 1996). As a general argument for such initial starbursts in all kind of galaxies we use the metal enrichment of the intergalactic medium in galaxy clusters, which finds its natural explanation by strong star formation driving galactic winds, which transports the enriched material into the galactic halos (Trentham 1994).

The prototype for a starburst galaxy is M82, which contains numerous supernova remnants in the disk, Hα-filaments in the halo and a large radio halo, with radial synchrotron filaments, which can be traced down to the supernova remnants in the very centre of M82 (Reuter et al. 1992). Thus, the radio halo of M82 is the result of relativistic electrons which are accelerated in supernova shock waves and which are transported into the halo via a galactic wind. The estimated wind velocity perpendicular to the disk is about $10^3 \, kms^{-1}$ (Reuter et al. 1992), i.e.

$$t_{\rm burst} \simeq 10^7 \text{years} \left[\frac{H}{8\,{\rm kpc}}\right]\left[\frac{v_{\rm wind}}{1200\,{\rm km\,s^{-1}}}\right]^{-1}, \tag{6}$$

A wind velocity of $1200 \, kms^{-1}$ is necessary to explain the spatial extent of the radio halo (8 kpc) with the assumption that the supernovae are the sources for the relativistic electrons in the halo and that the starburst lasts about 10^7 years. Since the action of a disk dynamo produces strong disk fields and weak halo fields and the equipartition field strength in the halo of M82 is about 50μG, whereas the disk field strength is only 10μG one can conclude that only the wind can be responsible for high field strength in the halo. A wind can indeed amplify magnetic fields to high field strengths. The necessary velocity responsible for the magnetic field amplification in the halo, with $B_{\mathrm{halo}} \sim 50\mu G$ is

$$v_{\mathrm{w}\perp} \simeq 400\,\mathrm{kms}^{-1} \left[\frac{B}{5\cdot 10^{-5}\,\mathrm{G}}\right]\left[\frac{\rho}{10^{-25}\,\mathrm{gcm}^{-3}}\right]^{-1/2} \tag{7},$$

which is approximately 1/3 of the wind speed.

In what follows we assume that a huge number of M82 analogues were present in the early phase of the Universe and that these objects magnetize their immediate surroundings via galactic winds. The spatial scale on which $50\mu G$ appears is about $8\,kpc$. Next we consider what maximum amount of magnetic energy density can be injected into spatial scales considerably larger than the halo size of 8 kpc. We have the following picture in mind:

The universe is filled with M82's which all produce their own magnetic field and are not causally connected with their neighbours. The magnetic field lines are expelled from the disk via galactic winds, and their typical injection scale $l_0 \sim 8\,kpc$. The galaxies produce magnetic bubbles. For simplicity we suppose that in each bubble magnetic field lines form loops confined to the bubble walls. When bubble walls collide the fields from each bubble are "stitched" to those of its neighbours by local magnetic reconnection (e.g. Lesch and Bender (1990) for an application of magnetic reconnection in galactic context). Because the magnetic amplification is uncorrelated on scales larger than the bubble radius l_0, the loop structure in different bubbles is uncorrelated, and an individual field instead of turning back in a closed loop on the scale l_0, soon loses any memory of where it started and executes an infinite Brownian walk in space with step length l_{b}. These lines traverse regions of space even on scales where no causal connection has yet occured. We define B_l as the flux of B field lines remaining after tangles below a scale l are smoothed out by viewing the system with spatial resolution l. Thus, B_l^2 is the kinetic energy density available from straightening tangles on scale l. To estimate the flux B_l we need to calculate the rms net flux through the "smeared" surfaces of radius l and effective half widths of order l. Since the orientation of the lines is random the rms net magnetic flux through such a fuzzy surface is $B \sim \sqrt{\eta}/l^2$, where η is the total number of lines which traverse the surface in any direction. If the lines were smooth on scale l or if the surfaces were sharp we would have $\eta \sim l^2$, but for Brownian walks the total effective length of each of the lines decreases like l^{-1}. Therefore this model leads to a spectrum $B_l \propto l^{-3/2}$.

It is worth noting that this acausal propagation is capable of producing large-scale magnetic fields that are just as strong as the causal superposition of dipole fields in vacuum. A simple way to derive this scaling is to imagine space filled with randomly placed, randomly oriented current loops of radius l_b and local field strength B_{l_b}; the magnetic dipole moment of these loops add linearly and so the mean field strength at a distance l due to $N = (l/l_b))^3$ randomly oriented loops is about

$$B_l \simeq \sqrt{N} B_{l_b} \left(\frac{l_b}{l}\right)^3 \simeq B_{l_b} \left(\frac{l}{l_b}\right)^{-3/2}, \qquad (8)$$

where $B_{l_b} \left(\frac{l_b}{l}\right)^3$ denotes the field strength contributed by each loop at a distance l. This example demonstrates that one does not need to generate currents with correlations extending to infinity in order to create a field of infinite random walks.

In our scenario l_b is about 8 kpc, the radius of the radio halo of M82. The magnetic field strength B_{l_b} is about $10\mu G$. If we use Eq.(8) we reach about $5 \cdot 10^{-9} G$ at a length scale of Mpc. Obviously the proposed mechanism is capable of magnetizing the intergalactic medium on time scales of about 0.5 Gyrs, providing that enough M82-like starbursters are present at a critical cosmic epoch.

We thank Charles Dyer and Judith Perry for helpful discussions, and Quentin Dufton for assistance.

References

Beck, R., Brandenburg, A., Moss, D., Shukurov, A., Sokoloff, D., 1996, ARA&A in press

Camenzind, M., Lesch, H., 1994, A&A 284, 411

Chiba, M., Lesch, H., 1994, A&A 284, 731

Katz, Gunn, 1991, ApJ 377, 365

Kronberg, P.P. Perry, J.J., and Zukowski, E.L.H., 1992, ApJ 387, 528

Kronberg, P.P., 1994, Rep Prog. Phys. 57, 325

Lanzetta, K.M., Yahil, A., Fernandez-Soto, A., 1996, Nature 381, 759

Larson, R.B., 1984, MNRAS 206, 197

Lesch, H., Crusius, A., Schlickeiser, R., Wielebinski, R., 1989, A&A 217, 99

Lesch, H., Bender, R., 1990, A&A 233, 417

Lesch, H., Chiba, M., 1995, A&A 297, 305

Lubow, S.H., Balbus, S.A., Cowie, L.L., 1986, ApJ 309, 496

Lynden-Bell, D., Kalnajs, A.J., 1972, MNRAS 157, 1

Trentham, N., 1994, Nature 372, 157

Rees, M., 1987, QJRAS 28, 197

Reuter, H.P., Klein, U., Lesch, H., Wielebinski, R., Kronberg, P.P., 1992, A&A 256, 10

Ruzmaikin, A.A., Shukurov, A.M., Sokoloff, D.D., 1988, *Magnetic Fields of Galaxies*, Reidel, Dordrecht

Addresses of the authors:

Philipp P. Kronberg, University of Toronto, 60 St. George Street, Toronto, Ontario M5S 1A7, Canada

Harald Lesch, University Observatory, Universität München, Scheinerstr.1 D-81679 München, Germany

Dust
and Dark Matter
in Halos

The Quest for Extragalactic Dust

A. Ferrara

1 Introduction

The presence of dust in the intergalactic space can have dramatic consequences for the interpretation of observations of the distant universe. Rudnicki (1986) describes in detail the evolution of our knowledge of extragalactic dust (EGD) either associated with (primeval) galaxies and their halos or truly in the intergalactic space; the author also reviews the attempts to derive lower limits to the amount of EGD. From a theoretical point of view, Heisler & Ostriker (1988) pointed out that dust obscuration is necessary to account for the apparent lack of high-redshift quasars which are assumed to exist, but obscured by intervening dust. High redshift dust has been detected from the statistical reddening of quasars with intervening damped Ly-α systems in the line of sight (for a review, see Fall & Pei 1995).

Perhaps the most important cosmological impact of EGD is on the determination of the deceleration parameter, q_0; indeed, this parameter is extremely sensitive to the assumptions concerning the extinction. In an important paper, Margolis & Schramm (1977) have estimated that the correction due to EGD to q_0 is

$$(\Delta q_0) = 2.8 \times 10^3 \left(\frac{A_v}{1 \ \mathrm{mag} \ \mathrm{Mpc}^{-1}} \right) h^{-1}, \tag{1}$$

where A_v is the visual extinction and $h = H_0/100$ km s^{-1}Mpc^{-1}. Thus, as little as 2×10^{-4} mag Mpc^{-1} absorption would cause a measurement error of $(\Delta q_0) \sim 1$. The conclusion drawn by these authors is that "uniform filling of the universe with dust is exceedingly unlikely whereas dust halos around galaxies and clusters of galaxies may be quite likely". The theoretical prediction by Margolis & Schramm seems to find fresh support in the results of a series of key observations aimed at the detection of dust in the extended disks of galaxies and/or their halos. Zaritsky (1994), studying the B and I colors of distant galaxies seen through the halo of two nearby spirals, concluded that background galaxies at smaller projected separations are statistically redder than those in the outer regions. This fact suggests the existence of an extended dust halo with a derived scale length of 31 ± 8 kpc. This figure is in remarkable agreement with the prediction of Heisler & Ostriker (1988), who required a scale length of 33 kpc to account for the observed quasar number counts. Similar results have been obtained recently also by Lequeux & Guelin (1996). Several evidences of extinction by dust in clusters of galaxies are present in the literature both from a deficiency of UV excess objects in the cluster background (Boyle etal. 1988, Hu 1992) or from diffuse FIR emission (Wyse etal. 1993) (see for a different view Ferguson 1993). Finally, optical observations (for a review, see Dettmar 1992) show that nearby edge-on galaxies are characterized by numerous dust filaments protruding from the plane and extending well into the galaxy halo.

The existence of dust at large galactocentric distances poses the question of the origin of the EGD. Two alternatives are currently possible: either at least part of the dust is primordial or it has been originated in galaxies. In this paper we will investigate the latter possibility and in particular on the viable physical mechanisms that could eject dust from the parent galaxy in the intergalactic space.

2   Dust Injection in Galactic Halos

There are two main mechanisms able to transport dust originally located in the main body of a galaxy into the halo: (i) a convective flow (i.e. a "galactic fountain", Shapiro & Field [1976] or a chimney, Norman & Ikeuchi [1989]) in which the gas heated by a supernova blastwave becomes buoyant and raises into the halo carrying along the dust grains that are coupled with the gas; (ii) a wind driven by the radiation pressure above the clusters of young OB stars. In order to work efficiently, the first mechanism requires the occurence of the so-called "blow-out" in which the gas of the shell created by the explosion is effectively reaccelerated and injected into the halo. However, recent studies have demonstrated that this phenomenon, at least in the Galaxy, should be much rarer than previously thought, either because the growth of the shell is inhibited by the presence of a magnetic field (Tomisaka 1990; MacLow & Norman 1992) or by effects related to non-coeval star formation (Shull & Saken 1995). Here we explore the second possibility, i.e. dust is injected into the intergalactic space by radiation pressure.

2.1 Dusty Chimneys

The ionizing flux above OB associations is very intense and can produce "HII chimneys" (Dove & Shull 1994) that are density bounded perpendicularly to the plane allowing Lyc photons to escape the disk and penetrate into the halo. The radiation pressure on dust grains located above these HII chimneys is particularly strong due to the intensity of the radiation field and it might devoid partially or entirely these regions of their solid particle content *before* the grains are hitted - and possibly destroyed - by the shock wave formed after the first supernova explosions in the association. To check this hypothesis we have performed a set of numerical simulations of the above "dusty chimneys" based on a mixed Monte Carlo/particle approach. A complete description of the calculation and of the detailed results will be presented in Ferrara & Shull (1997). Here we summarize the hypotheses of the model and the main findings.

We assume that dust grains are immersed in the time-dependent radiation field of a stellar association containing N OB stars (typically $N = 100$). The spectrum of the radiation field has been adapted from the results of Sutherland & Shull (1997) who calculate the evolution of the composite radiation spectrum from the most updated stellar models. We take the vertical distribution of the gas, $n_g(z)$, from Dickey & Lockman 1990; the dust distribution, $n_d(z)$, is assumed to be initially exponential with a scale height $z_d = 120$ pc. The initial position and radius of each (spherical) grain are extracted via a Monte Carlo procedure from the parent distributions, $n_d(z)$ and MRN (Mathis, Rumpl & Nordsiek 1977), respectively. We then follow the dynamical evolution of the grain ensemble as driven by radiation, gravity and drag (viscous + coulomb) forces; the grain charge is also calculated consistently solving the detailed balance equation in which both collisional and photoelectric charging rates are included. The preliminary results of the study are shown in Figs. 1-4 below. Fig. 1 illustrates the spatial distribution of the dust grains (all sizes $a = 10 - 250$ nm are shown) for a $N = 100$ OB stars association at the evolutionary time $t = 42$ Myr. At this time, the luminosity, L_ν, of the radiation field at 5 eV is roughly 3 orders of magnitude lower than the maximum intensity $L_\nu \sim 10^{24}$ erg s^{-1} Hz^{-1} sr^{-1}, reached at $t \sim 17$ Myr. From an inspection of Fig. 1 the presence of a central cavity completely devoided of dust swept by the very strong radiation pressure, is clearly seen. In addition, a relevant fraction of grains has been pushed well out of the main disk of the Galaxy, at $|z| > 500$ pc. These effects are even more dramatic for a larger association with $N = 500$, as reported in Fig. 2 at the same evolutionary time. The configuration of the region affected by the radiation field resembles now an

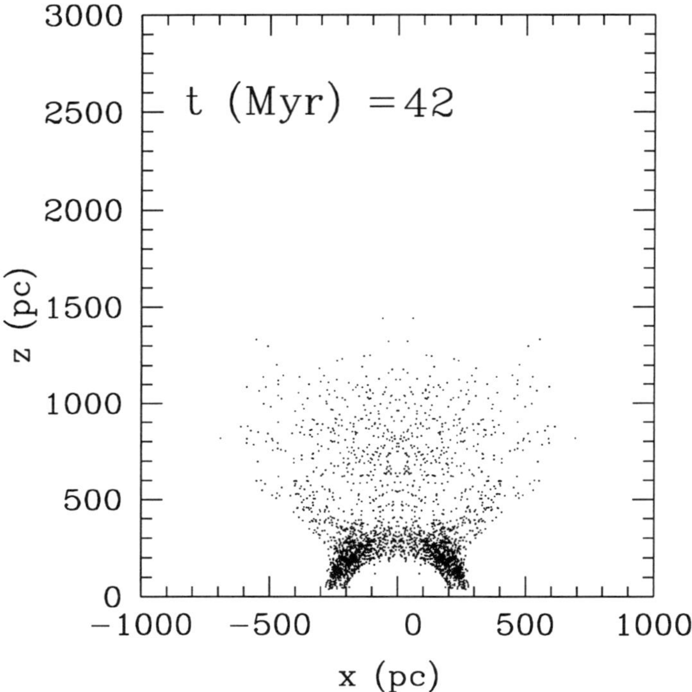

Figure 1: Spatial distribution of dust grains for an association (located at the origin of the axes) containing $N = 100$ OB stars at the evolutionary time $t = 42$ Myr

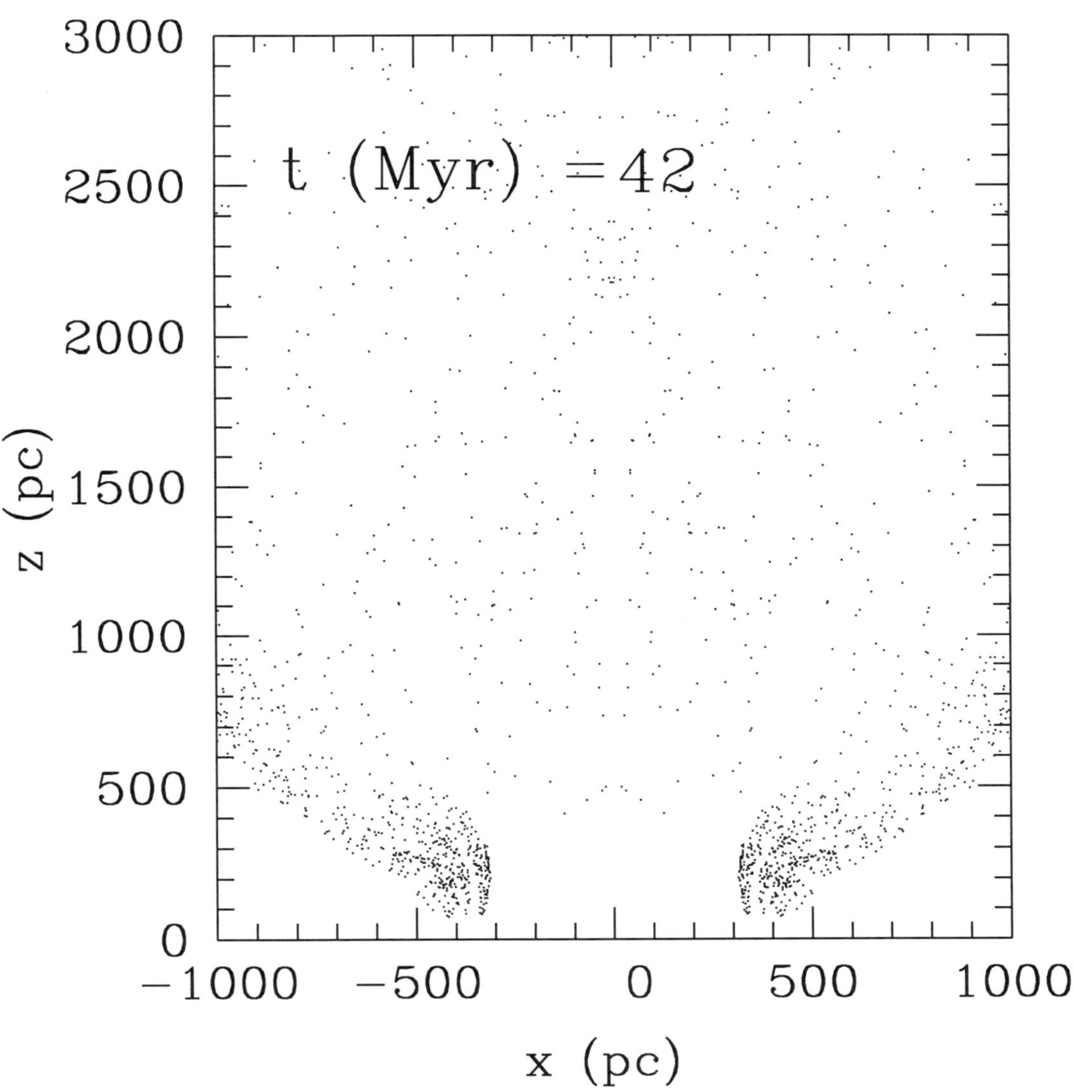

Figure 2: The same as Fig. 1 for $N = 500$

inflated bubble extending at least for 3 kpc above the midplane of the Galaxy; it is also possible to identify the filamentary structures that form the lateral walls of the dusty chimney. As it can be noted from Fig. 3, a vertical cut through the origin (i.e. above the center of the association) shows that the z-distribution of the dust becomes much flatter with time than the original exponential one, seen as the curve with a peak at $z = 0$. Thus, the dust is rather homogeneously distributed by the radiation pressure in the portions of the halo above the associations; moreover, this implies that the dust-to-gas ratio has to be larger in the upper halo than in the disk. Finally, since the drag force becomes vanishingly small because the gas density is rapidly decreasing with height, some grains can continue their journey for several kpc before they feel the gravitational pull of the Galaxy. The velocities achieved by the grains are rather high, as shown in Fig. 4 and they can exceed (for a $N = 500$ association) 100 km s^{-1}. Smaller grains tend to move faster since - neglecting drag forces - the ratio of the radiation-to-gravitational force is $\propto a^{-1}$.

3 Effects and Detection of EGD

If dust is present in galactic halos it could be responsible for several important detectable effects. In addition to the standard direct observations discussed in the Introduction and involving either the obscuration of distant object by the grains or their FIR emission, a number of indirect but nonetheless intriguing consequences of this component can be individuated and used to put constraints on any theory modelling the EGD.

3.1 Radio Emission from Spinning Grains

Grains are put into rotation by the collective effects of collisions with gas atoms, absorbed or emitted photons, photoelectric emission and molecule formation on the grain surface. If, in addition, grains are charged and with a non-zero dipole moment, they will emit radiation at easily accessible radio continuum wavelengths (Ferrara & Dettmar 1994). As a consequence, it is well into the possibilities of the actual instrumentation to search for a signature of EGD detecting this effect. In brief, grains are found to have substantial radio emission peaked at a cutoff frequency typically in the range 10-100 GHz, which should be larger than the free-free from the ionized *Reynolds layer* of the Galaxy by a factor of a few. The model can be used to determine the amount and size distribution of dust in galactic halos.

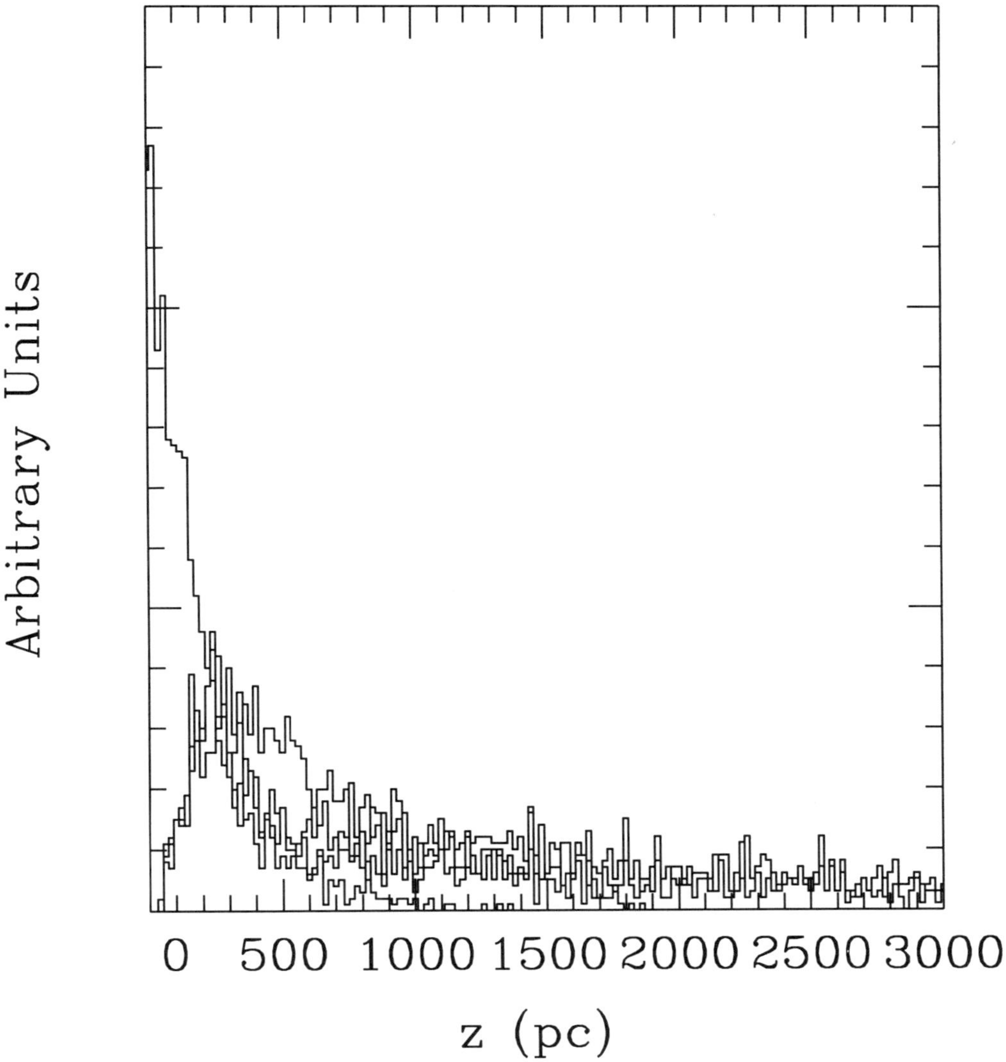

Figure 3: Vertical distribution of dust grains above an association containing $N = 500$ OB stars at four evolutionary times, $t = 0, 17, 29, 42$ Myr

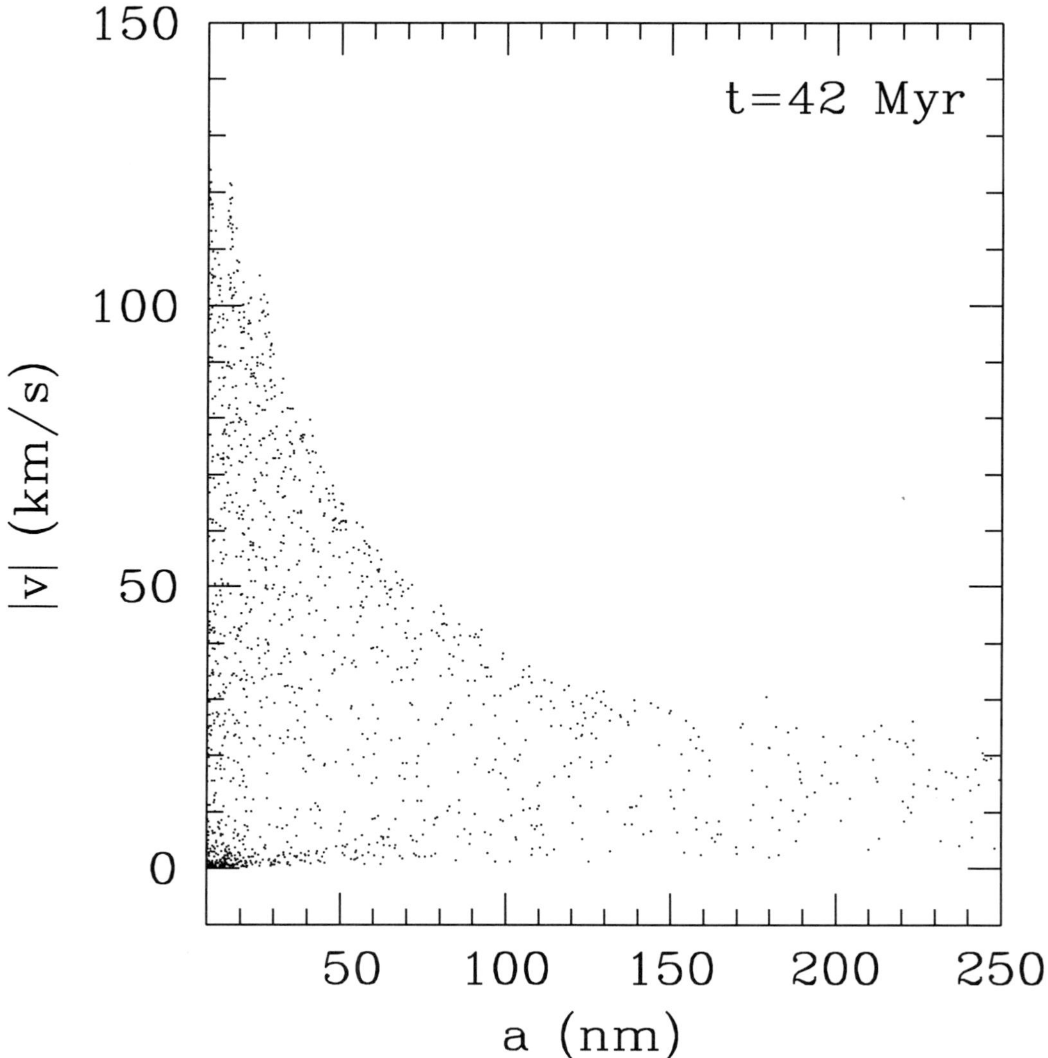

Figure 4: Velocity distribution of dust grains as a function of their size a for the case shown in Fig. 2

3.2 Opacity of Spiral Galaxies

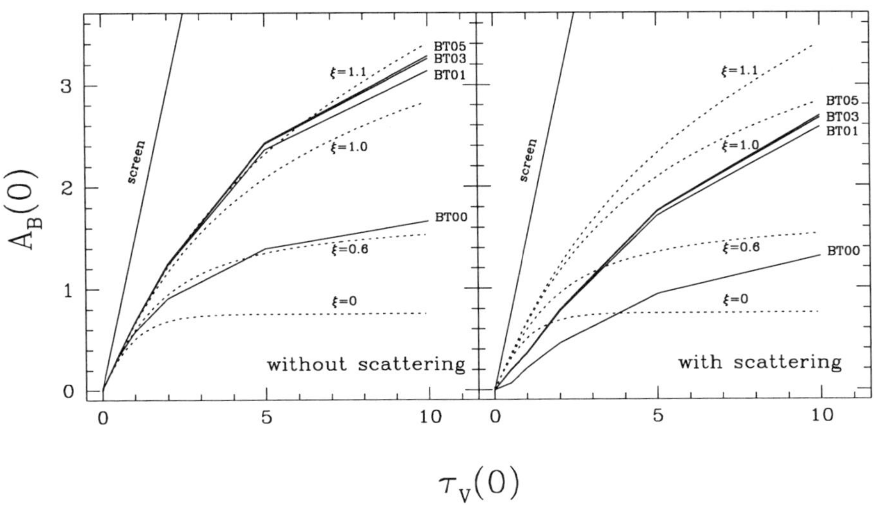

Figure 5: Central extinction vs. optical depth for B-band models with $i = 20°$ (solid lines); left panel is for models with absorption only, right panel for absorption + scattering models. Dotted lines refer to sandwich models with different dust-to-star thickness ratios (ξ). The straight solid line is for the screen model. BTxx labels indicate different bulge-to-total luminosity ratios (see text).

As mentioned in the Introduction, understanding the effects of EGD is crucial to cosmology. For example, extinction by EGD associated with galaxies may have profound effects on the interpretation of differential galaxy number counts in apparent magnitude, $N(m)$ and redshift $N(z)$ (Wang 1991, Im 1995). For the latter distribution the effects may be particularly important, since the high-redshift tail can be curtailed. As a consequence, the detection of high redshift galaxies can become more difficult.

For the above reasons it is important to develop suitable tools to investigate both the dust distribution and the opacity of galaxies. Bianchi, Ferrara & Giovanardi (1996, BFG) have presented a set of Monte Carlo simulations of dusty spiral galaxies, modelled as bulge + disk systems, aimed to the study of their extinction and polarization properties. The scattering properties are calculated from the Mie theory for dust grains taken according to the MRN model; in addition, BFG follow the transfer of the four Stokes parameters. Photometric and polarimetric maps of the galaxy for different optical depths, inclinations and bulge-to-total ratios have been produced. BFG find that the overall effect of scattering is to reduce substantially the extinction as a function of the optical depth, particularly in early-type spirals; scattering also reduces the reddening due to dust. These effects can be appreciated from Fig. 5, showing the extinction through the center of a galaxy in the B band (A_B) for an almost ($i = 20°$) face-on galaxy, as a function of the central optical depth. The straight line corresponds to the Holmberg screen model, while dotted ones to sandwich models (Disney etal. 1989) with different ξ, the dust-to-stellar thickness ratio. The left panel is for models with absorption only, while the right panel includes the scattering: screen and sandwiches are the same in both panels, and only include absorption. The models are labelled according to the different bulge-to-total luminosity ratios ranging from 0 (resembling a Sd galaxy, model BT00) to 0.5 (Sa, BT05).

Finally, the degree of linear polarization produced by scattering is usually of the order of a few percent, it increases with optical depth, and inclination. The polarization pattern is always perpendicular to the major axis.

3.3 Emission Line Ratios in the Diffuse Ionized Gas

The thick ($z \sim 1$ kpc) *Reynolds layer* of diffuse ionized gas (DIG) discovered in the Galaxy and in external ones poses some of the most challenging problems for our understanding of the large scale structure of the Galactic ISM (Reynolds 1995). One of the best ways to study this component is represented by spectroscopic observations of emission lines like $H\alpha$, $[NII](\lambda6583A\&A$) and $[SII](\lambda6716A\&A$). The different excitation conditions found in a given galaxy as a function of height above the plane pose a relevant question concerning the amount of light originating in the disk (where the most obvious ionization sources are located) and light scattered back from EGD. This aspect has been investigated recently by Ferrara etal. (1996), using Monte Carlo simulations similar to those described in the previous Section to calculate the radiation transfer of $H\alpha$ line emission, produced both by HII regions in the disk and in the diffuse ionized gas (DIG), through the dust layer of the galaxy NGC891.

The amount of light originating in the HII regions of the disk and scattered by EGD can be then compared with the emission produced by recombinations

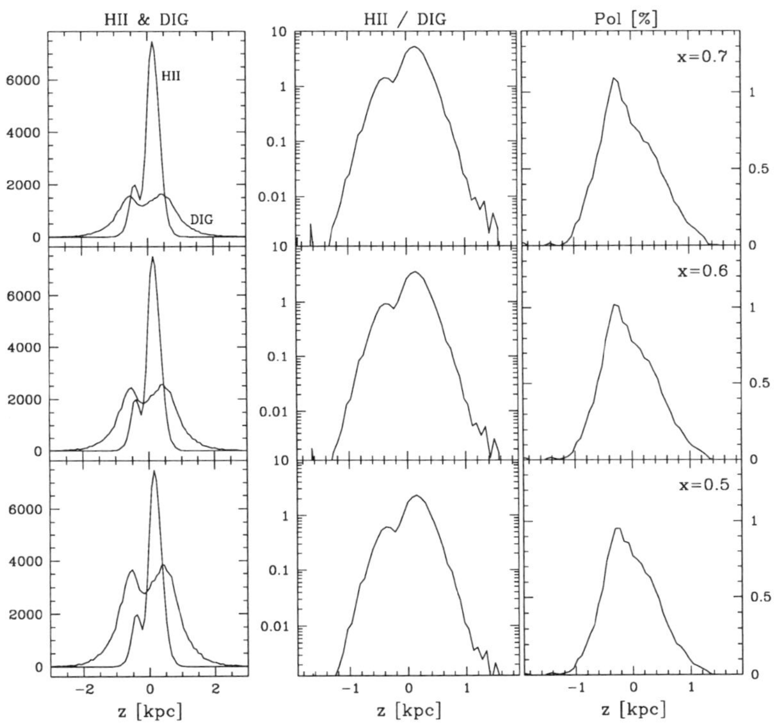

Figure 6: z-axis cuts through the center of NCG891 for (left column) $H\alpha$ luminosity profiles corresponding to contributions of HII regions and DIG component, (middle) ratio of the two components, and (right) profiles of the linear polarization degree. Plots refer to the cases of x=0.7, 0.6, 0.5 (from top to bottom) where x is the ratio between $H\alpha$ luminosity from HII regions to the total (DIG + HII regions) one.

in the DIG. The cuts of photometric and polarimetric maps along the z-axis show that scattered light from HII regions is still 10% of that of the DIG at $z \sim 600$ pc (Fig. 6), whereas the degree of linear polarization is small ($< 1\%$). This could explain the observed behavior of emission line ratios as a function of height (Ferrara etal. 1996).

4 Conclusions and Consequences

We have explored theoretically the possible mechanisms that could be responsible for the injection of dust, originally formed in the main body of galaxies, into their halos and possibly in the intergalactic space. We have shown that radiation pressure on dust grains initially located above the most powerful stellar associations could be the most efficient (and "gentle") mechanism to this aim as opposed to a "fountain"–type process in which grains are convected away from the galaxy together with the hot, shocked gas produced by supernova explosions. Detailed radiative transfer models of galaxies are likely to be the most powerful tool to assess the presence of such a EGD component associated with galactic halos. If such a component can be indeed proved to exist, it might have remarkable effects on the interpretation of high redshift observations, as for example QSO counts.

Several implications can be identified from the previous results, each one deserving more future study. In brief, they can be summarized as follows.

• A dusty flow from the disk into the halo (and possibly back) could have relevant consequences for the chemical evolution and metal content of the galaxy. In addition, it could modify the observed radial metallicity gradients of galactic disks.

• If a low-density, flat distribution of dust exists – as predicted by our models – and extends for several kpc, this would imply that the dust-to-gas ratio must be a function of z; this is a serious concern for any model addressing the thermal balance of the DIG via photoelectric heating by dust grains.

• One of the effects of injection of dust by radiation pressure is that grains are segregated by size, with the smaller grains becoming more and more abundant away from the plane. This segregation effect could become apparent from future IR observations of galactic halos.

• The filaments protruding from the plane of several edge-on galaxies and rooted in HII regions, can be explained as the lateral walls of the dusty chimneys studied here.

- If the grains are injected at in the halo with velocities of the order of 100 km s^{-1}, non-thermal sputtering can become important. This could be related to the evidence that the vertical distribution of refractory elements has an observed scale height well in excess of the one for neutral hydrogen (Edgar & Savage 1989).

- The presence of dust can be important for the thermal balance of the halo fountain gas and for the abundances of the high velocity clouds.

I would like to thank my collaborators in this project S. Bianchi, R.-J. Dettmar, C. Giovanardi, and M. Shull

References

Bianchi, S., Ferrara, A. & Giovanardi, C. 1996, ApJ, 465, 127 (BFG)

Boyle, B. J., Fong, R., & Shanks, T. 1988, MNRAS, 231, 897

Dettmar, R.-J. 1992, Fund. Cosm. Phys., 15, 143

Dove, J. B. & Shull, J. M. 1994, ApJ, 430, 222

Disney, M. J., Davies, J. I., & Phillipps, S. 1989, MNRAS, 239, 939

Edgar, R. J. & Savage, B. D. 1989, ApJ, 340, 762

Fall, S. M. & Pei, Y. C. 1995, in QSO Absorption Lines, ed. G. Meylan, (Springer: Verlag), 23

Ferguson, H. C. 1993, MNRAS, 263, 343

Ferrara, A. & Shull, J. M. 1997, in preparation

Ferrara, A. & Dettmar, R.-J. 1994, ApJ, 427, 155

Ferrara, A., Bianchi, S., Dettmar, R.-J., & Giovanardi, C. 1996, ApJL, 467, 69

Heisler, J. & Ostriker, J. P. 1988, ApJ, 332, 543

Hu, E. M. 1992, ApJ, 391, 608

Im, M. 1995, BAAS, 187, 3003

Lequeux, J. & Guelin, M. 1996, in New Extragalactic Perspectives in the New South Africa, ed. D. Block, (Kluwer: Dordrecht), in press

Mac Low, M. M. & Norman, M. L. 1992, BAAS, 181, 6706

Mathis, J. S., Rumpl, W. & Nordsiek, K. H. 1977, ApJ, 317, 637

Margolis, S. H. & Schramm, D. N. 1977, ApJ, 214, 339

Norman, C. A. & Ikeuchi S. 1989, ApJ, 345, 372

Reynolds, R. J. 1995, in The Physics of the Interstellar Medium and Intergalactic Medium, eds. A. Ferrara etal., ASP Conf. Series, 80, 388

Rudnicki, K. 1986, in Proc. Internat. School of Physics "Enrico Fermi", 86, 480

Shapiro, P. R. & Field, G. B. 1976, ApJ, 205, 762

Shull, J. M. & Saken, J. M. 1995, ApJ, 444, 663

Sutherland, R. S. & Shull, J. M. 1997, in preparation

Tomisaka, K. 1990, ApJL, 361, 5

Wang, B. 1991, ApJL, 383, 37

Wyse, M. W., O'Connell, R. W. , Bregman, J. N. & Roberts, M. S. 1993, ApJ, 405, 94

Zaritsky, D. 1994, AJ, 108, 1619

Address of the author:

A. FERRARA Osservatorio Astrofisico di Arcetri, Largo E. Fermi 5, 50125 Firenze, Italy

Future radio and infrared observations of dust in galactic halos

Uwe Herbstmeier

1 Why looking for dust in galactic halos?

That dust exists in the halos of galaxies is shown by the results of optical studies. Dust features at high distances from the disk plane, z, are seen in absorption of deep images of galaxies (Dettmar 1992). B and I band photometry of distant galaxies through halos of nearby ones (Zaritsky 1994) suggests that dust reddens the light of the background objects. It is much more difficult, however, to trace halo dust by its infrared and submillimeter emission. Analyses of he IRAS data on galaxies show that a possible emission from the halo regions plays no significant rôle (e.g. Telesco 1993).

Why looking for dust in halos of galaxies?
Zaritsky (1994) presents some arguments. Galactic halos are believed to comprise the major amount of a galaxy's mass, the dark matter. If of baryonic nature or not is still a highly debated question. That there is evidence for more baryonic mass in halos then directly visible in emission is shown by the QSO absorption line systems including also metal absorption lines. Zaritsky argues that if metal-enriched gas is present at large galactic radii, dust might also be present. Even the amount of halo dust is by far too small to represent the dark matter, the dust particles in the halo can serve as a tracer of the gravitational potential and baryonic mass (examples for such objects are proposed by Gerhard, this volume).

Other reasons to look for dust in galaxy halos are given by its rôle in the analysis of other phenomena at high z-distances. E.g. if dust is mixed with the diffuse ionised gas (DIG, e.g. Dettmar 1992, Domgörgen this volume) it could modify the UV radiation intensity which is considered to be an ionising agent for the DIG. Or the dust can change the spectral line ratios observed for the DIG which are used to characterise the DIG.

To analyse the dust in galaxy halos is also of interest by its own. Dust can be used as tracer for various parameters of the halo interstellar medium (ISM). It is also an important actor in the scenerios of the different halo models. This will be sketched in the following sections.

Why to look for dust in emission?
Absorption techniques are much more sensitive to low dust column densities compared to emission analyses. But, it is hard to find suitable background sources (galaxies nearly edge-on) or to distinguish between in-plane and halo dust (galaxies nearly face-on). Therefore the increased sensitivity of the infrared and sub-millimeter technology opens new possibilities to search for dust emission in galaxy halos. If this emission can be detected it also helps to analyse the composition of the dust.

In the following I will try to summarize what we expect for the dust emission of galaxy halos and how the various observational goals could be achieved in the future.

2 What determines the infrared and submillimeter dust emission?

To emit in the infrared, grains have to be heated. This is achieved by transfer of radiation energy into internal heat. Therefore, the intensity of the interstellar radiation field (ISRF), which is absorbed by the dust grains, is an important parameter in the calculation of the dust's infrared and submillimeter emission.

The way how the radiation energy is transfered determines the spectral characteristics of the emission. Various calculations exist to determine the influence of the grain size distribution and the dust composition on the heating process (e.g. Draine & Lee 1984, Désert et al. 1986): Big particles are heated to equilibrium temperatures, emitting blackbody spectra modified by the wavelength dependent emissivity. Small grains like the polycyclic aromatic hydrocarbons (PAHs) show transient behaviour between an intrinsically high excitation (temperature) state and an equilibrium temperature. Infrared emission of such small particles produce an increased intensity in the mid-infrared. Additionally, emission of PAHs shows characteristic bands. Based on these results various models of the dust composition and infrared emssion spectrum were calculated to fit the observational results obtained for our Milky Way.

As an example in Table 1 the calculated specific intensities for IRAS, L'($3.3\,\mu$m), zero width 200 and 800 μm bands are listed. The values are taken from the work of Désert et al. (1990). The authors derived that the emission in the

Table 1: Infrared emission of dust in the solar neighbourhood (taken from Désert et al. 1990)

λ (μm)	3.3	12	25	60	100	200	800
PAH	1.82(-4)	3.12(-2)	1.90(-2)	4.23(-3)	1.11(-3)	1.49(-4)	3.06(-4)
VSG	3.02(-6)	1.77(-3)	2.06(-2)	1.06(-1)	1.22(-1)	5.88(-2)	7.19(-3)
BG	0.00(0)	0.00(0)	2.11(-7)	6.62(-2)	7.45(-1)	1.48(0)	5.36(-2)
Total	1.83(-3)	3.30(-2)	3.96(-2)	1.76(-1)	8.68(-1)	1.54(0)	6.11(-2)

Units are: $MJy\,ster^{-1}\,10^{-20}\,cm^2$ (normalised to the hydrogen column density)

wavelength range $3 - 15\,\mu$m is dominated by the band and continuum emission of the PAHs. The far-infared / submillimeter intensity ($80 - 1000\,\mu$m) originates in the equilibrium temperature emission of the big grains (BGs). Between 15 and $80\,\mu$m the observations show higher intensities than derived from PAHs and big grains only. Therefore an additional component of very small grains (VSGs) is required to account for the brightness of dust clouds in this spectral range. The nature of the VSGs is, however, mysterious.

3   What do we expect for the halo dust emission?

3.1   Dust mixed with the diffuse ISM

According to Boulanger & Pérault (1988) the far-infrared (FIR) emission at $100\,\mu$m of the diffuse interstellar gas of the Milky Way is well correlated with the neutral atomic hydrogen column density. The authors derive a mean infrared emission to HI column density ratio of $0.8\,10^{-20}\,MJy\,ster^{-1}\,cm^2$. A similar relationship is also found for the emission in the range $100 - 1000\,\mu$m as observed by COBE (Puget et al. 1996 and references therein, see also Burton this volume). This indicates that the neutral diffuse interstallar gas is well mixed with dust.

Can this result be implied for the diffuse warm ionised matter (WIM)? Boulanger & Pérault (1988) didn't find any indication for dust mixed with the ionised component of the diffuse ISM. But, this result is not conclusive if the distributions of WIM and diffuse HI gas are significantly correlated as indicated by Reynolds (1995). Taking this into account Puget at al. (1996 and references therein) argue that about half of the warm ionised medium could show a similar infrared emission like the neutral component from interspearsed dust whereas the other half may be dust depleted.

Table 2: Central density n_0, scale heights H of the neutral and ionised diffuse ISM

Component	n_0 (cm^{-3})	H (pc)	$N_{>0.5\text{kpc}}$ (10^{19} cm^{-2})	$N_{>1\text{kpc}}$ (10^{19} cm^{-2})	References
HI comp. 2	0.249	180	both:		1
HI comp. 3	0.016	670	1.5	0.45	1
HI very broad	–	–	0	1	3
HII	0.025	900	7.2	2	2

References: 1: Lockman & Gehman (1991), 2: Reynolds (1993), 3: Westphalen, Kalberla (this volume)

What do we derive for the galactic halo?
The densities of the diffuse components decrease with z exponentially, but the scale heights are of the order of several hundreds of parsecs (see Table 2). Moreover, Westphalen and Kalberla (this volume) show evidence for the existence of an additional HI halo of the Milky Way. Summing up all contributions a column density of $6.5\,10^{19}$ cm^{-2} is derived for the disk-halo interface region (z-distance 0.5 – 1 kpc) and $3.5\,10^{19}$ cm^{-2} for the halo (z-distance > 1 kpc). With the FIR emissivity given above this results in a brightness of 0.5 and $0.3\,\text{MJy}\,\text{ster}^{-1}$ for both components. The estimate is very optimistic as the dominant component in the halo, the WIM, shows a reduced FIR emissivity and the ISRF may be shielded by gas clouds in the galactic plane. Also the nature of the newly postulated HI halo is unknown, therefore also its dust content.

For our Milky Way this very faint emission from a diffusely distributed dust component in the halo cannot be separated from the foreground emission significantly. Only looking in the halos of other galaxies with the most sensitive detectors may reveal the dust emission.

3.2 Clouds as probes of the disk-halo interface

The smooth functions described in the previous section approximate only the true distribution of the ISM. The matter is confined in distinct clouds which can be analysed in more detail. In the disk-halo interface, defined here as the z-range of $0.5 - 1$ kpc, the Intermediate Velocity Clouds (IVCs) are found (for a review of the distances of IVCs see de Boer, this volume). These are clouds, whose radial velocities cannot be described by the rotation curve but are still not as fast as the so-called High Velocity Clouds (HVCs).

What do we know about the dust properties of IVCs?
Deul & Burton (1990) have shown that IVCs are part of the FIR cirrus, which was mapped by IRAS. This result is confirmed by the analysis of specific IVCs by Heiles et al. (1988, a sample of cirrus clouds), Meyerdierks (1992, IVC $133 + 23 - 55$), Herter et al. (1990, IVC $9 + 51 - 20$), Herbstmeier et al. (1993, Draco Nebula), Herbstmeier (1990, IVC $86 + 39 - 44$) and others. Comparing these results one can derive that the the far-infrared emissivity (I_{100}/N_{H}) isn't unique for this cloud population. Values are found in the range from 0.5 to $2.0\,10^{-20}\,\mathrm{MJy\,ster^{-1}\,cm^2}$. Heiles at al. (1988) derived for their sample that the ratio between the IRAS intensities at 60 and $100\,\mu$m, I_{60}/I_{100}, is larger for IVCs with radial velocities $|v| > 30\,\mathrm{km\,s^{-1}}$ than for other cirrus clouds. Similar results are found for e.g. IVC $86 + 39 - 44$ (Herbstmeier 1990). Heiles and collaborators also determined that the ratio of I_{12}/I_{100} is marginally lower for IVCs then for the rest of the cirrus. Odenwald & Rickard (1987) found for the Draco nebula an even stronger reduction of the intensity ratio.

As an explanation of the observational results Heiles et al. (1988) proposed that strong shocks have sputtered the big grains in the IVCs and enhanced the population of small grains, which dominate the radiation at $60\,\mu$m. On the other hand the authors argue that very small particles are formed in slow shocks and destroyed in fast ones to explain the only slight decrease in the I_{12}/I_{100} ratio. Theoretical calculations by Borkowski & Dwek (1995) confirm at least that in strong shocks big grains are sputtered to small particles which become then a dominant population. But in total the number of grains is reduced.

Are the IVCs representative for this part of the Milky Way?
The database is still to incomplete to prove the proposed models on the IVC infrared emission. More IVCs have to be identified with their FIR counterpart in the cirrus images. The infrared spectrum of the IVCs has to be analysed for specific features of this population. E.g. looking for emission bands in the mid-infrared (MIR) and their dependencies from the location within the cloud (core-edge variations) will trace the dynamics for these clouds. I.e. are IVCs crashing on galactic clouds or are they collision products of HVCs with the Milky Way ISM? This analysis will help to judge whether these phenomena are typical for the disk-halo interface or not.

IS dust emission from IVCs also important in other glaxies?
The IVCs are part of the galactic cirrus. If the emission of a cirrus-like component contributes to the infrared brightness of other galaxies, IVCs could play

an important rôle. Analyses of the IRAS results for M81 and M101 by Devereux (1994) indicate, however, that no significant contribution of a cirrus-like component can be found.

3.3 High Velocity Clouds as probes of the halo

Also beyond 1 kpc clouds can be discerned. Despite their distances are still a matter of debate (see de Boer this volume) HVCs are generally assumed to populate the halo of the galaxies at z-distances of a couple of kiloparsecs (Wakker & van Woerden 1991). Originally only an arbitrarily defined limit in radial velocity of $|v| = 100\,\mathrm{km\,s^{-1}}$ distinguished HVCs from IVCs. But later it was shown that especially the FIR characteristics is different for both cloud populations. Whereas the FIR emission of IVCs is seen in the IRAS survey, a similar contribution from HVCs was unsuccessfully looked for by Wakker & Boulanger (1986, three clouds in Complexes A and M), Fong et al. (1987), and Mebold & Herbstmeier (1993, Magellanic Stream). Several explanation were proposed for the lack of FIR emission:

1. The clouds selected for these analyses are exceptional.

2. Most of the HVCs are underabundant in dust.

3. The big grains are missing in the HVCs. Only small grains are present.

4. The dust is too cold to radiate at MIR or FIR (reduced ISRF).

Explanation 1 may hold for the Magellanic Stream as this cloud population is associated with the Magellanic Clouds. We will come back to this in the next section. The other three clouds are taken from a sample representing most of the northern HVCs.

Why is it important to check the dust content of the HVCs?
It is important to check which of the afore-mentioned explanations can be verified, as Wolfire et al. (1995) model the origin of the HVCs related to their internal structure and to the influence of the environment. This analysis tries to explain why a large fraction of HVCs show a multi-phase structure, i.e. cores in an extended cloud halo. The authors calculated that the energetics of the cloud environment determines whether the cloud cores can evolve or not: The HVCs serve as probes for the filling fraction of the hot X-ray emitting gas in the halo (see e.g. Herbstmeier et al. 1995 for the HVC – X-ray corona relation). As

an important ingredient in their model the authors adopted the photoelectric heating of PAHs and very small grains mixed with the high velocity gas as a major source of internal energy. They follow the analysis of Bakes & Tielens (1994). A predominance of small dust particles is expected if the precursors of HVCs originate in the blow-out material of disk supernova remnants, which passed the strong shocks there (Borkowski & Dwek 1995). To prove the results of Wolfire et al. (1995) it is therefore important to look for the amount of small dust grains and PAHs locked in HVCs in the MIR.

Alternatively the clouds could also show a significant population of big grains which survive grain processing by shocks (see Ferrara this volume) or which are levitated by a galactic wind triggered by the radiation pressure (Ferrara et al. 1991). These grains can reach a far distance where they mix with HVCs. Heating of these grains is less efficient due to the dilution of the ISRF. Therefore the dust is very cold and radiates then only at submillimeter wavelengths (Ferrara private communication). This emission is expected to be very faint if present at all. The submillimeter maps from COBE don't hint on a major component associated with the HVCs (Puget at al. 1996 and references therein).

3.4 The Magellanic Stream as probe for an extended halo or galactic wind

Aside HVCs other components are expected in the galactic halo. They are all associated with dust or are at least supposed to be.

Breitschwerdt & Schmutzler (1994) propose a galactic wind which originates in the blow-out of superbubbles around OB associations. This wind is driven by hot gas and cosmic rays also carrying away dust particles along magnetic flux tubes.

Radiation pressure from the stellar disc exerts an upward force on dust grains. This force can photo-leviatate the dust and also, via dust-gas coupling, clouds in the halo or even expel the dust and gas in the intergalactic space (Ferrara et al. 1991).

Additionally, the very broad component found in the HI spectra of the Leiden / Dwingeloo survey (Hartmann & Burton 1996) shows no major variation with the line of sight (Westphalen and Kalberla this volume). This behaviour could be explained by a neutral halo component at far z-distance.

Can we observe the dust in these components?
If at least one of these components exists at large z-distances from the galacitc

plane ($30 - 50\,\mathrm{kpc}$) and is covering a large fraction of the disk surface, it is expected to interact with the Magellanic System, the Magellanic Clouds and the related Magellanic Stream. Mebold & Herbstmeier (1993) report on the search for dust in the Magellanic Stream. It was unsuccessful, but the authors found associations of the HVCs with several IVCs emitting in the FIR.

As an example cloud IV of the Magellanic Stream (Cohen 1982) is shown in Figure 1 observed with the 100m telescope at Effelsberg and its $100\,\mu$m image obtained from the IRAS database processed using the GEISHA analysis system (developed by the Laboratory for Space Research at Groningen, The Netherlands, funded partly by the University of Groningen, the Space Science Department of ESTEC, and the Air Force Geophysics Laboratory, USA). The horseshoeshaped HVC ($-250 < v < -160\,\mathrm{km\,s^{-1}}$) is coincident with a dusty low-velocity feature ("LVC") seen in the velocity range $-15 < v < 5\,\mathrm{km\,s^{-1}}$ (Herbstmeier & Westphalen in preparation). The low-velocity gas could be the result of a collision between the fast dusty wind and the infalling Magellanic Stream cloud. Both material flows are braked resulting in the features at low velocities. Due to this interaction the dust in the wind matter is heated and radiates in the infrared.

Recent results of Weiner & Williams (1996) support this explanation. These authors find H_α emission at the edges of the Stream clouds. Whereas Weiner & Williams (1996) suggest that the Magellanic Stream interacts with the hot ionised ISM of the galactic halo producing the H_α rim brightening of the clouds. But the HI and infrared results issue the question whether the outer halo contains a significant fraction of neutral gas mixed with dust.

4   Ways to look for dust in the halos of galaxies

The previous sections summarized some ideas on dust in the halo of galaxies. Here, some ways to follow up the scientific goals shall be denoted briefly.

- Future research using old and new survey data
  The new Leiden / Dwingeloo HI line survey (Hartmann & Burton 1996) will provide an upgraded database for the the comparison of the clouds' gas distribution with their infrared emission observed with IRAS. The cross-correlation of both data sets, similar to Deul & Burton (1990) will, add further examples to the set of IVCs identified in the infrared cirrus (see also Burton this volume). Based on this comparison a better

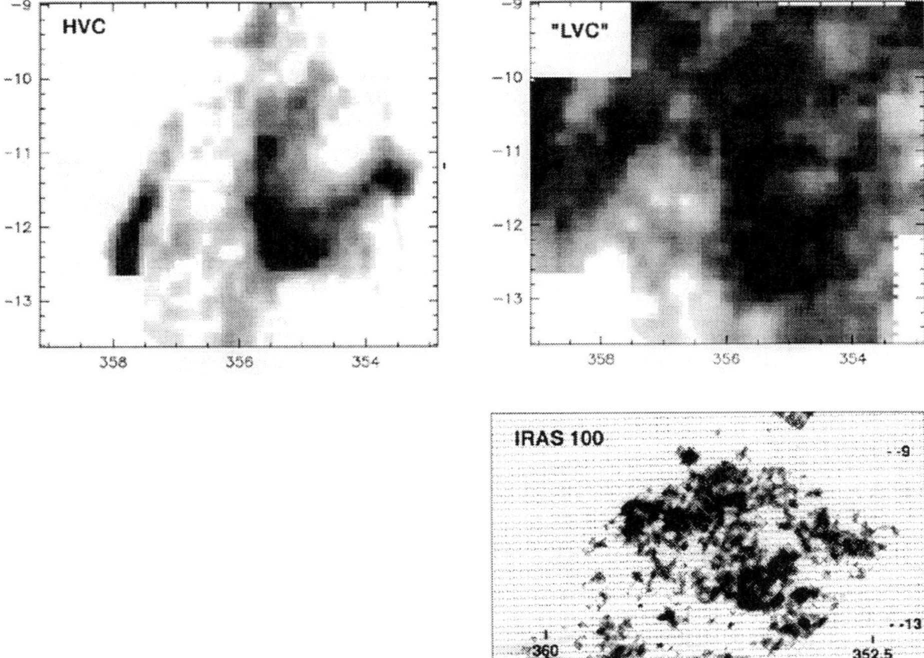

Figure 1: Magellanic Stream cloud IV in RA, DEC (1950): HVC: HI column density high-velocity gas, "LVC": HI column density low-velocity gas, IRAS100: IRAS 100 μm image (all relative units). For more details see text.

statistics for the IVC dust properties can be derived. Additionally, the search for dust in HVCs can be repeated with these new HI data and an IRAS database, which quality was improved during the last few years by improving the various corrections and calibration steps.

- ISO
  Several of the scientific goals will be followed up by ESA's Infrared Space Observatory (ISO, Kessler 1992). The nature of the diffuse interstellar dust will be analysed following several proposals from the central as well as from the open time programs (Bonnet 1994, Leech 1996: Mattila et al. – Diffuse IR emission of the galaxy, Boulanger et al. – Small particles in the ISM, Dust in the low density phases of the interstellar medium, Onaka et al. – Observations of diffuse ISM, Davies et al. – The diffuse FIR emission from spiral galaxies, Lemke et al. – Search for very cold dust outside the optical disk of galaxies, Rouan et al. – PAHs / very

small grains in edge-on galaxies, Reach et al. – Infrared emission from the warm ionised medium). These data will be used to characterise dust in the ISM extending also into the halo of galaxies. Distinct IVCs as well as HVCs will also be subject of ISO observations (Bonnet 1994, Leech 1996: Lemke et al. – Searches for cold dust in intergalactic and high velocity clouds, Reach et al. – Grain destruction in intermediate-velocity clouds, Engelmann et al. – Interaction of IVC $86 + 38 - 44$ with the hot plasma, Moritz et al. – Infrared photometry and spectroscopy of the Draco nebula, Kerp et al. – The high-velocity-cloud 90+42.5-130). These proposals intend to analyse the influence of the clouds' history on the dust composition and infrared emission. Moreover, the Serendipity Survey, which is performed with the ISOPHOT $200\,\mu$m camera during slews between to target positions (Bogun 1995), will provide information on the IVCs and HVCs crossed by the slew scans and may also hint on a possible cold dust component at high galactic latitudes.

- Mid-infrared cameras
  Small dust particles, like the PAHs, emit predominantly in the MIR. New technology for this wavelength range opens the possibility to look for these particles in HVCs. A proposal by Herbstmeier et al. is in preparation to use the new MPIA Mid-infrared Array Expandable Camera (MAX) on UKIRT looking for this emission in HVC complexes of extragalactic systems. This 128×128 pixel Si:As BIB aray is operated at wavelengths between 4 and $27\,\mu$m. On the UKIRT 3.8 m telescope the pixel scales according to 0.27"/pixel.

- FIR facilities
  Herbstmeier, Ferrara and Zylka intend to use the 19 channel bolometer of the Max-Planck-Institut für Radioastronomie to look for the 1.3 mm continuum emission of HVCs in other galaxies as well as in our own Milky Way. The field on the sky which can be covered is currently limited to about 8' on practical reasons. Recent results of Neininger et al. (1996) promise that the sensitivity of the instrument is sufficient to detect cold dust in extragalactic HVCs or extensions of high-z filaments or at least to derive significant upper limits. Currently it is also examined whether the upcomming SCUBA array of the Royal Observatory at Edinburgh can be used to provide additional measurements in the submillimeter regime.

Acknowledgement: I like to thank Anderea Ferrara very much for his invitation to visit the Osservatorio Astrofisico di Arcetri to discuss several of the topics raised by the search for dust in galaxy halos.

References

Bakes E.L.O., Tielens A.G.G.M., 1994, ApJ 427, 822
Bogun S., 1995, PhD Thesis, Max-Planck-Institut für Astronomie, Heidelberg
Bonnet R.M., 1994, ISO Call for Observing Proposals
Borkowski K.J., Dwek E., 1995, ApJ 454, 254
Breitschwerdt D., Schmutzler T., 1994, Nat 371, 774
Cohen R.J., 1982, MNRAS 199, 281
Désert F.X., Boulanger F., Shore S.N., 1986, A&A 160, 295
Désert F.X., Boulanger F., Puget J.L., 1990, A&A 237, 215
Dettmar R.-J., 1992, Fund. Cosmic Phys. 15, 148
Deul E.R., Burton W.B., 1990, A&A 230, 153
Devereux N., 1994, in The First Symposium on the Infrared Cirrus and Diffuse
    Interstellar Clouds, eds. R.M. Cutri et al., A.S.P. Conf. Ser. 58, p. 207
Draine B.T., Lee H.M., 1984, ApJ 285, 89S
Ferrara A., Ferrini F., Franco J., Barsella B., 1991, ApJ 381, 137
Fong R., Jones L.R., Shanks T., et al., 1987, MNRAS 224, 1059
Hartmann Dap, Burton W.B., 1996, Atlas of Galactic Neutral Hydrogen, Cam-
    bridge University Press, in press
Heiles C., Reach W.T., Koo B.-C., 1988, ApJ 332, 313
Herbstmeier U., 1990, PhD Thesis, Universität Bonn
Herbstmeier U., Heithausen A., Mebold U., 1993, A&A 272, 514
Herbstmeier U., Mebold U., Snowden S.L. et al., 1995, A&A 298, 606
Herter T., Shupe D.L., Chernoff D.F., 1990, ApJ 352, 149
Kessler M.F., 1992, in Infrared Astronomy with ISO, eds. T. Encrenaz et al.,
    Nova Science, p. 3
Leech K. (ed.), 1996, ESA ISO Info no. 6
Lockman F.J., Gehman C.S., 1991, ApJ 382, 182
Mebold U., Herbstmeier U., 1993, in: New Aspects on the Magellanic Cloud
    research, Lecture Notes in Physics 416, Springer Verlag, Berlin, Heidel-
    berg, New-York, p. 105
Meyerdierks H., 1992, A&A 253, 515
Neininger N., Guélin M., Garcia-Burillo S., Zylka R., Wielebinski R., 1996,
    A&A, in press
Odenwald S.F., Rickard L.J., 1987, ApJ 318, 702
Puget J.-L., Abergel A., Bernard J.-P., et al., 1996, A&A 308, L5

Reynolds R.J., 1993, in Back to the Galaxy, eds. S.S. Holt et al., AIP Conf. Proc. 278, 156

Reynolds R.J., Tufte S.L., Kung D.T., McCullough P.R., Heiles C., 1995, ApJ 448, 715

Telesco C.M., 1993, in Infrared Astronomy, Cambridge University Press

Wakker B.P., Boulanger F., 1986, A&A 170,84

Wakker B.P., van Woerden H., 1991, A&A 250, 509

Weiner B.J., Williams, T.B., 1996, AJ, in press

Wolfire M.G., McKee C.F., Hollenbach D., Tielens A.G.G.M., 1995, ApJ 453, 673

Zaritsky D., 1994, AJ 108, 1619

Address of the author:

UWE HERBSTMEIER Max-Planck-Institut für Astronomie, Königstuhl 17, D-69117 Heidelberg, Germany, uherbst@mpia-hd.mpg.de

Detection of Cold Dust in the Outer Parts of edge-on Galaxies

N. Neininger

1 Motivation of the study

Cold dust represents most of the interstellar dust in normal galaxies and may be used as a tracer of both molecular and atomic gas (see Cox & Mezger 1989 and references therein). Guélin et al. (1993, 1995) mapped the $\lambda\,1.2\,\mathrm{mm}$ continuum emission of two nearby spirals, NGC891 and M51. This emission there was found to correlate tightly with CO and poorly with HI . It was not even clearly detected beyond NGC891's molecular 'ring', in a region where HI emission is still strong. The mean dust temperatures derived from the λ $1.2\,\mathrm{mm}$ and FIR flux densities were found to be $\leq 20\,\mathrm{K}$.

In order to further study the properties of the ISM, we observed more edge-on galaxies of similar type, starting with NGC4565 (Neininger et al. 1996). It was chosen mainly because of its weak CO emission: the dust emissivity per H atom is on the average $2-4$ times larger for the molecular clouds than for the HI clouds. Thus, the dust associated with the atomic gas becomes predominant only when the H_2 column density becomes very small – which is the case in the outer parts of NGC4565. The edge-on geometry ($i \geq 85°$) ensures long lines of sight in the disk which helps to detect weak components of the ISM.

In a third step, we concentrated on the outer regions of different edge-on galaxies, of which I present NGC4013 here.

2 Observations

The observations of NGC891 were carried out in February 1993 with a 7-channel bolometer array (Kreysa 1992), those of NGC4565 with an instrument upgraded to 19 channels in March 1995 and the most recent data were taken in Spring 1996. The beamwidths are $11''$ (HPBW), and the spacing between two adjacent channels $20''$. The equivalent bandwidth Δ_ν and central frequency $\nu_\circ$

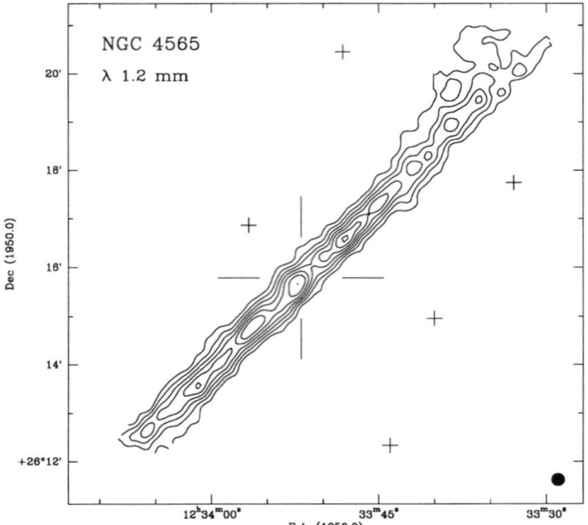

Figure 1: The λ 1.2 mm continuum emission of NGC4565 smoothed to a resolution of $20''$; contour levels are 3, 6, ... 21 mJy/beam.

should be close to 70 GHz and 245 GHz (1.2 mm), respectively (see Guélin et al. 1995).

The maps are mosaics of up to 26 overlapping submaps, each of the size of a few arcmin^2. During the observations, the subreflector was wobbled at 2 Hz in azimuth with a beam throw between $30''$ and $56''$. For each channel, a second order baseline was fitted to every azimuth scan, the scans were combined and restored into single beam maps and regridded in equatorial coordinates; after correcting the intensities for atmospheric absorption, they were calibrated with respect to the planets. Finally, the different channel maps were combined to yield a single map which reached a maximum sensitivity of $1 \ldots 1.5$ mJy per $12''$ beam for the observations of NGC4565.

3 Results and First Findings

Typically, the bulk of the molecular gas lies within a radius of 4–5 kpc from the center. The central component is relatively bright in NGC891 whereas the central 3 kpc region of NGC4565 hosts little interstellar matter, except for a compact 'ring' of molecular gas (of diameter $\simeq$ 1 kpc). In both galaxies a strong molecular ring is visible which contains most of the molecular gas. The bulk of the atomic gas is situated further out, forming a broad 'plateau' peaking at $R \simeq 9 - 15$ kpc and extending up to $R \simeq 15 - 20$ kpc. The outer plateau and molecular ring of NGC4565 show narrow density structures, which are probably spiral arms.

The λ 1.2 mm emission, which is the most reliable tracer of interstellar dust, generally follows closely the CO brightness distribution in the central region and in the molecular ring. In NGC4565, however, the λ 1.2 mm emission follows HI in the outer regions of the disk as soon as the CO emission becomes weak. This way, it extends significantly further out than the emission of the molecular gas and it is possible to derive the properties of the cold dust that is associated with the atomic gas.

The outer HI disks of the chosen galaxies are warped. Because of the lack of detected emission, no trace of it can be seen in the λ 1.2 mm maps of NGC891. However, the onset of NGC4565's warp is clearly visible in the 1.2 mm cold dust emission at the NW side. In the SE, the HI warp is by far less prominent, but a hint at it is seen just at the edge of our map.

Encouraged by this, we set out to explicitly observe the warped outer regions of other edge-on galaxies. Among those, NGC4013 seemed to be the most interesting object, showing a strong pronounced warp in HI (Bottema 1995). We mapped part of the disk and the western warp combining 9 coverages. The first look at the data shows λ 1.2 mm emission out to a radius of at least 15 kpc which is closely related to the HI . There are indications of a rising column density of the dust towards the edge of the disk which will be investigated in the ongoing data reduction. Moreover, the continuum map has features rising out of the plane that correspond equally well with the 'fingers' in the HI map of Bottema (1995). Altogether, it is clear that the cold dust may well exist far outside of the inner regions of galactic disks that are visible in molecular tracers. NGC4565 showed its presence in the warped outer disk and NGC4013 moreover hints at dust ejected out of the plane.

4   Derived Dust Properties

We detected λ 1.2 mm cold dust emission at galactocentric distances > 10 kpc in the HI ring and in the warp of NGC4565 and of NGC4013. In these outer regions, where the gas is mostly in the atomic form (this is indicated by the weakness of the CO emission), it is possible to measure the dust emissivity per H-atom. For the average dust temperature of 15 K (derived from a fit to 160 μm, 200 μm and 1.2 mm data) the comparison of the integrated HI line intensity with the observed λ 1.2 mm flux intensity yields an absorption cross section per H-atom $\sigma_\lambda^H = 5 \times 10^{-27}$ cm^2 (H-atom)$^{-1}$ and a mean dust absorption coefficient $\kappa = 0.002$ cm^{2}g^{-1} for NGC4565 (Neininger et al. 1996). These cross sections and temperatures are similar to those predicted in local diffuse clouds (see Draine & Lee 1984).

The mass of the molecular gas in NGC4565 is low compared to the mass of atomic gas. The molecular gas mass, inside the strip along the major axis covered by our CO observations, is found to be 1.0×10^9 M$_\odot$, when using the Galactic CO-to-H$_2$ conversion factor ($X = 2.3 \times 10^{20}$cm^{-2}K^{-1}km^{-1}s – Strong et al. 1988), and $\simeq 0.4 \times 10^9$ M$_\odot$ when using the values derived from the λ 1.2 mm emission. This corresponds respectively to $\simeq 1/2$ and 1/5 of the H$_I$ mass in the same area (2×10^9 M$_\odot$).

References

Cox, P. Mezger, P.G.M., 1989, A&A Review 1, 49

Bottema, R., 1995, A&A 295, 605

Draine, B.T., Lee, H.M., 1984, ApJ 285, 89

Dumke, M., Braine, J., Krause, M., Zylka, R., Wielebinski, R., Guélin, M., 1996, A&A, in preparation

Guélin, M., Zylka, R., Mezger, P.G., Haslam, C.G.T., Kreysa, E., 1995, A&A 298, L29

Guélin, M., Zylka, R., Mezger, P.G., Haslam, C.G.T., Kreysa, E., Lemke, R., Sievers, A., 1993, A&A 279, L37

Kreysa, E, 1992, in *ESA-Symposium on Photon Detectors for Space Instrumentation*, Noordwijk, p. 207 (ESA SP-356)

Neininger, N., Guélin, M., García-Burillo, S., Zylka, R., Wielebinski, R., 1996, A&A (in press)

Strong, A.W., et al., 1998, A&A 207, 1

Address of the author:

N. NEININGER Max-Planck-Institut für Radioastronomie, Auf dem Hügel 69, D-53121 Bonn, Germany

Cold Gas as Galactic Dark Matter

Ortwin Gerhard

In this talk I have discussed the possibility that some of the dark matter in the halos of galaxies may be baryonic and in the form of cold, dense gas clouds. The ideas summarized briefly below are described in more detail in a paper by Gerhard & Silk (Astrophys. J. 471, 1996). This also contains the many references to previous work which I have omitted below.

A number of observational and dynamical arguments point to the existence of sofar unobserved baryonic dark matter and possibly cold gas around galaxies. Cosmic nucleosynthesis predicts more baryons than are seen in luminous form, by a factor of several. Analysis of the Lyman alpha absorption lines detected towards high redshift quasars suggests that considerably more baryons may be present in the early Universe than are seen today in luminous galaxies. The shapes of the observed rotation curves in spiral galaxies are often well-predicted from scaling up the distribution of gas mass. From the several experiments searching for Macho events towards the Magellanic Clouds we have now evidence for a significant fraction of the Galactic halo consisting of $\sim 0.3 - 0.6\,\mathrm{M}_\odot$ objects, presumably non-luminous baryonic objects.

Dynamical stability arguments and analysis of recent gamma ray observations suggests that a substantial, sofar unseen gas component must reside in the outer halo. To avoid evaporation into the hot gas component of the Galactic halo, and, more importantly, collisional dissipation and settling of the gas into the disk plane, requires that such gas must be in the form of dense, cold clouds. Specifically, the column density must exceed $N_H \sim 10^{23}\,\mathrm{cm}^{-2}$ and the area filling factor must be $\lesssim 0.01$. For a typical temperature of $T \sim 10\,\mathrm{K}$, the condition that the clouds be self-gravitating determines their mass and radius. Typical values are $M_c \lesssim 1\,\mathrm{M}_\odot$ and $L_c \lesssim 0.02\,\mathrm{pc}$. The precise thermal state of these clouds is uncertain, both because of unknown heating rates in the outer halos of galaxies, and because modern calculations of molecular cooling rates are not available for the low element abundances expected in these clouds. However, observational constraints limit their temperature to values of order $10\,\mathrm{K}$ or less.

The critical theoretical question is why the gas clouds do not form stars even at low efficiency. This is a serious problem because any such gas clouds have presumably spent several Gyr in the Galactic halo – by contrast, any pristine gas now forming the first stars in dwarf galaxies like I Zw 18 may only recently have collapsed to high enough densities. We have considered two mechanisms for stabilizing cold halo clouds. The first relies on embedding the gas in a collisionless minicluster of particles which reduces the effective self-gravity of the gas. One finds that a polytropic gas cloud, which in isolation would be unstable, can be stabilized by a surrounding minicluster when the gas mass fraction is $\lesssim 30\%$. Such miniclusters might form if originally the gas clouds are gravitationally unstable to continuing fragmentation. There would then result clusters of Jupiters near the fragmentation limit ($\sim 10^{-2} - 10^{-3}\,M_\odot$). Because these low-mass objects do not inject as much energy into their surroundings as do normal main sequence stars, the residual gas can remain near the Jupiters and settle in the common potential well.

However, such objects now seem unlikely to make up the main constituent of the dark matter halo, because from the microlensing experiments the fraction of the halo in Machos in this mass range must be $\lesssim 10\%$. It is just possible that the miniclusters consist of lower-mass Machos: there is a small window in the mass range below $\sim 10^{-7}\,M_\odot$. In addition, there is an alternative possibility that the gas clouds are stabilized by miniclusters of cold dark matter particles.

An entirely different stabilizing mechanism relies on the assumption that the clouds contain dust grains whose temperature is fixed by the galactic and inter-galactic radiation field at a larger value than the gas temperature and cosmic background radiation temperature. Then the dominant heating mechanism for the gas is by collisions with the grains, and this may under certain conditions lead to a stable equation of state. Local overdensities in the gas would suffer increased heating from the dust, and hence be stabilized against collapse. This mechanism may indeed be relevant in some Bok globules in the interstellar medium.

There are many observational probes with which cold gas clouds in the outer halos of galaxies or in our own Galactic halo can in principle be detected. Perhaps surprisingly, none of these is as yet particularly restrictive. Some of the most promising are: Obscuration of background objects; weak, broad molecular line emission; diffuse gamma ray emission; FIR emission from cold dust. Attempts at detection would usefully be directed towards clouds at large galactic radii, because in our model one predicts that inner halos should by now be largely depleted of gas – substantial gas fractions should have survived only in the outer halos.

If indeed a significant fraction of the mass of galactic halos is or once was in the form of cold gas, then these halos would have played a much more active role in the build-up of galactic disks and their chemical evolution than is commonly supposed. Moreover, these processes would then naturally occur fastest in massive galaxies, with much of the mass in dwarf systems remaining in dark halos even today.

Address of the author:

ORTWIN GERHARD, Astronomisches Institut, Universität Basel, Venusstr. 7, CH-4102 Binningen, Schweiz.

Theory
of the Hot Phase

Evolutionary Scenarios of Gaseous Halos

G. Hensler, M. Samland, O. Michaelis, I. Severing

1 Introduction

If we wish to discuss the gaseous structure of a galaxy halo, at first, we have to specify what we denote as a halo. This seems to be an easy task, because it can refer to everything found outside the main luminous stellar body of a galaxy. The steep decrease of radial stellar brightness distributions supports this impression. Nevertheless, from the first viewpoint the stellar distributions of the different morphological galaxy types might lead to some confusion. While in ellipticals the overall spheroid represents the main luminous body, in disk galaxies the outermost spheroid is faint and only an innermost spheroid, the bulge, dominates in accordance to the Hubble type with the additional brightness of the characteristic disk population. Since the formation scenarios of spiral and elliptical galaxies are reasonably assumed to have occurred in different manners, for the giant ellipticals preferably by mergers, also the gaseous halos will have evolved differently. Because content and size of this paper does not allow the dedication of a general review, here we wish to restrict ourselves to some issues about the evolution of gaseous halos of disk galaxies from our studies.

From our present-day picture about the structure of our Milky Way Galaxy (MWG) the part which does neither belong to the galactic disk nor to the bulge is defined as the *halo*. One can distinguish a stellar halo component from the stellar population characteristics of metallicity, kinematics, and age, which also show a local separation because of the high velocity dispersion of halo stars and their low local number density in the solar vicinity (see e.g. Gilmore 1989). While we can determine by this to some certainty its inner boundary as the height below which the disk population dominates, the radial extent of the stellar halo is almost impossible to discern. Globular clusters (GCs) are seen to be gravitationally bound still at galactocentric distances larger than those of the Magellanic Clouds and even the tidal evaporation of losely bound stars to the intergalactic space are invoked.

While brightness and number density of stars in the disk are decreasing vertically almost exponentially (and by this show a quite regularly shaped edge-on disk), the situation is more complex and much more irregular for the interstellar medium (ISM). Within our MWG the vertical stratifications of various ISM phases belonging to the galactic disk can be deduced from observations in different spectral ranges. Here we refer to the cool molecular cloud phase of low scaleheight, the Lockman layer (see recent results within this volume), the Reynolds layer (see also herein). Other components like infalling high-velocity clouds (HVC) to low-velocity clouds (LVC) and the gaseous corona detected in UV absorption lines of high ionisation stages of O, C, and N are attributed to the halo, although their origin is not yet completely clear. A variety of velocities are visible for infalling HI clouds and complexes (Wakker & Schwarz 1991, Danly et al. 1992) but only the LVC observations allow to locate them closely above the galactic disk, while for the HVCs distances are uncertain. A picture that combines the infalling clouds and embeds them into a hot gaseous halo where they approach the disk has been proposed by Kerp et al. (1996).

The present-day gaseous halos of disk galaxies show a multi-variate interplay between infalling galactic or intergalactic matter with outflowing galactic gas. The disk gas is stirred up by means of energetical events, is swept up by expanding superstructures like supershells and superbubbles, and is probably ejected into the halo region through funnels of vertically opening magnetic fields and forming a synchroton-radiating hot halo. Although the outflowing disk gas is not directly observed except as expanding H_α shells in starburst dwarf galaxies (e.g. Heckman et al. 1995), in external galaxies signatures are detectable like structured H_α filaments, supershells, HI holes, clumps, and X-ray plumes, which are well interpreted as results of this outflowing process. Several symposia have been dedicated to the *disk-halo connection* (like e.g. IAU Symp. No. 144 in 1990) in order to improve our insight. Although we do not really know to what extent these dynamical and energetical activities lead to the present-day halo formation or support its maintenance, it is obvious that gaseous galactic halos must have exist already during the early stages of galaxy evolution as an inseparable part of each galaxy.

2 The Early Formation of Gaseous Halos

2.1 The Early Galactic Collapse and Supernova Activity

Once a gravitationally bound protogalactic gaseous system is able to cool and to collapse the formation of a galaxy commences. Although the issues of processes acting during this early epoch can be traced in old stellar populations as well as in high-z gas properties, it is still one of the white areas in astrophysical research. Our ignorance is based on the two major problems: 1) We cannot observe the very primordial state of the ISM and, therefore, do not know about its scales and clumpiness; 2) It is still far from being understood whether the first generation of stars could be limited solely to massive ones because of the absence of cooling agents in the gas. On the other hand, the smallest aggregates which are supposed to be produced by the dominance of primordial cooling processes, the GCs with masses around $10^6 \mathcal{M}_\odot$ (Fall & Rees 1985), consist of stars with abundances in excess of $10^{-3} Z_\odot$. Thus they must have formed from small but already metal-enriched clouds.

There is now much evidence that metal-absorption line systems in QSO spectra arise from the gaseous halos of parent galaxies. While they have been recognized at first by Mg II and C IV lines according to their redshift towards the optical window they have recently also been detected in S IV, N V, and O VI (Tripp et al. 1996). The lack of knowledge of their dynamical properties and their origin has led to different theoretical approaches under particular assumptions. As a working hypothesis it can be well assumed that the early hot halo gas is produced by supernovae typeII (SNeII), because the analyzed abundance ratios agree well with SNII yields (Reimers et al. 1992). Li & Ikeuchi (1992) considered this hot gas as formed with high overpressure within galaxies. For different internal and surrounding energy densities the hot gas expands out to different galactic radii, but even for moderate conditions it reachs 30 kpc and more achieving constant C IV, Si IV, N V, and O VI column densities. Mo & Miralda-Escudé (1996) approach to a more realistic description of the ISM with two gas phases, where cool clouds fall through a hot gaseous halo, while they stay in pressure equilibrium but are photoionized if they survive evaporation. The hot halo is formed in their scenario by shock heating of infalling gas.

A global evolutionary picture of the formation of disk galaxies which takes a three-phase ISM into account and treats two phases dynamically independently while allowing for star-gas interactions has been investigated by us (Samland 1994; Samland & Hensler 1996, Samland et al. 1996: hereafter SHT) taking advantage of the chemo-dynamical description. In one of the models which resembles the MWG at best (SHT), the early star formation with high rates of up to 50 $\mathcal{M}_\odot$ yr^{-1} is distributed over the whole galaxy but concentrates to smaller radii later during the collapse of the protogalaxy and leads to an excess of SNeII. This released hot gas cannot be kept bound because of the low external ISM pressure in a still flat gravitational potential but expands from all regions even from the inner bulge and streams outwards to distances of 20 kpc and more. While in the inner part the state of the hot thin intercloud medium (ICM) can only interact with the cooler embedded clouds (CM) by evaporation, the condition changes outwards due to cooling. Therefore condensation of hot metal-rich SNII gas onto the clouds' surfaces dominates at larger radii.

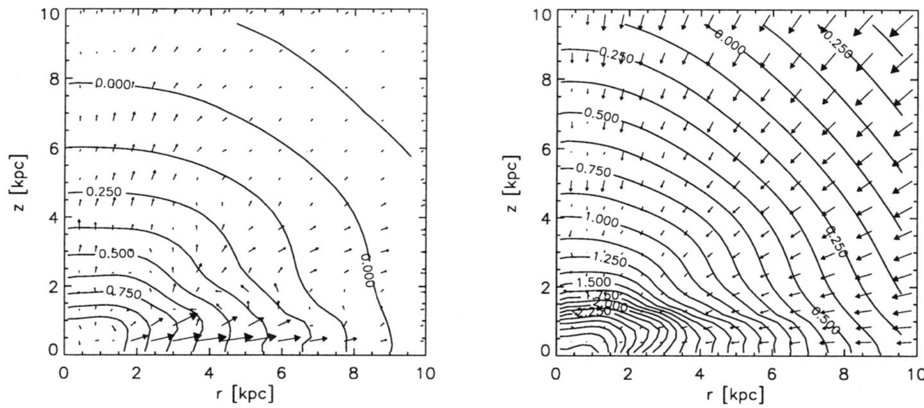

Figure 1: Surface densities and velocities of hot intercloud medium (left) and cloudy medium (right) 7 Gyrs after the onset of the galactic collapse. The ICM in the galactic plane has velocities up to 230 km/s, while the velocities of the infalling CM are less than 60 km/s. (from SHT)

While fig.1 demonstrates the dichotomy of the gas-phase dynamics even after 7 billion years, figs. 2 represents the change in the evaporation/condensation behaviour from 1.5 Gyrs to 7 Gyrs after the onset of the collapse. There the spatial distribution of a characteristic parameter $\beta = (\dot{\rho}_{cond} - \dot{\rho}_{evap})/(\dot{\rho}_{cond} + \dot{\rho}_{evap})$ is shown at two epochs. Negative values of β represent areas with dominating evaporation, at positive labels condensation exceeds. As one can discern at smaller radii evaporation of cloudy material dominates while condensation of hot gas onto the clouds enrich them in the outermost regime. The evaporation zone has shrinked radially at 7 Gyrs because of the mentioned pressure enhancement. The more complex structure at 7 Gyrs decovers the formation and growth of large cloudy aggregates by means of condensation. This state of condensation is strikingly perceivible in fig.3 where the main aggregate from fig.2 is shown with a smaller companion both embedded in the hot outflowing supernova gas. In the vicinity of the giant clouds the arrows of the ICM flow direction are clearly bending towards the clouds' surfaces. The clouds can achieve masses of up to $10^8 \mathcal{M}_\odot$ and fall further in slowly.

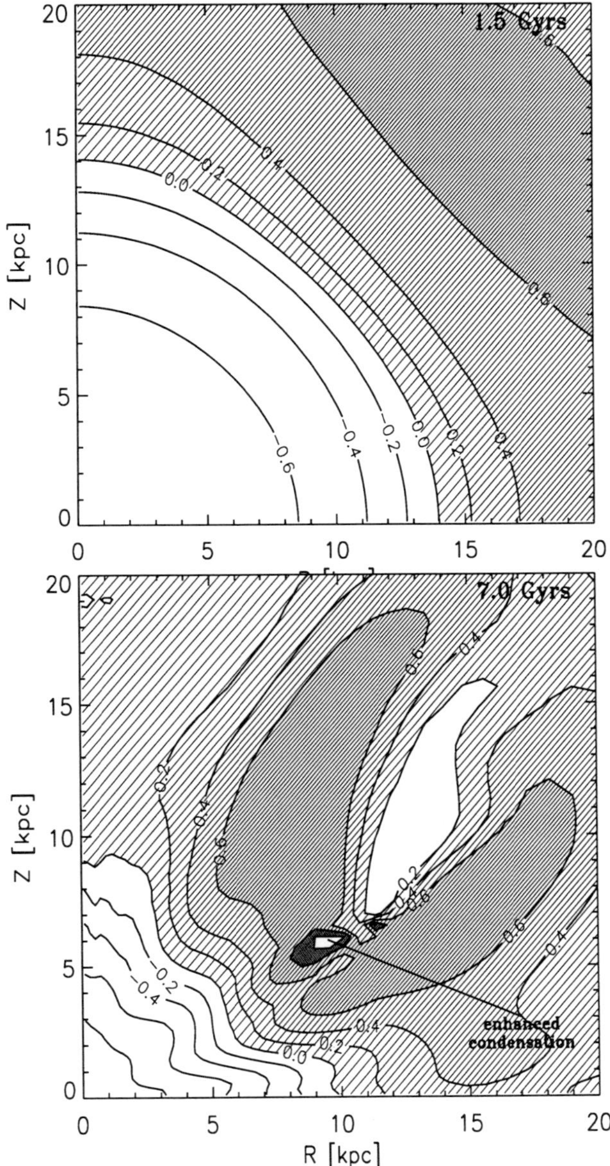

Figure 2: The distribution of the evaporation/condensation parameter $\beta = (\dot{\rho}_{cond} - \dot{\rho}_{evap})/(\dot{\rho}_{cond} + \dot{\rho}_{evap})$ at 1.5 and 7.0 Gyrs after the commencement of the collapse (from Samland 1994). See the discussion of the different regimes in the text!

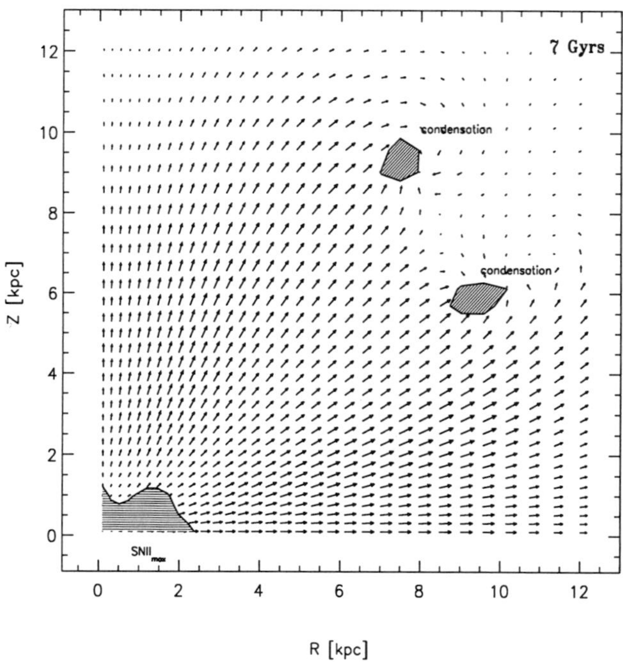

Figure 3: Giant clouds in the innermost halo are formed by condensation (compare with fig.2) of hot outflowing gas (arrows; largest velocities shown amount to less than 250 km s^{-1}). The central hatched area represents the zone of highest supernova typeII activity.

The evolution of these condensation seeds are not yet treated in full detail of physical processes and internal structures but they give the impression and allow to speculate that they represent the precursors of an almost coherent population of lately formed GCs.

2.2   Clouds in a Hot Gas Flow

Concerning the evolution of these CM aggregates which are comparable with giant molecular clouds in the present ISM the problem of their evolution in the hot streaming gas has to be studied. Three at the moment unsolved questions have to be addressed accordingly:

1. To what extent are the clouds dynamically coupled to the hot gas motion? This is the question for the drag.

2. Concerning the drag, the effective cross section of the clouds are of interest. By that, how is the shape of the cloud influenced by the hot gas stream?
3. How does the mass of the clouds evolve? For this, investigations have to figure out, what amounts of mass are either stripped off from the clouds or are sticking on the surface by instabilities?

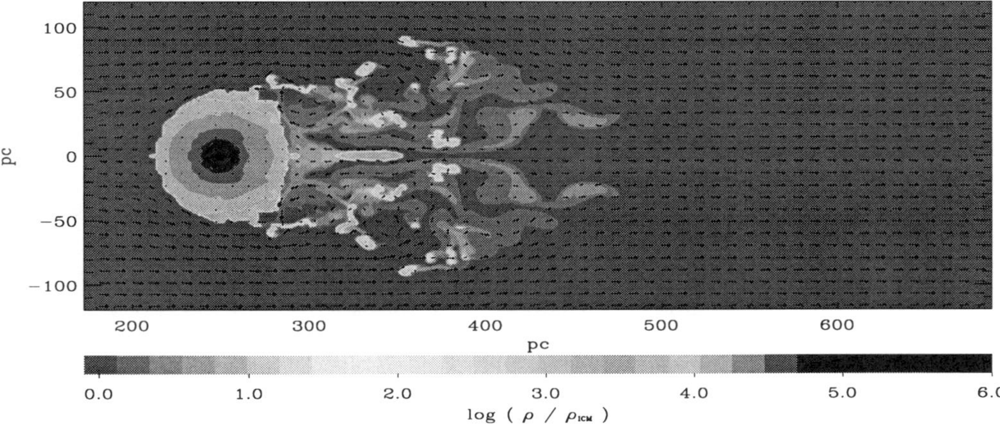

Figure 4: Density of cloudy material in a Mach=0.3 streaming hot gas at 27 Myrs after the start of the simulation (from Severing 1995).

For this purpose we have performed hydrodynamical simulations (Severing 1995, Severing et al. 1996). Because of their high mass these clouds are self-gravitating; since their infall velocities does not reach more than almost 200 km s^{-1} they can be treated subsonically in a several 10^7 K hot ICM. In order to stabilize the clouds in an undisturbed case we introduce functional heating and cooling processes into the energy equation. As a short resume of the results from a model grid with different Mach number we can conclude that the shape of the clouds is slightly stretched perpendicularly to the streaming direction, that the total mass is reduced by a few up to ten and more percent, while an accretion of up to 20% of streaming-in hot gas within the cloud's cross section is reached. It is suggestive that this amount of gas can be homogeneously distributed within the cloud and, by this, enhance the metal abundances, or can polute smaller star-forming fragments inhomogeneously. The homogeneous enrichment has to be preferred for GCs in order to explain their small intrinsic abundance scatter.

3   Disk-Halo Connection by Expanding Superbubbles

Hot gas is observed in X-rays of a handful edge-on spiral galaxies (see e.g. Junkes & Hensler 1996). It also exists in the halo of our MWG. The question of its origin seems to be clarified by observed signatures of vertically escaping gas from the disks of edge-on galaxies.

Because of the local concentration of massive stars in OB assiciations and due to their comparable short lifetimes their SNeII explosions are suggested to produce superbubbles as an accumulated highly energetical effect. 100 to 1000 explosions of massive stars contribute successively to the hot bubble which expands within the stratification of a galaxy disk preferably in vertical direction. If a hot vertically elongated cavity is formed, shocks of succeeding SN gas can easily propagate through it and deposite kinetic energy in the outer shell of the superbubble, by this transporting energy and preferably hot gas from the disk into the halo.

Although this process sounds very plausible in order to account for the existence of hot highly-ionized gas in the halo and for sufficient sweep-up of disk gas to account for the LVC (Norman & Ikeuchi 1989), problems emerge in detailed studies. Analytical studies can consider an instantaneous explosion of a specific number of massive stars or a constant energetic wind (e.g. Castor et al. 1975) expanding into a vertically stratified gaseous neighbourhood, while hydrodynamical simulations refined the models taking sequential explosions (Tomisaka & Ikeuchi 1986) and different external components of the ISM into account. Most of the models show the formation of a cool dense HI shell (MacLow & McCray 1988) and the break-out of the superbubble from the disk into the halo (MacLow et al. 1989; see also recent review by MacLow 1996).

These results can be suspected because they neglect the tension of horizontal magnetic fileds as well as the external pressure of an already existing hot gaseous halo. Although sequential SNeII which explode discretely in accordance to the lifetimes of simultaneously formed massive stars provide an enhanced superbubble expansion in comparison to the wind-driven case, already the additional contribution of an external halo pressure prevents the blow-out of a multiple supernova remnants from 100 explosions (Michaelis 1995, Michaelis et al. 1995). Only if one considers OB associations formed up to 80 pc above the equatorial plane one can make the break-out easier (see fig.5). Because until now heat conductivity and the evaporation of Rayleigh-Taylor

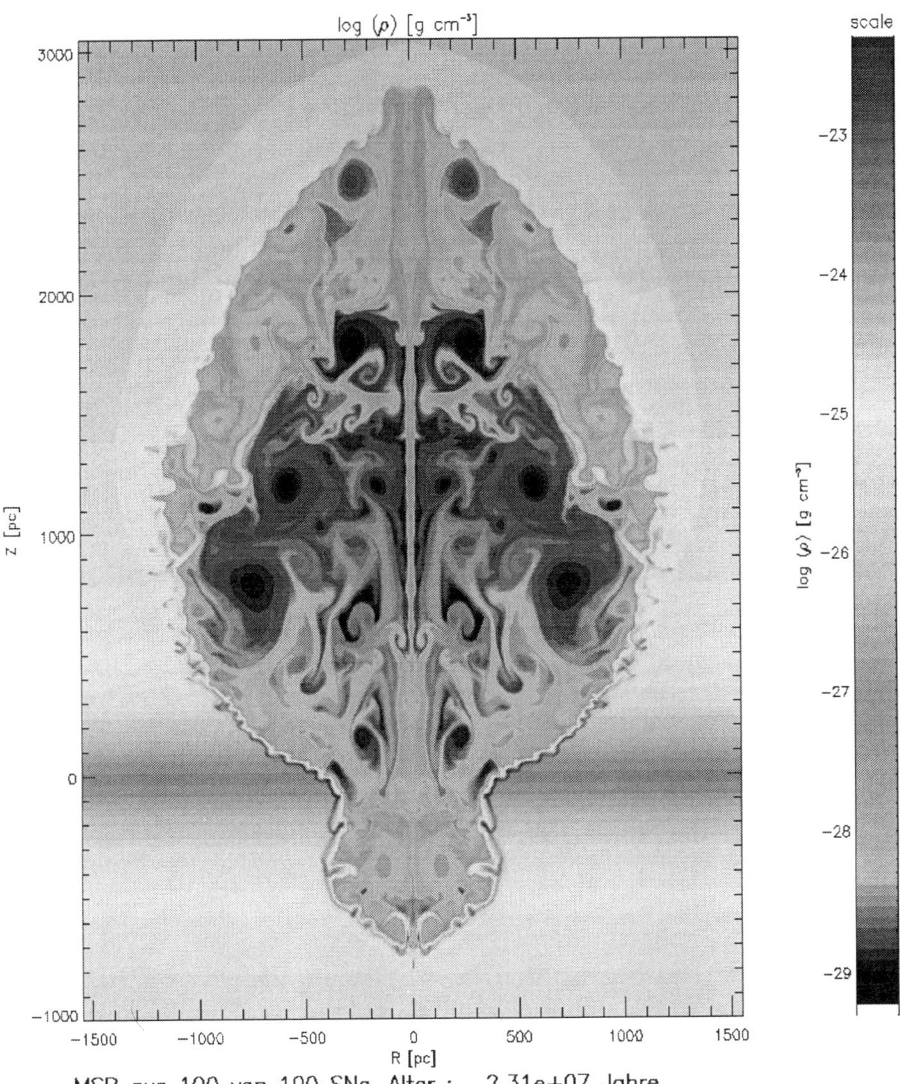

Figure 5: Expanding superbubble formed from 100 discrete supernova explosions of massive stars in an OB association which is located 80 pc above the equatorial plane. The snapshot is taken 23 Myrs after the first explosion of the most massive 100 $\mathcal{M}_\odot$ star.

instability fingers sticking into the hot gas or of embedded interstellar clouds (Silich et al. 1996) are neglected, the superbubble temperatures are extremely high and the bubbles' X-ray luminosities exceed the observed ones by far. As a consequence, evaporating dense gas would decrease the bubble temperature (Silich et al. 1996), by this, diminish the internal pressure and problably prevent the bubble blow-out into the halo. In contrast, two possibilities could be supposed to alter this pessimistic conclusion and to facilitate the disk-halo flow again: After the horizontal expansion of the bubble has stopped, it could be reversed so that the bubble closes at its bottom and moves upwards by buoyancy; magnetic fields can break up and form a vertical chimney wherein cosmic-ray diffusion accelerates the gas (Breitschwerdt et al. 1993). By this, the halo support by hot disk gas as concluded from observational signatures can be realized at present days.

4   Conclusions

The evolution of the gaseous halos of disk galaxies has to be divided into two epochs and processes: the early formation stage and the time of maintenance.

The early state of the halo during the formation stage of the parent galaxy is characterized by the dichotomy of the nature of its components. While one gas phase stems from the protogalactic cloud which is cooling and contracting in the gravitationally bound system, a hot gas phase produced by SNeII of the early but strong star-formation epoch expands, but carries already nucleosynthesis products throughout the whole galaxy. The protogalactic system does not collapse radially on short free-fall and also not on dissipation timescales but is determined by cooling and phase transition processes. Phase transitions tend to mix the metal-rich gas from the hot phase to the cooler clouds where modestly ionized and dense transition layers become visible as QALS with metallicities of $[Z_D] > -1.5$. Infalling cloud aggregates are growing immensely by condensation and could account for the formation of a later GC population.

Hot gas also exists in present-day halos of galaxies. However, its origin is still suggestive. Obviously, strong star-formation rates produce bright X-ray halos by means of gas blow-out from the disk. For moderate and smoothly distributed star-formation activity in disks the situation is still discussed controversally. Although plausible, detailed simulations of multiple supernova remnants are still lacking the treatment of important physical processes and components, like e.g. magnetic fields, heat conductivity, buoyancy, differential rotation, reliable external stratifications, etc. First models give only hints until now but do not allow an decisive conclusion about this disk-halo connection.

References

Breitschwerdt D., McKenzie J.F., Voelk H.J., 1993, A&A 269, 54

Castro J., McCray R., Weaver R., 1975, ApJ 200, L107

Danly L., Lockman F.J., Meade M.R., Savage B.D., 1993, ApJS 81, 125

Fall M., Rees M.J., 1985, ApJ 298, 18

Gilmore G., 1989, in *The Milky Way as a Galaxy*, 19th Adv. Course Swiss Soc. of Astrophys. & Astron., Saas Fee, p. 227

Heckman T.M., et al., 1995, ApJ 448, 98

Junkes N., Hensler G., 1996, Proc. of the international X-ray Conference *Röntgenstrahlung from the Universe*, eds. H.U. Zimmermann et al., MPE Report No. **263**, p. 459

Kerp J., et al., 1996, A&A, in press

Li F., Ikeuchi S., 1992, ApJ 390, 40

MacLow M.-M., 1996, Proc. 11th IAP Meeting, *The Interplay between Massive Star Formation, the ISM and Galaxy Evolution*, ed. D. Kunth, Editions Frontières, Gif sur Yvette, in press

MacLow M.-M., McCray R., 1988, ApJ 324, 776

MacLow M.-M., McCray R., Norman M.L., 1989, ApJ 409, 663

Michaelis O., 1995, diploma thesis, University of Kiel

Michaelis O., Hensler G., Samland M., 1995, ApJ, submitted

Mo H.J., Miralda-Escudé J., 1996, ApJ 469, 589

Norman C.A., Ikeuchi S., 1989, ApJ 345, 372

Pettini M., 1994, ApJ 426, 79

Reimers D., Vogel S., Hagen H.-J., et al., 1992, Nature, 360, 561

Samland M., 1994, PhD thesis, University of Kiel

Samland M., Hensler G., 1996, Rev. Mod. Astron. 9, 277

Samland M., Hensler G., Theis C., 1995, ApJ, submitted

Severing I., 1995, diploma thesis, University of Kiel

Severing I., Hensler G., Samland M., 1996, in preparation

Silich S.A., et al., 1996, A&A, submitted

Tripp T.M., Lu L., Savage B.D., 1996, ApJS 102, 239

Tomisaka K., Ikeuchi S., 1986, PASJ 38, 697

Wakker B.P., Schwarz U.J., 1991, A&A 150, 484

Addresses of the authors:

G. Hensler, O. Michaelis, I. Severing
Institut für Astronomie und Astrophysik, Universität Kiel, Olshausenstr. 40,
D-24098 Kiel, Germany
M. Samland
Astronomisches Institut der Universität Basel, Venusstr. 7, CH-4102 Binningen, Swizerland

Non-equilibrium plasmas in galactic halos

Dieter Breitschwerdt

1 Introduction

Already 40 years ago, Spitzer (1956) postulated the existence of a hot ($\sim 10^6$ K) Galactic corona in order to confine diffuse gas clouds in the halo. There was clear evidence for the existence of extraplanar HI gas with velocities in excess of 20 km/s with respect to the local standard of rest (Münch and Zirin, 1961). Of course, this idea was no more than a convenient speculation, because the detection of a hot phase of the Interstellar Medium (ISM) had still to await the launch of the COPERNICUS satellite in 1972. Important pieces of evidence came from the ubiquitous O VI absorption lines (Jenkins & Meloy, 1974). Thus the existence of a widely distributed hot intercloud medium (HIM) seemed compelling. The detection of highly ionized Fe IV in the Cygnus Loop (Woodgate et al., 1974) argued for a thermal origin of the X-ray emitting plasma.

At first glance it seemed tempting to identify the earlier discovered (Bowyer et al., 1968) diffuse soft X-ray background (SXRB) by thermal emission of a tenuous and widespread HIM, even more so because nonthermal mechanisms failed to reproduce it. A purely extragalactic origin was dubious, because the absolute intensity of the measured flux exceeded the downward extrapolation of the spectrum above 2 keV (Henry et al., 1968). In particular the ultrasoft 1/4 keV energy band, required a local contribution, because unity optical depth is already achieved for a column density of $N(H) \sim 10^{20}$ cm^{-2}. However, with a typical temperature of the HIM of $\sim 5 \times 10^5$ K (McKee and Ostriker, 1977), the correct ratios of the WISCONSIN C-band (160 - 284 eV), B-band (130 - 188 eV) and Be-band (77 - 111 eV) count rates within the framework of a Raymond and Smith (1977) equilibrium plasma model is not possible (cf. McCammon and Sanders, 1990). The spectra are systematically too soft and a somewhat higher equilibrium temperature of $\sim 10^6$ K is required. The origin of the diffuse SXRB is still under debate. On the other hand, estimating the energy input into the ISM, it became soon clear that supernovae in concert with early type stellar winds were responsible for the existence of the HIM.

In addition, the scale height of such a hot plasma, which is of the order of a few kpc, by far exceeds that of the neutral gaseous disk and hence should fill a hot galactic corona. There is even a possibility that some fraction of it may leave the Galaxy in the form of a thermal galactic wind (Mathews and Baker, 1971; McKee and Ostriker, 1977). Including centrifugal forces, the escape temperature at the solar circle should be larger than 4×10^6 K (Habe and Ikeuchi, 1980). Therefore, the bulk of the hot gas should not have enough energy to escape and will become thermally unstable due to radiative losses, falling back onto the disk like in a "galactic fountain" (Shapiro and Field, 1976; Bregman, 1980). The existence of so-called intermediate velocity clouds has been interpreted as a fountain return flow.

In any case, a dynamical model describing the evolution of the ISM, seems appropriate for understanding the mass, momentum and energy transfer between the various ISM phases. The most comprehensive model, based on a detailed balancing of mass and energy in a 3-phase ISM has been put forward by McKee and Ostriker (1977). As a result, they obtained a fairly high volume filling factor for the HIM of $f_{\mathrm{HIM}}^{\mathrm{V}} \sim 0.7 - 0.8$. However, the existence of an overall pervasive HIM is questionable. HI observations of supershells in our Galaxy (Heiles, 1984) as well as HI holes in external galaxies (Brinks and Shane, 1984) point towards $f_{\mathrm{HIM}}^{\mathrm{V}} \geq 0.2$, assuming a strict anticorrelation between neutral and hot material. Taken at face value, an obvious explanation is the spatial and temporal correlation of type II supernovae as a result of their coeval origin and evolution.

The debate, whether the flow will stall at a height of less than a few kpc, or whether it can reach considerable distances, from which it can ultimately fall back onto the disk in a "clustered" galactic fountain type return flow ("chimney" model; Norman and Ikeuchi, 1989) or, even leave the galaxy as a wind, is still going on. Numerical simulations are inconclusive, because they depend on the boundary conditions chosen. Although the break-out of a superbubble can be inhibited by an obstructing ambient magnetic field of $\sim 5\,\mu$G perpendicular to the shock normal, the hot gas may still reach the halo region ($|z| \geq 1\,\mathrm{kpc}$) during the bubble lifetime (Tomisaka, 1991). This is also true if the shock has to penetrate an extended HI layer with a scale height of 500 pc, given the initial energy is sufficiently large. Assuming on the other hand that the field is deformed into loops due to Parker instability, so that there is a substantial regular component parallel to the flow, the acceleration of the shock propagating down a density gradient could even lead to Rayleigh-Taylor instability and break-up of the outer shell, ejecting hot gas into the halo (Tomisaka, 1991).

In conclusion, I find it hard to imagine that the magnetic field preserves a perfect topology throughout the superbubble evolution, in order to confine a hot plasma expanding into a highly turbulent ambient medium. At least terrestrial plasmas do unfortunately not follow such a behaviour. Observations of the SXRB, and recent ROSAT PSPC observations of nearby edge-on galaxies show extended X-ray halos, connected to their galactic disks (cf. Section 2.1). It is therefore highly probable that *outflow does occur* in normal spiral galaxies. This point of view will be adopted throughout the paper.

2   Observations

The most important observations that should be explained by any serious model of the galactic halo are discussed below.

2.1   Soft X-ray background and X-ray halos

Diffuse emission in the soft energy bands has been first reported by the Wisconsin Survey (McCammon & Sanders, 1990) between $0.08 - 0.3\,\mathrm{keV}$ and $0.5 - 1.5\,\mathrm{keV}$ (M - I-band). With much better spatial resolution, the global distribution of this emission has been confirmed by the ROSAT All Sky Survey (RASS). Above 0.5 keV the background is fairly isotropic, although the photon mean free path in gas with an average density of $1\,\mathrm{cm}^{-3}$ is only about 1 kpc. As much as 50% of the emission in the M-bands may be due to a soft excess of extragalactic point sources (Hasinger et al., 1993). However, a purely extragalactic origin would still fail to reproduce the measured count rates in the direction of the galactic plane due to absorption. Conversely, a purely galactic origin of stellar sources could only contribute about 15% at higher galactic latitudes (Schmitt and Snowden, 1990).

Since deep pointed ROSAT PSPC observations towards heavy absorbers (e.g. Draco nebula), the so-called shadowing experiments (e.g. Snowden et al., 1991; Kerp et al., 1993), were carried out, it has been unequivocally demonstrated that also a substantial fraction of the SXRB is diffuse, thermal emission from a plasma. At the ultrasoft energy bands ($E \leq 0.3\,\mathrm{keV}$), unity optical depth corresponds to a column density of a few $10^{20}\,\mathrm{cm}^{-2}$. HI observations and Ly_α absorption measurements towards background stars, show that this corresponds to a distance of about 100 pc. The X-ray intensity increases with latitude by a factor of 2-3.

The standard interpretation explains the SXRB as emission from a static hot plasma in collisional ionization equilibrium (CIE). Temperatures are derived by fitting Raymond and Smith (1977) emission models to broad band spectra, and densities are derived from the emission measure and extension of the emitting volume. Thus according to the so-called displacement model (Snowden et al., 1990) the ultrasoft emission should originate in an anisotropic "Local Hot Bubble" with a temperature $T \sim 10^6$ K, density $n_e \sim 4.5 \times 10^{-3}$ cm^{-3} and an extension between 30 pc in the plane and 300 pc perpendicular to it. The emission above 0.5 keV is attributed to an extended diffuse source in the halo with a temperature of $T \sim 2.2 \times 10^6$ K (Wang and McCray, 1993)

However, the observations do not support a spatial separation of different energy bands. There are several discrepancies with the above explanation: (i) temperatures in the bubble are lower according to some absorption line widths, which are *upper limits* (e.g. Jenkins and Meloy, 1974); (ii) shadowing experiments have revealed substantial C-band emission from *outside* the Local Bubble; note that the halo source at a temperature of 2.2×10^6 K would only contribute to C-band emission by a few percent; (iii) shadowing towards a heavy absorber like the molecular cloud MBM12 (Snowden et al., 1993) is compatible with 20% of the emission above 0.5 keV coming from *inside* the Local Bubble; (iv) the non-detection of EUV emission lines, that should be highly abundant in a 10^6 K plasma, by the EUVE satellite places severe constraints on any model; the emission measure is a factor of 5 to 10 below the standard CIE model value (Jelinsky et al., 1995); (v) from the dispersion measure of pulsar PSR 0950+08 (Reynolds, 1990), whose distance is known from parallax to be about 130 pc, an average electron density of $\langle n_e \rangle \approx 2.3 \times 10^{-2}$ cm^{-3} is inferred, a factor of 5 higher than in the Local Hot Bubble Model, if this pulsar is still inside the bubble; (vi) the pressure in the bubble is at least a factor of 3 above the pressure in the Local Cloud, which therefore needs substantial support by a magnetic field, in order to survive inside the hot cavity.

Deep ROSAT PSPC observations of edge-on spiral galaxies allow a spatial separation of the disk and halo contribution, due to the high spatial resolution of the instrument. Observations of the galaxy NGC891 are plagued by high Galactic foreground absorption ($N_{\mathrm{H}} \sim 7 \times 10^{20}$ cm^{-2}). Nonetheless there has been a 3σ-detection of an X-ray halo with a radial extension of ~ 6.5 kpc and a vertical extension of ~ 4.8 kpc (Bregman and Pildis, 1994), as well as an X-ray jet protruding about 6 kpc from the disk. Spectral modelling requires a soft halo component with $T_{\mathrm{sX}} \sim 3 \times 10^6$ K.

The galaxy NGC4631 is at high latitude and therefore foreground absorption effects are small. In the 0.3 kev band a huge X-ray halo with a vertical extension of about 12 kpc has been detected (Wang et al., 1995). Interestingly, the spectral modelling by a standard Raymond and Smith (1977) plasma model rejects a single temperature fit at the 90% confidence level. Instead at least 2 temperatures, i.e. a *harder* component of $T \sim 3 \times 10^6$ K and a *softer* component of $T \leq 6 \times 10^5$ K are needed for a satisfactory fit. This of course leaves open the physical explanation, why there should be two coexisting plasmas at different temperatures in the halo.

2.2   Diffuse ionized gas and cosmic rays in galactic halos

The Galaxy, NGC891 and NGC4631 are comparable with respect to their star formation activity, according to their similar FIR luminosities. Therefore it is not surprising that all of them exhibit radio halos with a scale height of $1 - 2$ kpc and also a thick disk of diffuse ionized gas (DIG) with a scale height of about 1 kpc (Dettmar, 1992). A causal connection is strongly supported by the observational evidence that the polarized radio intensity and the H_α emission are spatially correlated (cf. Dettmar, 1992) in some regions, where star formation in the disk below is high. This is a hint that cosmic ray electrons may be advected by a general outflow, since pure diffusion would smear out the radio synchrotron emission to some extent.

Since the X-ray emission is also causally related to the star formation activity, it seems compelling that a successful galactic halo model must be capable of explaining both the extension of X-ray halos and their spectral characteristics and the extension of radio halos and the observed flat behaviour of the radio spectral index in the edge-on galaxies (cf. Hummel, 1991). Such a model will be presented in the following section, in which non-equilibrium cooling of halo plasmas will play a crucial rôle.

3   Dynamical Halo Model

3.1   Brief description of the model

The transfer of mass, momentum and energy from the galactic disk to the halo is described in a hydrodynamical model. Proceeding from the observational fact that there is rough energy equipartition between thermal gas, cosmic rays and magnetic fields in our local environment, it is clear that this not sheer co-incidence. Instead, it is a strong argument in favour of dynamical interaction between these different ISM components, with Parker's instability just being one example. Since the escape of cosmic rays from the Galaxy along magnetic field lines leads to a resonant generation of MHD waves due to the so-called streaming instability (Lerche, 1967), we have included the momentum transfer from cosmic rays to the thermal plasma via these self-excited waves as a mediator into the model (Breitschwerdt et al., 1987; 1991; 1993). The treatment is essentially one-dimensional and the regular magnetic field defines a flux tube geometry perpendicular to the galactic disk (cf. Fig. 1). Such a field structure

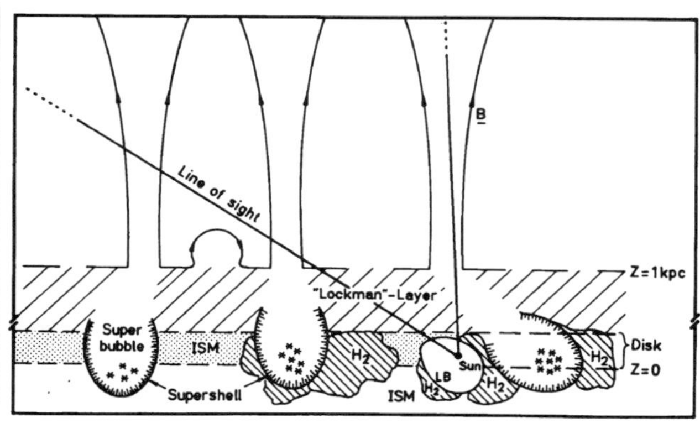

Figure 1: Schematic edge-on view, not drawn to scale, of the galactic disk and halo (Breitschwerdt and Schmutzler, 1994).

has been observed in NGC4631 (Hummel et al., 1988), although its spatial extension is not typical for normal spiral galaxies. However, all we require is some "locally open field lines" in a galactic disk where blow-out of hot gas can occur. At some distance from the disk the flux tubes open up. The cosmic rays diffuse through the ISM, but reaching a height of 1 kpc in the halo, their transport is dominated by advection, since the wave field is no longer isotropic and they drift through the gas at essentially the Alfvén speed. The combined forces due to gas, cosmic ray and wave pressure lead to an acceleration $u(du/dz)$ of the gas against the gravitational acceleration $-g_{\text{eff}}$ (including a bulge, disk and dark matter halo component for the gravitational potential), where u is its velocity and z the coordinate perpendicular to the disk. We have shown (e.g. Breitschwerdt et al., 1991) that a steady state flow can become transsonic at a distance of the order of ten kpc, reaching a terminal velocity of a few 100 km/s. The overall mass loss rate is of the order of $1\,M_\odot/\text{yr}$, comfortably lower than the average star formation rate, but still high enough to be significant for the dynamical and chemical evolution of the ISM.

3.2   X-ray emission model

The X-ray emission from a fast adiabatically expanding flow, such as a galactic wind is radically different from the static case. The time scales for atomic processes (e.g. ionization, recombination) and the dynamical time scale are quite different in general. Therefore the emission spectrum is essentially *never in equilibrium*. It is worth pointing out, that the assumption of collisional ionization equilibrium (CIE) is based on artificial boundary conditions, because collisional ionization and radiative recombination are *not inverse processes*. Therefore an external source is needed to compensate for radiative losses in an optically thin plasma.

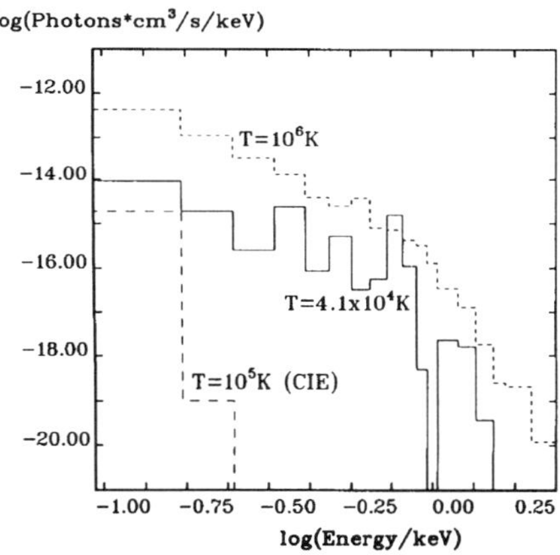

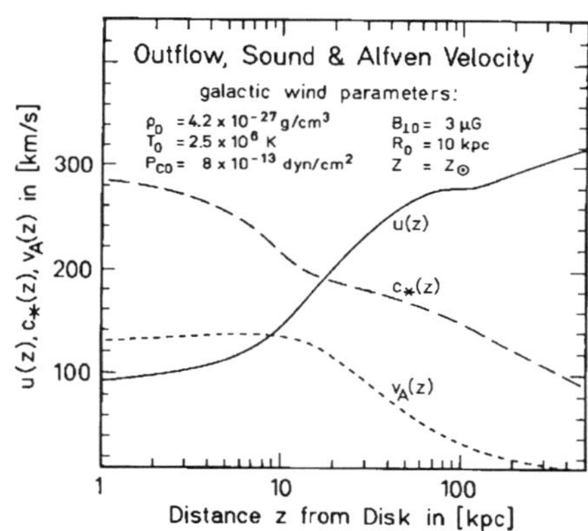

Figure 2: Non-equilibrium emission spectrum (normalized to n_e^2) of a fast adiabatically cooling (model parameters s. Fig. 3). The short dashed line shows the spectrum at $T = 10^6$ K, the solid line at $T = 4.1 \times 10^4$ K. The long dashed line is a 10^5 K CIE spectrum for comparison.

Figure 3: Dynamically and thermally self-consistent local galactic wind flow driven by disk superbubbles (Breitschwerdt, 1994). The solid line is the outflow velocity, the long dashed line the compound sound speed (including waves and cosmic rays) and the short dashed line is the Alfvén speed.

Moreover, we have shown (Breitschwerdt and Schmutzler, 1994), that due to a proper treatment of the internal energy balance, consisting of a kinetic part (translational energy of plasma particles) and a potential part (energy stored in various stages of ionization of hydrogen and 10 most abundant heavy elements), the emission processes depend on the thermodynamic path, or in other words, the thermal history of the plasma. This in turn is intimately related to its dynamical evolution. Therefore a self-consistent thermal and dynamical treatment of the flow is the only valid solution. Since we start with a temperature of $\geq 10^6$ K, the assumption of CIE is not too bad, but for lower temperatures it becomes progressively worse. As can be seen from a model calculation in Fig. 2, the similarity between the non-equilibrium spectra at 10^6 K and at 4.1×10^4 K is striking. For comparison, a 10^5 K-CIE spectrum has a completely different spectral appearance. A high resolution spectrum at 4.1×10^4 K shows, that it is dominated by *recombination continuum* (Breitschwerdt and Schmutzler, 1994). The local galactic wind solution from which this spectrum is calculated is shown in Fig. 3. It is therefore a general conclusion, that the existence of only one or a few emission lines and the form of the spectrum do not allow to derive a unique temperature.

3.3 New Interpretation of the soft X-ray background

We have calculated a dynamically and thermally self-consistent outflow (s. Fig. 3) and have integrated the contributions of the local emission towards the Galactic North Pole, including the minimum absorption by the Local Cloud ($N(H) \approx 3 \times 10^{18}$ cm^{-2}) and an extended HI region ("Lockman layer") with $N(H) \approx 1.5 \times 10^{20}$ cm^{-2} (s. Fig. 1). It is evident from Fig. 4, that such a non-equilibrium emission model can explain the count rates of the WISCONSIN Survey (cf. McCammon and Sanders, 1990) from the I- down to the B-band. Thus we can state the following results: (i) since we have no spatial separation between the ultrasoft and the soft energy bands, we can explain C-band emission from outside the Local Bubble, as it has been observed; (ii) the spectrum is a sum of different "local temperatures", and therefore the result that Wang et al. (1995) obtained for the spectrum of NGC4631 is naturally explained by our model. In fact we have a continuous non-equilibrium temperature distribution. The observed isotropy above 0.5 keV can be obtained by considering the contributions of other nearby flux tubes (s. Fig. 1).

Of course, we also need a local component in order to model the highly absorbed ultrasoft part of the SXRB. We have sketched elsewhere (Breitschwerdt and Schmutzler, 1994; Breitschwerdt et al., 1996), an alternative model for the Local Bubble. The basic assumptions are, that initially a superbubble expands into a fairly dense environment (molecular cloud) and subsequently undergoes fast adiabatic expansion due to a large density gradient in the ambient medium. This also leads to delayed recombination and the detailed spectrum is shown in Fig. 5. The results are briefly summarized: (i) the present temperature can be fairly low $\sim 4.2 \times 10^4$ K; (ii) the density can be as high as 2.4×10^{-2} cm^{-3} in order to accommodate the dispersion measure of pulsar PSR 0950+08; (iii) yet the overall pressure is in agreement with that of the Local Cloud; (iv) the spectrum in Fig. 5 shows a deficiency in EUV lines in contrast to the CIE spectrum and in agreement with the EUVE observations; (v) the bubble spectrum shows also emission in the M-bands. It should be emphasized, that the spectra

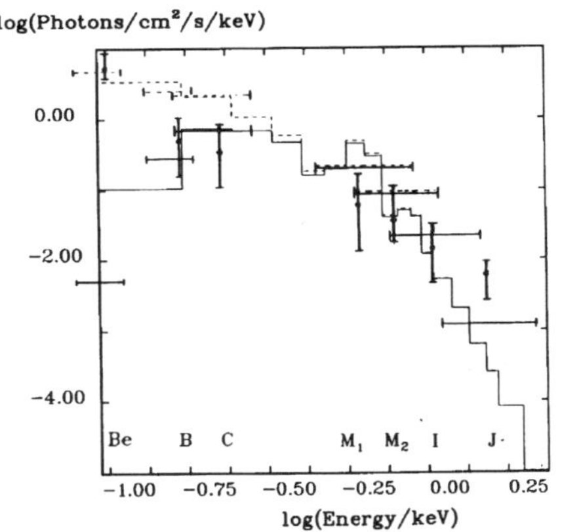

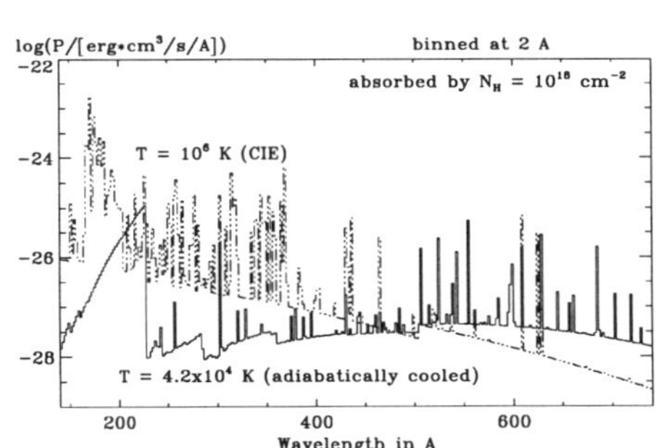

Figure 4: Integrated photon flux spectrum towards the North Galactic Pole. The points are the WISCONSIN Survey data and the vertical bars are measure of the spatial variation of the flux. The horizontal solid and dashed lines give the calculated band averaged fluxes for an absorption column density $N(H) \approx 3 \times 10^{18}$ cm^{-2} and $N(H) \approx 1.5 \times 10^{20}$ cm^{-2}, respectively.

Figure 5: Non-equilibrium EUV emission spectrum (normalized to n_e^2) of the Local Bubble in comparison to the standard CIE spectrum; N_H denotes the HI column density of a local absorber.

of the X-ray halos of NGC891 and NGC4631 could be modelled in a similar way, given a sufficient photon statistics that allows to derive enough spectral information. In principle, the model can be further constrained. Due to the information of the spectral index behaviour along the minor axis, we can calculate a wind model in which relativistic electrons are advected appropriately (cf. Sect. 3.4). Then with the same model we should try to model the X-ray spectrum. This is still work in progress.

3.4 Spectral index modelling

Here is a brief summary of the modelling of the spectral index in the halo of edge-on galaxies (Breitschwerdt, 1994); NGC4631 is taken as an example. Relativistic electrons in halo magnetic fields experience severe energy losses above 100 MeV due to synchrotron emission ($\propto E^2$), inverse Compton effect ($\propto E^2$) and adiabatic losses ($\propto E$). Therefore the flat spectral index behaviour as well as the extension of the radio halo in NGC4631 (s. Fig. 6) are rather surprising.

In the framework of a galactic wind model, in which advection dominates diffusion, an elegant explanation can be given. This even holds in the case when the halo magnetic field remains constant over a large distance, as the observations for NGC4631 (Hummel et al., 1988) suggest. We treat the electrons as test particles, because due to their lower abundance in cosmic rays, they do not contribute significantly to the cosmic ray pressure. Next we calculate a galactic wind flow appropriate for NGC4631. The mass is taken from Weliachew et al., (1978) and the input parameters of the best model fit (M5 in Fig. 6) are: $\rho_0 = 3.34 \times 10^{-27}$ g/cm^3, $T_0 = 2 \times 10^6$ K, $P_{c0} = 0.5$eV/cm^{-3}, $B_0 = 2.0\,\mu G$ and the electron spectral index was taken to be $\gamma_0 = 2.6$, for a disk source function of $Q(E,z) = K_0 E^{-\gamma_0} h_g\, \delta(z)$, where K_0 is a constant, h_g is the height of the source distribution and $\delta(z)$ is the Dirac delta function. The solution of the diffusion-advection transport equation for the number density N of electrons

$$-\frac{\partial}{\partial z}\left(D(E,z)\frac{\partial}{\partial z}N - u(z)N\right) - \frac{\partial}{\partial E}\left(\frac{1}{3}\frac{du(z)}{dz}E\,N\right.$$
$$\left. - \frac{dE}{dt}N\right) = Q(E,z)$$

is solved for dominant advection $((D(E,z) = 0)$ and with the appropriate loss terms dE/dt. The results are shown in Fig. 6. We see that model M5 can explain both the large extension of the cosmic ray halo and the flat spectral index behaviour. It turns out that the associated mass flux is fairly large, but in the central region the star formation activity is also fairly high as can be

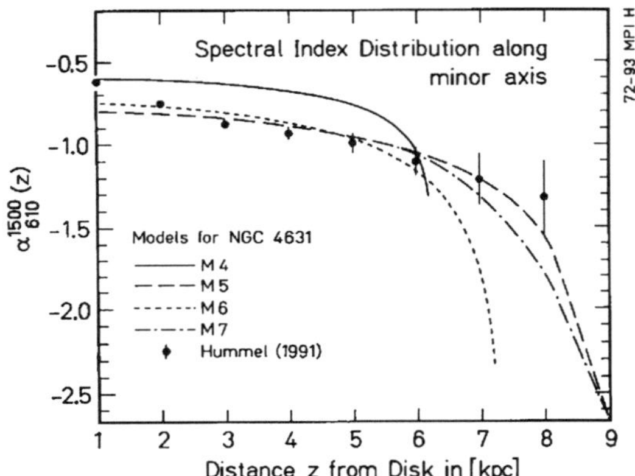

Figure 6: Variation of the spectral index α_{610}^{1490} along the minor axis of NGC4631, derived from VLA radio measurements at 1490 MHz and 610 MHz by Hummel (1991). The long dashed line (M5) is the best model fit (Breitschwerdt, 1994).

seen from ROSAT PSPC observations (Wang et al., 1995; Vogler and Pietsch, 1996). The physical explanation for our result is fairly simple: A dynamical halo model with dominant cosmic ray advection and *increasing acceleration* of the flow can partially compensate for z-increasing energy losses (Breitschwerdt, 1994).

4   Conclusions

Any modelling of the ISM in normal spiral galaxies should take the observational fact of a strong physical connection between the galactic disk and halo into account. The driving mechanism behind mass, momentum and energy transfer is clearly the supernova activity. The difficulty in the theoretical description lies in the essential non-linearity due to the feedback via star formation. Observationally, the existence of *diffuse* X-ray and optical (DIG) as well as radio continuum emission in disk and halo of spirals has been established in many cases. Therefore, in a first attempt of modelling the ISM in the halo, one should include the dynamical coupling between its major constituents, i.e. gas, cosmic rays and magnetic fields. Other phenomena, such as high velocity clouds, can be regarded as tracers, which also give valuable information on the halo conditions, but are unlikely to be primary elements of a model. One of the direct consequences of an outflow picture is the reduction of the volume filling factor of the HIM. Whether the outflow just recycles matter back in a galactic fountain or chimney, or develops into a galactic wind, depends mainly on the coupling between cosmic rays and plasma. Under a wide range of conditions in normal spirals, the cosmic rays, which according to their isotopic composition have to leave the galaxy eventually, can advect $\sim 1\,\mathrm{M_\odot/yr}$ of gas to infinity.

Such an outflow has a distinct spectral signature in the X-rays, because non-equilibrium cooling "freezes in" the high ionization stages. Thus the spectrum is characterized by delayed recombination and line emission. The results are in agreement with observations of the soft X-ray background. At present, we may speculate, whether the ionization problem of the DIG may be solved by assuming that supernova shock waves, propagating into the lower halo, heat the gas, but leave it for some time "underionized". Thus the spectral signature of the ambient warm ionized medium is retained for some time. Clearly other possibilities remain (cf. H. Domgörgen, this volume).

Another signature of outflow is the flatness of the radio halo spectral index. There is a clear hint for such a mechanism, if we take starburst galaxies as an extreme form of star formation activity. Modelling of e.g. M82 has shown (Breitschwerdt, 1994), that both the spectral index and the X-ray emission can be successfully explained by a galactic wind.

High spectral resolution and high sensitivity future observations in various wavelengths will guide our understanding towards what the "normal" state and evolution of the ISM in spiral galaxies is. The future detection of individual spectral lines in the halo will be an essential diagnostic tool for determining the thermal and hence the dynamical state of the gas.

Acknowledgements

Financial support by the *Deutsche Forschungsgemeinschaft* (DFG) by a Heisenberg Fellowship is gratefully acknowledged.

References

Bowyer, C.S., Field, G.B., Mack, J.E., 1968, NATURE **217** ,32

Bregman, J.N., 1980, ApJ **236**, 577

Bregman, J.N., Pildis, R.A., 1994, ApJ **420**, 570

Breitschwerdt, D., 1994, *Habilitationsschrift*, Universität Heidelberg, 158p

Breitschwerdt, D., McKenzie, J.F., Völk, H.J., 1987 in *Interstellar Magnetic Fields*, eds. R. Beck und R. Gräve, Springer, p. 131

Breitschwerdt, D., McKenzie, J.F., Völk, H.J., 1991, A&A **245**, 79

Breitschwerdt, D., McKenzie, J.F., Völk, H.J., 1993, A&A **269**, 54

Breitschwerdt, D., Schmutzler, T., 1994, NATURE **371**, 774

Breitschwerdt et al., 1996, (rapporteur paper), 1st ISSI workshop, eds. R. Steiger, R. Lallement, M. Lee, (in press)

Brinks, E., Shane, W.W., 1984, ApJ Supp. **55**, 179

Dettmar, R.-J., 1992, Fund. of Cos. Phys. 15, 143

Habe, A., Ikeuchi, S., 1980, Prog. Th. Phys. **64**, 1995

Hasinger, G. et al., 1993, A&A **275**, 1

Heiles, C., 1984, ApJ Supp. **55**, 585

Henry, R.C., et al., 1968, ApJ **163**, L11

Hummel, E., 1991 in IAU Symp. 144, ed. H. Bloemen, Kluwer, p. 257

Hummel, E., Lesch, H., Wielebinski, R., Schlickeiser, R., 1988, A&A **197**, L29

Jelinsky, P., Vallerga, J.V., Edelstein, J., 1995, ApJ **442**, 653

Jenkins, E.B., Meloy, D.A., 1974, ApJ **193**, L121

Kerp, J., Herbstmeier, U., Mebold, U., 1993, A&A **268**, L21

Lerche, I., 1967, ApJ **147**, 689

Mathews, W.G., Baker, J.C., 1971, ApJ **170**, 241

McCammon, D., Sanders, W.T., 1990, Ann. Rev. Astron. Astrophys. **28**, 657

McKee, C.F., Ostriker, J.P., 1977, ApJ **218**, 148

Münch, G., Zirin, H., 1961, ApJ **133**, 11

Norman, C.A., Ikeuchi, S., 1989, ApJ **345**, 372

Raymond, J.C., Smith, B.W., 1977, ApJ Supp. **35**, 419

Reynolds, R.J., 1990, ApJ **348**, 153

Schmitt, J.H.M.M., Snowden, S.L., 1990, ApJ **361**, 207

Shapiro, P.R., Field, G.B., 1976, ApJ **205**, 762

Snowden, S.L., Cox, D.P., McCammon, D., Sanders, W.T., 1990, ApJ **354**, 211

Snowden, S.L., McCammon, D., Verter, F., 1993, ApJ **409**, L21

Snowden, S.L., *et al.*, 1991 *Science*, **252**, 1529

Spitzer, L.jr., 1956, ApJ **124**, 20

Tomisaka, K., 1991 in IAU Symp. 144, ed. H. Bloemen, Kluwer, p. 407

Vogler, A., Pietsch, W., 1996, A&A (in press)

Wang, Q.D., McCray, R., 1993, ApJ **409**, L37

Wang, Q.D., Walterbos, R.A.M., Steakley, M.F., Norman, C.A., Braun, R., 1995, ApJ **439**, 176

Weliachew, L., Sancisi, R., Guélin, M., 1978, A&A **65**, 37

Woodgate, H.E., et al., ApJ **188**, L79

Address of the author:

D. BREITSCHWERDT, Max-Planck-Institut für Extraterrestrische Physik, Giessenbachstraße, D-85740 Garching, Germany.

A Multiphase, Multiscale Halo Gas

Yu. Shchekinov

1 Introduction

Extraplanar gas in the Milky Way shows spatial structure consisting of several layers with scale heights varying from $\sim 0.5 - 1$ kpc for HI gas (Lockman et.al., 1986), up to $\sim 2 - 5$ kpc for CIV, SiIV and NV ions (Sembach & Savage,1992). Apparently, halo gas has multiphase composition: Savage & Sembach (1994) have identified two types of highly ionized gas with different ratios of CIV and NV column densities; Shull & Slavin (1994) have shown from a comparison of emission and absorption lines that CIV gas has density $0.01 \leq n \leq 0.02$ cm^{-3} considerably higher than the mean value estimated from observed column density and scale height; Sembach & Savage (1992) have found fluctuations of column density of CIV ions with a factor of 10. Patchy (and possibly multiphase) structure of extraplanar gas is also found in edge-on spiral galaxies (Dettmar, 1992). Thus, adequate description of galactic halos should incorporate their multiphase nature. In this contribution, I give a short description of thermal properties of a two-phase steady state halo, and its vertical structure. Detailed description is given elsewhere (Shchekinov, 1996).

2 Equation of State of a Multiphase Gas

In the framework of the chimney theory ejection of clouds into a hot halo can be connected with fragmentation of polar parts of supershells generated by clustered SNe explosions at stages preceding formation of chimney (Cioffi, 1986). Ejected clouds form an atmosphere around the galactic disk with vertical density distribution depending on the initial kinetic energy, number density in the surrounding intercloud gas, and other characteristics regulating their motion. Evaporation of clouds influences considerably energetics of the intercloud gas. In particular, Begelman & McKee (1990) stress that evaporation stabilizes thermal instability. Since hot intercloud component of the halo is expected to have temperature $T \sim 10^5 - 10^6$ K, e.g. within the interval favourable for thermal instability, cloud evaporation is *required* for the halo to be stable against global (with wave length comparable with typical scale height) thermal instability (Shchekinov, 1996).

In a simplest case the equation of state of a hot halo gas affected by evaporation effects can be described by the equation of energy balance

$$\Gamma_{ch} + \frac{1}{2}\sigma(T)v_0^2 = \Lambda(T)n^2 + \sigma(T)\frac{\gamma}{\gamma-1}\frac{P}{\rho}, \tag{1}$$

where Γ_{ch} is the bulk heating due to ejection of hot gas by chimneys, $\sigma(T)$, the mass ejection rate per unit volume due to cloud evaporation, v_0, the characteristic relative velocity of clouds with respect to the intercloud gas, the relative velocity field of clouds is assumed in (1) to be isotropic; $\sigma(T)$ is proportional to the number density of clouds $\mathcal{N}$. The last term in the l.h.s. of (1) is the energy released in course of deceleration of evaporated material, while the last term in the r.h.s. is the energy which hot intercloud gas wastes to heat evaporated gas up to temperature T. In the limit $n \to 0$ the energy balance of the intercloud component can be supported at non-zero level, even if the chimney heating is negligible $\Gamma_{ch} \to 0$; the limiting value of temperature in this case can be found as

$$T = \frac{\gamma-1}{2\gamma}\frac{\mu}{R}v_0^2. \tag{2}$$

In general, solution of (1) consists of two branches, high-temperature thermally stable solution with cooling dominated by the last term in the r.h.s. of equation (1), and low-temperature thermally unstable branch where radiative cooling dominates. On the high-temperature branch temperature weakly depends on density

$$T \simeq T(\Gamma_{ch}, v_0, \mathcal{N}), \tag{3}$$

while on the low-temperature branch

$$T = \left(\frac{\lambda_0 n^2}{\Gamma_{ch}}\right)^{1/\alpha}, \tag{4}$$

for a power-law cooling function $\Lambda(T) = \lambda_0 T^{-\alpha}$, where $\alpha \simeq 0.6$ for the recombination cooling dominating in temperature interval $T = 10^5 - 10^6$ K. Equation (4) describes thermally unstable solution.

3 Spatial Structure of a Steady Halo

On the high-temperature branch z-dependence of temperature of the hot halo component is rather weak: in regions dominated by the bulk chimney heating equation (1) gives

$$T \propto \left(\frac{\Gamma_{\mathrm{ch}}}{\mathcal{N}}\right)^{2/7}, \tag{5}$$

and thus, within the heights occupied by the gas ejected by chimneys the intercloud gas is approximately isothermal. For the heating rate $\Gamma_{\mathrm{ch}} \sim 10^{-26}$ erg cm^{-3} s^{-1}, corresponding to the energy ejected by approximately 0.15 chimneys per 1 kpc^2, and distributed in vertical direction with a scale of 3 kpc, and for clouds with $\mathcal{N} \sim 10^{-3}$ pc^{-3} and radius $R \sim 3$ pc, temperature can be found from (1) as $T \sim 10^6$ K. Within this layer density and pressure of the intercloud gas have scale height $z_{\mathrm{i}} \sim 3$ kpc.

Cloud component forms a distribution with a scale z_{cl} which depends on initial velocities, sizes and densities of formed clouds and their interaction with hot intercloud gas. Since clouds form at stages preceding blowout of a supershell, and therefore their initial velocity is larger than thermal velocity of a blowing out gas, those clouds whose dynamics is weakly affected by an ambient medium (ballistic clouds) have $z_{\mathrm{cl}} \geq z_{\mathrm{ch}}$ (z_{ch} is the scale height of a hot gas ejected by chimneys). However, in most cases ram pressure influences cloud dynamics considerably, resulting in $z_{\mathrm{cl}} < z_{\mathrm{ch}}$. Observed scale heights of highly-charged ions arising in clouds and cloud-intercloud interfaces are determined by a combination of z_{cl} and z_{i} since radius of clouds (and volume filling factor) increases with z in a barometric environment.

4 Conclusions

Evaporation of clouds affects considerably physical state of the halo, supporting energy balance and stabilizing halo against global thermal instability. In halo models with a domination of cloud evaporation hot intercloud gas has vertical distribution close to a barometric one with temperature $T \sim 10^6$ K for typical parameters of chimneys in the Galaxy. Vertical distribution of clouds differs qualitatively from the distribution of hot gas, and has different scale height. Assuming CIV, SiIV ions and Hα gas to be associated with clouds, and OVI and NV ions with hot intercloud gas and partly (NV) with interfaces between clouds and intercloud gas, a natural explanation of different vertical structure of these constituents can be found in this model.

Acknowledgments. I thank the organizers, and particularly R.-J. Dettmar, for enabling me to attend the seminar. This work was supported in part by the Russian Foundation of Basics Researches under grant 94-02-05016-a.

References

Begelman M.C., McKee C.F., 1990, ApJ, 358, 375

Cioffi D, 1986, in Proc. NRAO Conference on Gaseous Halos of Galaxies, eds. J.N. Bregman, F.J. Lockman, p. 45

Dettmar R.-J., 1992, Fundam. Cosmic Phys., 15, 148

Lockman F.J., Hobbs L.M., Shull J.M., 1986, ApJ, 301, 380

Savage B.D., Sembach K.R., 1994, ApJ, 434, 145

Sembach K.R., Savage B.D., 1992, ApJS, 83, 147

Shchekinov Yu., 1996, A&A, to be submitted

Addresses of the author:

YU. SHCHEKINOV, Osservatorio Astrofisico di Arcetri, Largo E. Fermi, 5, 50125, Firenze, Italy; Department of Physics, Rostov State University, Sorge 5, 344090, Rostov on Don, Russia.

Structure of the Galactic Halo and Disk

Wolfgang Kundt

Abstract

The neutral hydrogen clouds in the Galactic halo are thought to owe their existence to at least the following two processes: (1) Condensation on the (channel walls of the) Galactic Twin-Jet (which was more strongly fed in the past), and (2) high-velocity (filamentary) Supernova Ejecta. The condensations of type (1) form distinct chains whereas the SN ejecta form a large, isotropic, quasi corotating background. Replenishment (of the 'falling clouds', from the Disk) takes place through several hundred hot Galactic 'chimneys'.

1   Ill-understood Halo properties

Among the ill-understood observed objects in the Galactic halo are the • unevenly distributed high-velocity neutral-hydrogen clouds (HVCs), almost all of which 'rain down' into the disk at velocities intermediate between free-fall from infinity and being supported (against the Galaxy's gravity); • a number of yet cooler intermediate-velocity clouds (IVCs) which show (roughly) local galactic abundances, and which tend to be associated with HVCs, and to connect continuously to them in velocity space; and • (3) diffuse, isotropically distributed very-high-velocity neutral hydrogen seen in the form of faint line wings. All this in a region of space whose kinetic temperature should exceed 10^7K in order not to collapse under gravity, because of the thermal scale-height formula $h = kT/m_H g = 30 Kpc\, T_7$, ($T_7 := T/10^7 K$ [as always]; g := gravity acceleration $= 10^{-8} cm/s^2$).

In 1966, Oort gave the HVCs a careful consideration, but found them difficult to understand. In particular, he more or less dismissed (a) nearby SN shells, (b) coronal condensations, (c) nuclear ejections, (d) disk ejections in the form of cool clouds, (e) intergalactic accretion, and (f) galactic satellites. Among the reasons for dismissing explanations (a) through (f) were • (i) their moderate distances, of order Kpc (based on Münch & Zirin, 1961; cf. K.S.de Boer, these proceedings), • (ii) the high involved mass rate, in excess of one $M_\odot/yr$, • (iii) their (high) velocity dispersion (of order ±12 Km/s), signaling an age of less than $10^7 yr$, • (iv) their highly asymmetric distribution on the sky, in the form of disjoint chains of small surface-filling factor (which has survived modern scrutiny: Ed Murphy, this meeting), and • (v) their velocities – of order 120 Km/s – half-way between the local < 20 Km/s and the free-fall velocity from intergalactic space, which must exceed $\sqrt{2}$ times the Kepler velocity in the disk (of order 220 Km/s), and which Oort assessed between 380 and 496 Km/s. The HVCs look like having condensed recently in the halo along certain preferred strings.

At the time of Oort's writing, it was not yet known that active galaxies at large redshifts have cigar shape aligned with their radio axis, and that isolated star formation along the galaxy's twin jet is even observed in the nearest radio galaxy Cen A, out to 40 Kpc from its center: the (relativistic) jet interacts, at its surface, with the ambient medium, compresses it, whence it cools, contracts, and gets unstable to star formation. Cloud condensation on extragalactic jets is therefore demonstrated (cf. Kundt, 1996a). For these reasons, I have proposed in 1987 that most of the HVCs in the upper Galactic hemisphere can be understood as condensations on the Galactic twin-jet, a fossil remnant of the last Seyfert stage – analogous to water vapour condensations in a Wilson chamber, or (over night) on grass stalks – see Fig 1.

This proposal has not conquered the market. Had I accepted the chairman's offer of 60 min – instead of the LOC's 7 min – I might have succeeded this time. But let me make another effort: To begin with, there is the Magellanic stream which needs an independent explanation, though perhaps not a very independent one: galaxies are thought to pass repeatedly through active stages, whenever their center has been refuelled, with the earliest activities being the most energetic ones. The Magellanic stream may well be due to an earlier active stage of our Galaxy.

Is it possible – as is often considered – that the HVCs have been slowed down by collisions with coronal material, or magnetic fields, from some 400 down to the present 120 Km/s ? I don't think so. For coronal deceleration to $v = 120$ Km/s, its mass density ρ_{co} would have to be intolerably high: ram pressure balance during free fall yields $v = (Dg\rho_{cl}/\rho_{co})^{1/2} = 120 \ (\text{Km/s})(D_{19.5}(\rho_{cl}/\rho_{co})_{-2.8})^{1/2}$, so that the average coronal density would have to fall short by no more than some 600 of the clouds' density, whereas the scale-height argument wants its temperature to be above 10^7K, some 10^5 times that of the clouds. There is thus a density discrepancy (in pressure equilibrium) by a factor in excess of 10^2! The situation can be compared to airplanes in the troposphere, without propagation, whose crash is to be prevented by air friction. Note that magnetic fields cannot be a way out because they are almost weightless.

This argument against intergalactic accretion of the HVCs can even be made more convincing by encorporating the IVCs, which tend to associate with some HVCs, and to connect to them in LOS velocity (velocity 'bridges' in the l-v plot) even though both are tiny spots in the sky! To me, they look like the braking pads of Earth's orbiters during re-entry: the HVCs are transiently – and gently – braked from their typical 120 Km/s to less than 20 Km/s, at the positions of the IVCs, at angles of order 50° w.r.t. the normal to the Galactic plane (Kundt, 1992), i.e. normal to the jet's direction, see Fig 1. Apparently, not a quasi-static corona but a relativistic jet can brake the HVCs' free fall, soon after detachment.

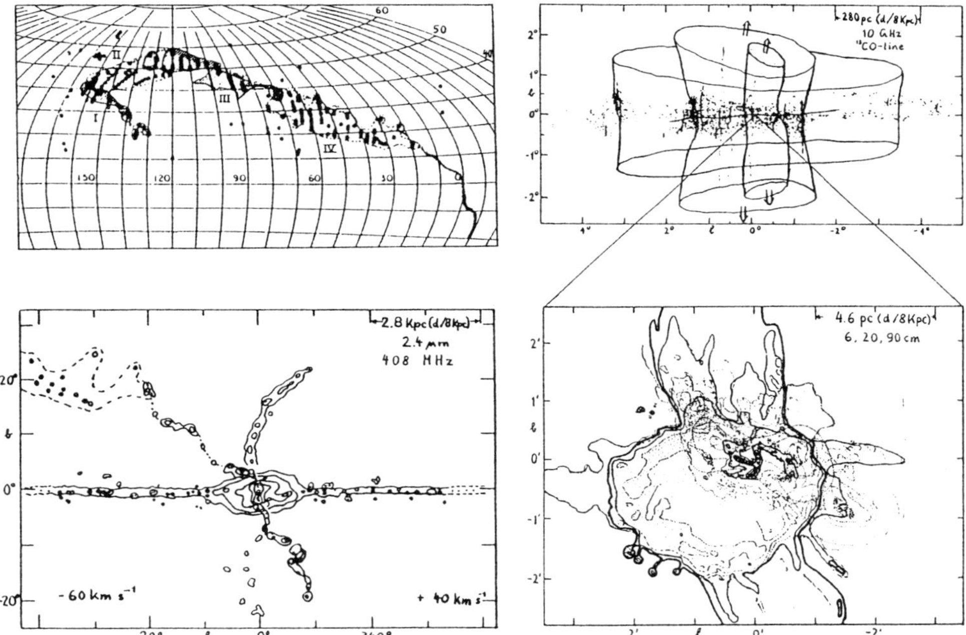

Figure 1: Multifrequency (radio to IR) view of the Galactic Center on four different angular (length) scales. The first two maps (left column), on scales of 10 and 5 Kpc respectively, show the reported and/or proposed jets from Sgr A. The third and fourth map – on scales of 1 Kpc and 10 pc respectively – highlight their apparent origin: Sgr A*. The distance to the Center has been assumed 8.0 Kpc, according to Reid (1993). After Kundt (1987, 1990, 1996a).

A final concern of Oort's is the replenishment of the infalling halo mass; do we need episodic super-explosions ? For some reason, he does not consider the way atmospheric clouds are continually replenished: by steady, invisible evaporation. The galactic analog would be hot, ionized interstellar medium (ISM) escaping from the disk at high velocities. Such a continual feeding of the halo through Galactic 'chimneys' may indeed happen, dragged along by cosmic rays as the driver gas (Kundt & Müller, 1987); which leads us to the concern of the next section.

Before, however, we must not forget the very-high-velocity diffuse neutral hydrogen, analyzed by Gernot Westphalen at this workshop; how does it get up there ? It may well be the 'tail' of the velocity-distribution of supernova ejecta (of all types) which has not been slowed down sufficiently by collisions with disk material. Note that it is my understanding that SNe are thick-walled (splinter) explosions – not to be described by strong shock waves – with a velocity distribution of the (filamentary, more than 10^3) splinters between a few 10^2 and some $10^{4.6}$ Km/s, (Kundt, 1996b). The splinters illuminate the CSM via crashes, and some splinters crash less than others. Many of them will escape from the Galactic disk.

2 Ill-understood Disk properties

This section would not belong to the present workshop were it not an important boundary condition on the corona: how to replenish the falling clouds, and how to inflate the halo? In my 1987 discussion of 'Stockert's chimney' with Peter Müller, we argue that Galactic chimneys are very different from stellar winds: they are forced open by the cosmic rays, both hadronic and leptonic, which can escape from an over-pressurized HII-region at almost the speed of light, dragging along with them a considerable amount of hot CSM. They are the long searched-for leak in the Galaxy's 'leaky-box' model for cosmic-ray escape. In 1992 I added to this that quite likely, the leptonic component of the cosmic rays has a comparable energy density to the hadronic one, and that it may well fill most of the volume of the Galactic disk, and even suffice to fill the whole halo of our Galaxy throughout its lifetime. This conclusion further reduces the average density of the coronal gas.

At the same time, this conclusion points at an additional source of ionizing photons for Ron Reynolds' warm ionized medium (Reynolds, 1990): electrons and positrons of Lorentz factor a few hundred can ionize collisionally – at least in its boundary layers – and in addition, their inverse-Compton radiation on the 2.73 K background falls into the (strongly ionizing) UV band.

Finally, if none of the standard hot, warm, and cool components of the disk fill its volume, the 'Local Hot Bubble' (LHB) may well be an illusion: Rather than provoking the Copernican principle, this hot neighbourhood may look like a forest from inside, with the trees represented by hydrogen clouds: we are locally embedded in a rather transparent medium – relativistic pair plasma plus hot stellar wind zones – with neutral hydrogen covering a small fraction of the sphere of seeing. With growing distance ($\gtrsim 10^2$ pc), the probability of hitting neutral hydrogen grows towards unity. We do not see a uniform cold component in our immediate vicinity because its distribution is patchy. Every typical Galactic observer may conceive herself inside an LHB.

Acknowledgements: I am thankful to Bülent Uyanıker for a discussion.

References

Kundt W., 1987, Ap&SS 129, 195

Kundt W., 1990, Ap&SS 172, 109

Kundt W., 1992, Ap&SS 195, 331

Kundt W., 1996a, 'Jets from Stars and Galactic Nuclei', Lecture Notes in Physics 471, Springer

Kundt W., 1996b, in 'Multifrequency Behaviour of High-Energy Cosmic Sources', ed. F. Giovannelli, in press

Kundt W., Müller P., 1987, Ap&SS 136, 281

Münch G., Zirin H., 1961, ApJ 133, 11

Oort J.H., 1966, Bull. Astr. Inst. Netherlands 18, 421

Reid M.J., 1993, ARA&A 31, 345

Reynolds R., 1990, in 'Galactic and Extragalactic Background Radiation', IAU Symp. 139, eds. S.Boyer & C.Leinert, Kluwer, 157

Address of the author:

W. KUNDT Institut für Astrophysik der Uni Bonn, Auf dem Hügel 71 D-53121 Bonn

Gamma-Rays
and High Energy Processes

γ-ray evidence for distributed stochastic acceleration of galactic cosmic ray electrons

R. Schlickeiser

1   Introduction

The possible existence of a strong flux of low-energy (< 10 MeV) cosmic ray electrons in the Galaxy has profound consequences for the heating, ionization and pressure balance of the interstellar matter. Evidence for such a component has been scarce in the past, since direct cosmic ray measurements are contaminated by solar modulation and Jovian electrons, while the nonthermal cyclotron and synchrotron radiation of these electrons in the galactic magnetic field is strongly affected by interstellar free-free absorption. However, recent observations made with the OSSE (Kurfess 1995) and COMPTEL (Strong et al. 1994) instruments on board of the *Compton Gamma Ray Observatory* provide evidence that the diffuse galactic gamma ray continuum emission extends down to photon energies below 100 keV. In particular, a significant spectral upturn somewhere below 1 MeV has been established. As the gamma-ray emission in this energy region is most likely electron bremsstrahlung in the interstellar medium, this may imply a corresponding sharp upturn of the source energy spectrum of the radiating electrons. Skibo and Ramaty (1993) as well as Skibo et al. (1995) estimate that integrated over the whole Galaxy a source power of $\sim 4 \cdot 10^{41}$ erg s^{-1} is required to maintain these electrons against the severe Coulomb and ionization losses. This power exceeds the power supplied to the nuclear cosmic ray component by at least an order of magnitude, and represents a serious problem for the general problem of the origin of cosmic rays. It is the purpose of this work to provide an alternative interpretation of the Compton observatory measurements. We show that the spectral upturn in the diffuse gamma-ray bremsstrahlung spectrum can be explained by the existence of in-situ acceleration in the Galaxy due to the presence of interstellar electromagnetic turbulence. We demonstrate that this interpretation requires much less severe conditions for the global energetics of cosmic ray sources.

The bremsstrahlung photon intensity is calculated by the line-of-sight integral of the bremsstrahlung emissivity over the field of view scanned by the OSSE and COMPTEL instruments, and is influenced by the spatial distribution of the interstellar gas and cosmic ray electron density. Assuming that the shape of the energy spectrum of the radiating electrons is the same throughout the galaxy, the calculated bremsstrahlung photon energy spectrum is

$$\frac{dI_b}{dtd\epsilon}(\epsilon) = N \int_0^\infty dE\, J_e(E)\frac{d\sigma_b}{d\epsilon}(\epsilon, E) \tag{1},$$

where $\frac{d\sigma_b}{d\epsilon}(\epsilon, E)$ denotes the (known) differential bremsstrahlung cross section, $J_e(E)$ the isotropic differential cosmic ray electron intensity as a function of the electron kinetic energy E, and the constant factor N accounts for the line-of-sight and solid angle integration. In order to determine the energy spectrum of the bremsstrahlung intensity (1) we have to consider the equilibrium spectrum of the radiating electrons, $J_e(E)$.

2   Interstellar dynamics of cosmic ray electrons

Quite generally, the dynamics of energetic charged particles (cosmic rays) in cosmic plasmas is determined by their mutual interaction and interactions with ambient electromagnetic, photon and matter fields. Among these by far quickest is the particle-wave interaction with electromagnetic fields which very often can be separated into a leading field structure F_o and superposed fluctuating fields δF. Theoretical descriptions of the transport and acceleration of cosmic rays in cosmic plasmas are usually based on transport equations which are derived from the Boltzmann-Vlasov equation into which the electromagnetic fields of the medium enter by the Lorentz force term (see e.g. Schlickeiser 1994). The quasilinear approach to wave-particle interaction is a second-order perturbation approach in the ratio $q_L \equiv (\delta F/F_o)^2$ and requires smallness of this ratio with respect to unity. In most cosmic plasmas this is well satisfied as has been either established by direct in-situ electromagnetic turbulence measurements in interplanetary plasmas, or by saturation effects in the growth of fluctuating fields. Nonlinear wave-wave interaction rates and/or nonlinear Landau damping set in only at appreciable levels of $(\delta F)^2$ and thus limit the value of $q_L \leq 1$. Since the electrical conductivity of the interstellar medium is very large, any large-scale steady electric fields are absent. We then consider the behaviour of energetic cosmic ray electrons in a uniform magnetic field with superposed small-amplitude $(\delta B)^2 << B_o^2$ plasma turbulence $(\delta\mathbf{E}, \delta\mathbf{B})$ by calculating the quasilinear particle acceleration rates and transport parameters. This is by no means trivial since especially for the interaction of non-relativistic charged electrons with ion- and electron-cyclotron waves thermal resonance broadening effects are particularly important (Schlickeiser and Achatz 1993, Schlickeiser 1994). The acceleration rates and spatial transport parameters are then used in the kinetic diffusion-convection equation for the isotropic part of the phase space density of charged particles $F(z, p, t)$ which for non-relativistic bulk speed $u << c$ reads

$$\frac{\partial F}{\partial t} - S_o = \frac{\partial}{\partial z}[\kappa \frac{\partial F}{\partial z}] - u \frac{\partial F}{\partial z}$$

$$+\frac{p}{3}\frac{\partial u}{\partial z}\frac{\partial F}{\partial p} + \frac{1}{p^2}\frac{\partial}{\partial p}[p^2 A \frac{\partial F}{\partial p} - p^2 \dot{p}_{loss} F] \qquad (2).$$

Here z denotes the spatial coordinate along the ordered magnetic field, p the cosmic ray particle momentum, κ is the spatial diffusion coefficient, A the momentum diffusion coefficient, and S_o denotes the "Stossterm" describing the mutual interaction of the charged particles and their injection.

With respect to the generation of energetic charged particles, stochastic acceleration of particles, characterized by the acceleration time scale $t_A = p^2/A$, competes with continous energy loss processes $\dot{p}_{loss}$, characterized by energy loss time scales $t_L = p/|\dot{p}_{loss}|$. As shown e.g. by Schlickeiser (1981) and Lerche and Schlickeiser (1982) for cosmic ray electron energies below 400 MeV Coulomb and ionization losses dominate bremsstrahlung, adiabatic expansion and radiation losses in all phases of the interstellar medium, i.e. $t_L \simeq t_C$. Moreover, the associated Coulomb and ionization loss time scale t_C at all momenta of interest is much smaller than the escape time scale from the Galaxy, associated with both the spatial diffusion and convection of particles, so that the Galaxy at these particle momenta is a thick target for the cosmic ray electrons (Pohl 1993). We may therefore simplify the transport equation (2) in the steady-state case to

$$\frac{1}{p^2}\frac{d}{dp}[p^2 A \frac{dF(\mathbf{r}, \mathbf{p})}{dp} - p^2 \dot{p}_C F(\mathbf{r}, \mathbf{p})] + \mathbf{S_0}(\mathbf{r}, \mathbf{p}) \simeq \mathbf{0} \qquad (3).$$

The equilibrium momentum spectrum of cosmic ray electrons at the position $\mathbf{r}$ in the Galaxy $N(\mathbf{r}, \mathbf{p}) = 4\pi \mathbf{p^2}\mathbf{F}(\mathbf{r}, \mathbf{p})$ and the equilibrium differential electron intensity $J_e(\mathbf{r}, \mathbf{p}) = \mathbf{v}\mathbf{p^2}\mathbf{F}(\mathbf{r}, \mathbf{p})$ follow from the solution of Eq. (3) for given source injection spectrum $S_0(\mathbf{r}, \mathbf{p})$, momentum diffusion coefficient $A(\mathbf{r}, \mathbf{p})$, and Coulomb and ionization loss rate $\dot{p}_C(\mathbf{r}, \mathbf{p})$. We consider each input quantity in turn.

2.1   Sources of cosmic ray electrons

It is well known that inelastic nuclear interactions and knock-on collisions of cosmic ray nuclei with ambient interstellar gas particles generate secondary electrons (Ramaty 1974), which may represent a considerable fraction of the total electron flux in the Galaxy (Schlickeiser 1982). The corresponding source function of secondary electrons depends on the input momentum spectrum of the colliding cosmic ray nuclei and the many details of the pion, muon and neutron decay (Ramaty 1974); however, at electron momenta below 10 MeV/c the production by the knock-on process dominates. According to Abraham et al. (1966) the production rate of secondary electrons per unit electron momentum interval of

$$S_k(\mathbf{r}, \mathbf{p}) = \frac{4.9 \mathbf{n_g}(\mathbf{r})\mathbf{p}}{4\pi \mathbf{m_e^2 c^2}[1 + (\frac{\mathbf{p}}{\mathbf{m_e c}})^2]^{3/2}} [\sqrt{1 + (\frac{\mathbf{p}}{\mathbf{m_e c}})^2} - 1]^{-2.76}$$

$$\simeq \frac{n_g(\mathbf{r})}{4\pi \mathbf{m_e c}} \begin{cases} 33.2(\frac{p}{m_e c})^{-4.52} & \text{for } 0.21 m_e c \leq p \leq m_e c \\ 4.9(\frac{p}{m_e c})^{-4.76} & \text{for } m_e c \leq p \leq 10 m_e c \end{cases} cm^{-3} \, s^{-1} (eV/c)^{-1} \quad (4).$$

An additional source of cosmic ray electrons is the acceleration of primary electrons in supernova shock fronts that also gives rise to a power law source spectrum in momentum.

2.2   Coulomb and ionization interactions

After injection the energetic electrons Coulomb interact with the fully ionized coronal interstellar plasma which gives rise to both, friction ($\dot{p}_C(\mathbf{r}, \mathbf{p})$) and diffusion ($A_c(\mathbf{r}, \mathbf{p})$) terms in the kinetic equation (2) (for review see Hinton 1983). Moreover, the energetic electrons lose energy ($\dot{p}_I(\mathbf{r}, \mathbf{p})$) by ionizing the neutral interstellar gas. In the following we will neglect the Coulomb diffusion term A_c with respect to the momentum diffusion term A_w associated with stochastic acceleration off interstellar plasma turbulence, and only calculate the two loss rates.

2.2.1   Interactions in neutral matter

The energy loss of particles traversing neutral matter was first calculated by Bohr using the classical theory. In quantum mechanics the problem has been investigated by several authors including Moeller, Bethe, Williams and Bloch in a satisfactory way using various atomic models. We follow the treatment of Heitler (1954).

The kinetic energy ($E = (\gamma - 1)m_e c^2 = \beta^2 m_e c^2/2$) loss rate of a test particle of charge Z traversing neutral matter consisting of s sorts of atoms and molecules with concentrations n_s and atomic numbers z_s is (Heitler 1954, Ch.37)

$$-(\frac{dE}{dt})(\beta \geq \beta_o) = \frac{3\, c\sigma_T\, Z^2\, (m_e c^2)}{4\beta} \sum_s n_s[B_s + B'_{el}], \; eV/s \qquad (5)$$

with

$$B_s = [\, \ln(\frac{m_e c^2 \beta^2 E}{I_s^2(1 - \beta^2)}) - 2\beta^2]\, , \quad B'_{el} = -[\beta^2 + 2\sqrt{1 - \beta^2}]\ln 2 + 1 + \beta^2 \quad (6).$$

I_s denotes the geometrical mean of all ionization and excitation potential of the absorbing atom or molecule s. For most elements I_s in eV is roughly $13z_s$; for the for astronomical applications important very light elements hydrogen and helium the experimental values are $I_H = 19eV$ and $I_{He} = 44$ eV, respectively. Equation (5) can be used for particle velocities high compared to the characteristic velocity of the medium's electrons $\beta \geq \beta_0 = 1.4e^2(\hbar c)^{-1} = 0.01$, corresponding in the case of atomic hydrogen to kinetic energies $E > E_0 = 49$ keV.

Adopting an interstellar medium consisting of hydrogen and helium with a total number abundance ratio of 10:1 we obtain a reasonably accurate approximation for the momentum loss rate

$$\dot{p}_I = \frac{m_e c}{\beta}\dot{\gamma} \simeq -\frac{1.8\, c\, \sigma_T m_e c}{\beta^2}[10.1 + \ln(\gamma - 1)]\, n_G(\mathbf{r}) \quad (\mathbf{eV/c})\, \mathbf{s^{-1}} \qquad (7).$$

Although approximation (7) strictly holds only for non-relativistic electron energies, it also provides a very accurate approximation if used at relativistic energies.

2.2.2 Interactions in fully ionized plasma

The rate of energy loss of a fast test electron in a completely ionized thermal plasma consisting of s species of respective mass m_s, charge $Z_s e$, density n_s and temperature T_s density n_e has been calculated by Butler and Buckingham (1962), Sivukhin (1965) and Gould (1972) as

$$-(\frac{dE}{dt}) = \frac{3\, c\, \sigma_T\, (m_e c^2)}{2\beta} \sum_s \frac{m_e}{m_s} Z_s^2 n_s\, W_s(\frac{\beta}{\beta_s})\, eV/s \qquad (8),$$

where $\beta_s \equiv \sqrt{(2kT_s/(m_sc^2)}$ and

$$W_s(t) \equiv \frac{2}{\pi^{1/2}}[\int_0^t dy \exp(-y^2) - 2t \exp(-t^2)] \tag{9}.$$

Because of the small electron-proton mass ratio $m_e/m_p = 1/1836$ it is easy to see that for all values of the metallicity of the plasma the Coulomb collisions are dominated by scattering off the thermal electrons, i.e.

$$-(\frac{dE}{dt}) \simeq \frac{30\ c\ \sigma_T(m_ec^2)}{\beta} n_e\ W_e(\frac{\beta}{\beta_e})\ eV/s \tag{10},$$

where numerically $\beta_e = 0.018T_6^{1/2}$. For values of $\beta \leq \sqrt{15}\beta_e/5 = 0.77\beta_e$ we find $W_e < 0$, indicating that at low particle velocities Coulomb collisions accelerate $((dE/dt) > 0)$ the test electrons up to velocities comparable to the thermal velocity β_e. In the range $\beta \geq 0.77\beta_e$ we may approximate Eq. (9) as

$$W_e(\beta \geq \beta_c(A, T_e)) = \frac{\beta^3}{x_m^3 + \beta^3} \tag{11},$$

with $x_m \equiv (3\pi^{1/2}/4)^{1/3}\beta_e = 1.10\beta_e = 0.020T_6^{1/2}$. Using the interpolation formula (11) in Eqs. (10) and (7a) we obtain for the momentum loss rate

$$\dot{p}_C(\mathbf{r}, \mathbf{p}) = -6.0 \cdot 10^{-13}(\mathbf{m_e}c)\ \mathbf{n_c}(\mathbf{r})\frac{\beta}{\mathbf{x_m^3} + \beta^3}\ \mathbf{eV\ c^{-1}\ s^{-1}} \tag{12},$$

which we use for all momenta above $0.77p_M$. Below the Coulomb barrier $p_M = 0.87\beta_e = 0.0160T_6^{1/2}m_ec$ (and above $0.77p_M$) the momentum loss rate varies linearly proportional to p, whereas beyond the barrier it decreases $\propto p^{-2}$ for non-relativistic electrons, while it becomes constant for relativistic electrons.

2.3   Stochastic acceleration

Stochastic acceleration rates in weakly turbulent plasmas for general plasma modes have been calculated by Schlickeiser and Achatz (1993). Cosmic ray electrons with Lorentz factors below 1836 interact with weakly-damped Alfvén waves and with weakly-damped right-handed circularly polarized Whistler waves. The momentum diffusion coefficient of cosmic ray electrons then consists of two parts: the first, A_A, resulting from the resonant interaction with the Alfvén waves, and secondly, A_W resulting from the interaction with the

Whistler-electron cyclotron plasma waves. Assuming only parallel propagating waves we use for the former (A_A) the calculation by Dung and Schlickeiser (1990), and for A_W the calculation by Steinacker and Miller (1992). In both cases we adopt a Kolmogorov-type magnetic turbulence power spectrum $I(k_\parallel) = I_0 k_\parallel^{-q}$, with $q > 1$, above $k_\parallel \geq k_{min}$. In this case

$$I_0 = (q-1)(\delta B)^2 k_{min}^{q-1} \tag{13}$$

determines the energy density in magnetic field fluctuations. Assuming further vanishing cross and magnetic helicities for the Alfvén waves we derive

$$A_A(\mathbf{r},\mathbf{p}) = \mathbf{A_0}(\mathbf{r},\mathbf{p})\mathbf{H}[\mathbf{p} - \mathbf{m_p V_A}][1 - (\frac{\mathbf{m_p V_A}}{\mathbf{p}})^{\mathbf{q}}] \tag{14},$$

where $H(x) > 0(< 0)$ for $x > 0(< 0)$ denotes the Heaviside step-function, and

$$A_0(\mathbf{r},\mathbf{p}) = \frac{\pi}{4}\frac{\mathbf{q-1}}{\mathbf{q}}\frac{(\delta\mathbf{B})^2}{\mathbf{B_0^2}}\frac{\mathbf{\Omega_e}}{\gamma}\frac{\mathbf{V_A^2 p^2}}{\mathbf{v^2}}(\mathbf{k_{min}R_e})^{\mathbf{q-1}} \tag{15}.$$

$V_A = \beta_A c = 2.18 \cdot 10^{11}(B_0/G)(n_e/1cm^{-3})^{-1/2}$ cm/s denotes the Alfvén velocity, $\Omega_e = |e|B_0/m_e c$ the nonrelativistic electron gyrofrequency, and $R_e = pc/|e|B_0 = \beta\gamma c/\Omega_e$ the electron gyroradius.

Assuming the same intensity of forward and backward propagating Whistler waves, we use Eqs. (6c), (31c), (35) of Schlickeiser and Achatz (1993) to derive for the Whistler part of the momentum diffusion coefficient

$$A_W(\mathbf{r},\mathbf{p}) = \frac{\pi(\mathbf{q-1})}{8}\mathbf{M^3}\frac{\mathbf{\Omega_e m_e^2 V_A^3}}{\mathbf{v}}\frac{(\delta\mathbf{B})^2}{\mathbf{B_0^2}}(\frac{\mathbf{k_{min}}}{\mathbf{k_c}})^{\mathbf{q-1}}$$

$$\int_1^{M^{1/2}} dk \frac{k^{1-q}}{(k^2+M)^2}\int_0^1 d\mu(1-\mu^2)\delta(\mu-\mu_c(k)) \tag{16},$$

where $M = m_p/m_e = 1836$, $k = k_\parallel/k_c$, $k_c = \Omega_p/V_A$, and $\mu_c(k) = \frac{MV_A}{vk}|\frac{1}{\gamma} - \frac{k^2}{M+k^2}|$. We calculate Eq. (16) for the Kolmogorov-type turbulence spectrum (13). Performing the μ-integration we only obtain a non-zero acceleration rate for values of $\mu_c \leq 1$, yielding for electron Lorentz factors $\gamma < 1836$

$$A_W(\mathbf{r},\mathbf{p}) = \mathbf{A_0}(\mathbf{r},\mathbf{p})\frac{\mathbf{q}}{2}(\frac{\mathbf{p}}{\mathbf{m_p V_A}})^{-\mathbf{q}}\mathbf{J_W(p)} \tag{17},$$

where the slowly varying function $J_W(p)$ is

$$J_W = 0 \quad \text{for } k_2 \leq 1 \text{ and } k_1 \geq M^{1/2} = 43 \tag{18a},$$

$$J_W = \frac{L_2^{2-q} - L_1^{2-q}}{2-q}[1 + 2\frac{V_A}{v}\frac{m_p V_A}{p}]$$

$$-(\frac{m_p V_A}{p})^2[\frac{L_1^{-q} - L_2^{-q}}{q} + \frac{L_2^{4-q} - L_1^{4-q}}{(4-q)k_e^4}] \tag{18b},$$

and

$$k_e = \sqrt{\frac{M}{\gamma}}, \quad k_{1,2} = \frac{v}{2V_A}[\sqrt{1 + (4\frac{V_A}{v}\frac{m_p V_A}{p})} \mp 1],$$

$$L_1 = \max[k_1, 1], \quad L_2 = \min[k_2, M^{1/2}] \tag{19}.$$

Comparing Eqs. (14) and (17) we note the markedly different momentum dependence of the diffusion coefficients; apart from cut-off effects the leading momentum dependence is

$$\frac{A_W(p)}{A_A(p)} \simeq J_W(p)(\frac{p}{m_p V_A})^{-q} \tag{20}.$$

2.4   Loss and acceleration time scales

It is convenient to introduce the dimensionless momentum variable $x = p/(m_e c)$, and to express the momentum diffusion terms (14) and (17) and the momentum loss rates (7) and (12) as

$$A_{A,W}(p) = \frac{m_e^2 c^2 x^2}{t_{A,W}(x)}, \quad \dot{p}_{I,C} = -\frac{m_e c x}{t_{I,C}(x)} \tag{21},$$

in terms of the characteristic acceleration $(t_{A,W}(x))$ and loss $(t_{I,C}(x))$ time scales. Evidently, the total momentum diffusion coefficient then is

$$A(p) = A_A(p) + A_W(p) = \frac{m_e^2 c^2 x^2}{t_S(x)}, \quad t_S = \frac{t_A t_W}{t_A + t_W} \tag{22}.$$

We treat the interstellar medium as a composite medium consisting of the well-intermixed coronal phase and atomic hydrogen layer. We therefore use the following parameters typical for the interstellar medium: $n_g = 0.3n_3$ cm^{-3}, $B_0 = 3B_3 \cdot 10^{-6}$ G, $(B_0/\delta B)^2 = 10$, q=5/3, $k_{min} = 2\pi/100 l_{100}$ pc $= 2 \cdot 10^{-20} l_{100}^{-1}$ cm^{-1}, and $T_6 = 0.1T_5$. For the momentum loss we assume Eq. (12) to hold with $n_c = n_g$. ¿From the preceeding sections we immediately derive

$$t_C(x > 0.012T_6^{1/2}) = \frac{1.7 \cdot 10^{12}}{n_g(\mathbf{r})}[\frac{x^3 + (0.02T_6^{1/2})^3(1 + x^2)^{3/2}}{(1 + x^2)}] \text{ s} \tag{23},$$

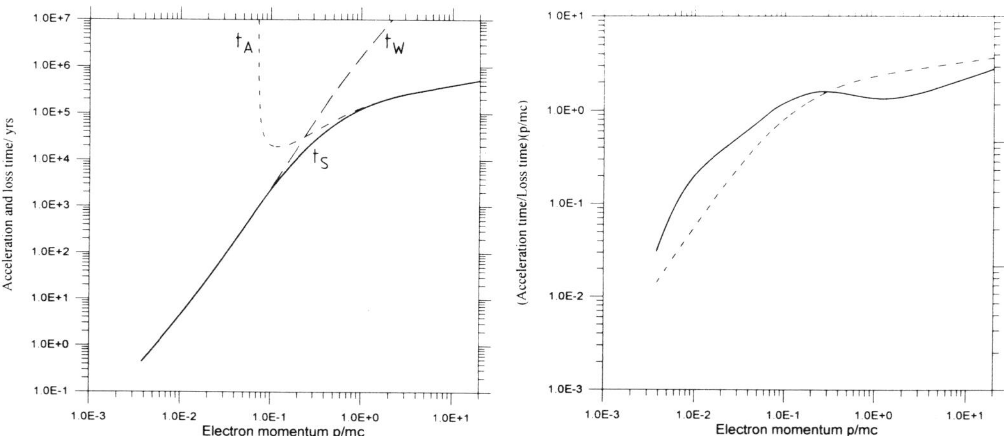

Figure 1: (a) left: Total (t_S) and individual acceleration time scales due to resonant interactions with Alfvén (t_A) and Whistler (t_W) waves; (b) right: Ratio of total acceleration and loss time multiplied with the normalized momentum $r(x)$ (full curve) and its approximation (dashed curve).

$$t_A(x > M\beta_A) = \frac{t_0(x)}{[1 - (\frac{M\beta_A}{x})^q]} \ , \quad t_W(x) = \frac{2t_0(x)}{qJ_W(x)}\left(\frac{x}{M\beta_A}\right)^q \qquad (24),$$

with

$$t_0(x) = 1.4 \cdot 10^{-9}\frac{q}{q-1}\frac{n_c(\mathbf{r})}{\mathbf{B_0^3}}\left(\frac{B_0}{\delta B}\right)^2\frac{x^{3-q}}{\sqrt{1+x^2}}\left[\frac{1705k_{min}}{B_0}\right]^{1-q}\text{ s}$$

$$= 7.5 \cdot 10^{14}\ n_3 B_3^{-7/3}\left(\frac{B_0}{\delta B}\right)^2 l_{100}^{2/3}\frac{x^{4/3}}{\sqrt{1+x^2}}\text{ s} \qquad (25).$$

Fig. 1a shows the resulting acceleration time scales due to Alfvén and Whistler waves, respectively, as well as the total acceleration time scale t_S according to Eq. (22). As a consequence of the leading momentum dependence (20) we find for the acceleration time ratio apart from the slow variation of the function $J_W(p)$ that $t_W(p)/t_A(p) \propto (\frac{p}{m_p V_A})^q$, indicating that the stochastic acceleration of cosmic ray electrons is dominated by Whistler waves at small semirelativistic momentum values ($p << p_W$), and by Alfvén waves at relativistic momentum values ($p >> p_W$). This behaviour is clearly apparent in Fig. 1a, which shows that in the interstellar medium p_W is of order $\sim 0.2m_ec = 0.1$ MeV/c. Since according to Eq. (25) $t_0 \propto p^{3-q}/\gamma$ we note that the momentum dependence of the Alfvén wave acceleration time $t_A(p) \propto t_0 \propto p^{3-q}/\gamma$ is determined by the turbulence spectral index q, while the Whistler wave acceleration time

$t_W(p) \propto p^q t_0 \propto p^3/\gamma$, at least to leading order, is independent from the value of q.

In Fig. 1b we show the variation of the ratio of the total acceleration time $t_S(x)$ and the loss time $t_C(x)$ multiplied with the normalized momentum x, i.e.

$$r(x) = [x t_S(x)]/t_C(x) \tag{26}.$$

We find that the behaviour of $r(x)$ can be reasonably well reproduced by the approximation

$$r_A(x \geq x_T) \simeq 1.8(\frac{x}{x+0.1})^{3/2}[1 + 0.5x^{2-q}] \tag{27},$$

which is also shown in Fig. 1b for comparison.

3 Electron Equilibrium Spectrum

With the substitutions (22) and the new variable x implying $\mathbf{F(r,p)d^3p} = \mathbf{f(r,x)d^3x}$, $\mathbf{S_0(r,p)d^3p} = \mathbf{q_0(r,x)d^3x}$ the steady-state electron transport equation (3) reads

$$\frac{1}{x^2}\frac{d}{dx}[\frac{x^4}{t_S(x)}\frac{df}{dx} + \frac{x^3 f}{t_C(x)}] = -q_0(x) \tag{28}.$$

Substituting (note Eq. (26))

$$f(x) = u(x)\exp[h(x)] , \quad \text{with } h(x) = \int_x^\infty dy \frac{t_S(y)}{y t_C(y)} = \int_x^\infty dy \frac{r(y)}{y^2} \tag{29},$$

we reduce Eq. (28) to self-adjoint form

$$\frac{d}{dx}[D(x)\frac{du}{dx}] = -x^2 q_0(x) , \quad \text{where } D(x) = \frac{x^4 \exp[h(x)]}{t_S(x)} \tag{30}.$$

We can readily solve Eq. (30a) as

$$u(x) = \int_0^\infty dx_0 x_0^2 q_0(x_0) G(x, x_0) \tag{31},$$

where the Green's function $G(x, x_0)$ obeys

$$\frac{d}{dx}[D(x)\frac{dG}{dx}] = -\delta(x - x_0) \tag{32},$$

and can be constructed from the homogeneous solutions of Eq. (28) and standard methods (e.g. Arfken 1970). It is instructive to investigate first special cases of the transport equation (28).

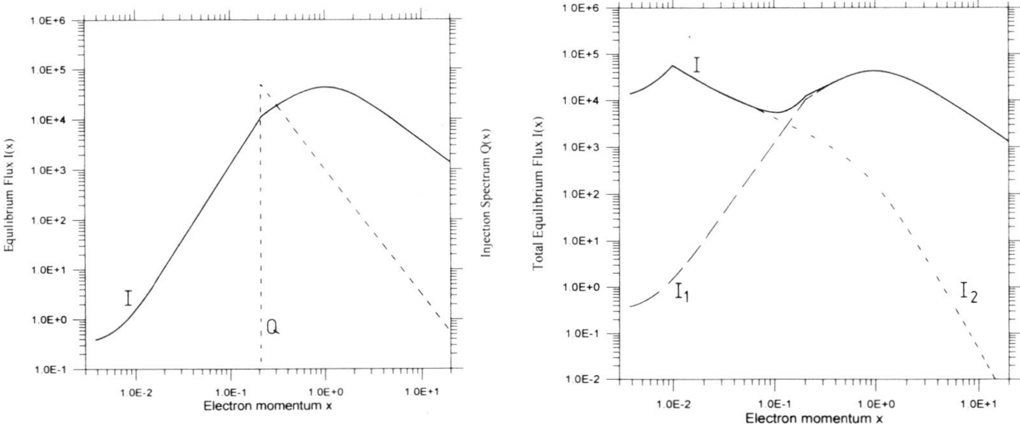

Figure 2: (a) left: Equilibrium differential electron flux in the case of no ($t_s = \infty$) stochastic acceleration for single power law injection (34) above $x_i = 0.21$ with spectral index $\alpha = 4.5$; (b) right: Equilibrium differential electron flux in the case of no ($t_s = \infty$) stochastic acceleration for two power law source components (36) with $\alpha_1 = 4.5$, $x_{i1} = 0.21$ and $\alpha_2 = 7$, $x_{i2} = 0.01$.

3.1   No stochastic acceleration $t_s = \infty$

We first consider the case of no stochastic acceleration discussed already by Skibo et al. (1995). With $t_S = \infty$ Eq. (28) readily solves as

$$N(x) = 4\pi x^2 f(x) = \frac{4\pi t_C(x)}{x} \int_x^\infty dx_0 x_0^2 q_0(x) \qquad (33).$$

Adopting as Skibo et al. (1995) power law injection above some x_i,

$$q_0(x) = q_s x^{-\alpha} H[x - x_i] \qquad (34),$$

with $\alpha > 4$, we obtain for Eq. (33) in this case

$$N(x) = 4\pi x^2 f(x) = \frac{4\pi q_s t_C(x)}{x(\alpha - 3)} (\max[x, x_i])^{3-\alpha} \qquad (35),$$

with the associated equilibrium electron flux

$$I(x) = \frac{v}{4\pi} N(x) = \frac{cx^3 f(x)}{\sqrt{1 + x^2}} = \frac{cq_s t_C(x)}{(\alpha - 3)\sqrt{1 + x^2}} (\max[x, x_i])^{3-\alpha} \qquad (36).$$

In Fig. 2a we show the resulting momentum variation (36) of the electron flux $I(x)$ in comparison with the input source spectrum (34) for values of $\alpha = 4.5$

and $x_i = 0.21$, which represents either an input source spectrum from shock wave acceleration in supernova remnants and/or the injection by the knock-on process (compare with Eq. (4)). Such a source spectrum reproduces well the known electron spectrum at very high energies > 10 MeV. However, as has been noted before also by Skibo et al. (1995), the heavy Coulomb losses momenta dramatically attenuate the source spectrum at low energies and cause a sharp turnover at nonrelativistic energies. In particular, it is impossible with such a source distribution to account for the spectral turn up below 200-500 keV visible in the bremsstrahlung gamma-ray spectrum. Skibo et al. have therefore required the existence of a dramatic turn up in the electron source spectrum below a few MeV, which may be due to a second, much steeper, low-energy source component. E.g. Skibo and Dermer (1995) have argued that these may originate from acceleration by weak shocks. To illustrate the presence of such a second population of electrons we calculate the electron equilibrium intensity resulting from a second power law injection spectrum (34) with values of $\alpha_2 = 7$ and $x_{i2} = 0.01$. The maximum intensity is now shifted to considerable lower momentum values. In Fig. 2b we show the individual (from both source populations) and the total equilibrium intensities. As one can clearly see by proper arrangement of the relative intensities, the second source component can indeed account for the required flattening of the electron intensity at MeV energies. However, as calculated by Skibo et al. (1995) in order to do so, this second source population then contains a total power of $4 \cdot 10^{41}$ erg s^{-1} if integrated over the whole Galaxy, which is about a factor 20 more than the power supplied to the nuclear cosmic rays. Besides the obvious problem that such enormous power requirement has for potential galactic cosmic ray sources, we will argue here in favour of an alternative explanation, namely the influence of finite, $t_s \neq \infty$, stochastic acceleration of electrons in the galaxy.

3.2   The influence of in-situ stochastic acceleration $t_s \neq \infty$

Allowing for in-situ stochastic acceleration the solution of Eq. (32) is

$$G(x, x_0) = \begin{cases} \int_{x_0}^{\infty} \frac{ds}{D(s)} & \text{for } 0 \leq x \leq x_0 \\ \int_{x}^{\infty} \frac{ds}{D(s)} & \text{for } x_0 \leq x \leq \infty \end{cases} \tag{37}.$$

In determining this Green's function we have required the boundary conditions (see e. g. Park and Petrosian 1995) that $f(x = 0)$ and $f(x = \infty)$ are finite. With Eqs. (29) and (31) for general source spectra $q_0(x)$ the solution $f(x)$

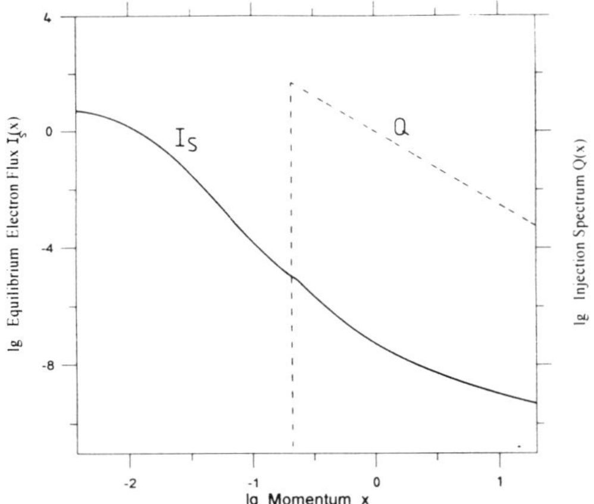

Figure 3: Equilibrium differential electron flux in the case of finite $t_s \neq \infty$ stochastic acceleration for single power law injection (34) for $q = 1.5$.

then is

$$f(x) = \exp[h(x)](\int_0^x dx_0 x_0^2 q_0(x_0) \int_x^\infty \frac{ds}{D(s)} +$$

$$\int_x^\infty dx_0 x_0^2 q_0(x_0) \int_{x_0}^\infty \frac{ds}{D(s)}) \tag{38}.$$

Inserting the cosmic ray source function (34) we derive after partial integration

$$f_s(x) = \frac{q_s \, e^{h(x)}}{\alpha - 3}(x_i^{3-\alpha} \int_m^\infty \frac{ds}{D(s)} - \int_m^\infty \frac{ds \, s^{3-\alpha}}{D(s)}) \tag{39}$$

with $m = \max[x_i, x]$. ¿From Eqs. (26) and (30b) we note that

$$\frac{1}{D(s)} = \frac{t_C(s) \, r(s) \, e^{-h(s)}}{s^5} = \frac{t_C(s)}{s^3} \frac{d}{ds}[e^{-h(s)}] \tag{40}.$$

Using Eq. (40) in Eq. (39) we readily obtain after partial integration

$$f_s(x) = \frac{q_s \, e^{h(x)}}{\alpha - 3}\Big[\frac{t_C(m)}{m^\alpha} e^{-h(m)}[1 - (\frac{m}{x_i})^{\alpha-3}]$$

$$+ \int_m^\infty ds \, e^{-h(s)} \frac{d}{ds}[\frac{t_C(s)}{s^\alpha}[1 - (\frac{s}{x_i})^{\alpha-3}]]\Big] \tag{41}.$$

For small momentum values $x \leq x_i$ implying $m = x_i$ the first term in the bracket of Eq. (41) vanishes, so that the momentum variation is simply given by $f_s(x \leq x_i) \propto \exp[h(x)]$. This behaviour is significantly different from the solution (35) at $x \leq x_i$ in case of no stochastic acceleration: the existence of finite interstellar stochastic reacceleration provides many more low-energy electrons below the minimum injection momentum x_i in the electron equilibrium spectrum. This behaviour can clearly be seen by comparing Fig. 2a with Fig. 3, where we calculate the associated equilibrium electron flux $I_s(x) = cx^3 f_s(x)(1 + x^2)^{-1/2}$ from Eqs. (39) and (41) for a value of $q = 1.5$.

At large momentum values $x > x_i$ both terms in the bracket of Eq. (41) contribute. For the equilibrium electron flux we find

$$I_s(x > x_i) = \frac{cq_s}{(\alpha - 3)\sqrt{1 + x^2}} \Big[t_C(x)[x^{3-\alpha} - x_i^{3-\alpha}]$$

$$+ x^3 e^{h(x)} \int_x^\infty ds \; e^{-h(s)} \frac{d}{ds} \big[\frac{t_C(s)}{s^\alpha}[1 - (\frac{s}{x_i})^{\alpha-3}]\big] \Big] \tag{42}.$$

As Fig. 3 demonstrates in this momentum range the effect of finite stochastic acceleration does not change the equilibrium spectrum significantly.

4 Summary and conclusions

In this work we have investigated again the implications of the OSSE and COMPTEL measurements of the diffuse galactic gamma ray bremsstrahlung continuum emission for interstellar dynamics of cosmic ray electrons. We show that the previous interpretation of the bremsstrahlung spectral upturn by Skibo and co-workers, as being due to the existence of a second electron source component at mildly relativistic electron energies with a total source power of $4 \cdot 10^{41}$ erg/s, is not unique. As an alternative we demonstrate that the spectral upturn can be explained by the existence of in-situ stochastic acceleration by the interstellar plasma turbulence in the Galaxy. Our interpretation avoids the requirement of additional source power in cosmic ray electrons.

References

Abraham, P. B., Brunstein, K. A., Cline, T. L., 1966, Phys. Rev. 150, 1088.

Arfken, G., 1970, Mathematical Methods for Physicists, Academic Press, New York.

Butler, S. T., Birmingham, M. J., 1962, Phys. Rev. 126, 1.

Dung, R., Schlickeiser, R., 1990, Astr. Ap. 240, 537.

Gould, R. J., 1972, Physica 60, 145.

Heitler, W., 1954, The Quantum Theory of Radiation, Dover, New York.

Hinton, F. L., 1983, in: Basic Plasma Physics I, eds. A. A. Galeev and R. N. Sudan, North-Holland, Amsterdam, p. 147.

Kurfess, J. D., 1995, in: Proc. 17th Texas Symposium on Relativistic Astrophysics, eds. H. Böhringer, G. E. Morfill, J. E. Trümper, Annals of the New York Academy of Science, Vol. 759, p.236.

Lerche, I., Schlickeiser, R., 1982, Astr. Ap. 116, 10.

Park, B., Petrosian, V., 1995, Ap. J. 446, 699.

Pohl, M., 1993, Astr. Ap. 270, 91.

Ramaty, R., 1974, in: High-Energy Partcles and Quanta in Astrophysics, eds. F. B. McDonald, C. E. Fichtel, MIT Press, Cambridge, p. 122.

Schlickeiser, R., 1981, Fortschr. d. Physik 29, 95.

Schlickeiser, R., 1982, Astr. Ap. 106, L5.

Schlickeiser, R., 1994, Ap. J. Suppl. 90, 929.

Schlickeiser, R., Achatz, U., 1993, J. Plasma Phys. 49, 63.

Sivukhin, D. V., 1965, in: Reviews of Plasma Physics, ed. M. Leontovich, Consultants Bureau, New York, Vol. 4, p. 93.

Skibo, J. G., Dermer, C. D., 1995, Proc. 24th Intern. Cosmic Ray Conf. (Roma), Vol. 3, p. 52.

Skibo, J. G., Ramaty, R., 1993, Astr. Ap. Suppl. 97, 145.

Skibo, J. G., Ramaty, R., Purcell, W. R., 1995, Proc. 24th Intern. Cosmic Ray Conf. (Roma), Vol. 2, p. 219.

Steinacker, J., Miller, J. A., 1992, ApJ 393, 764.

Strong, A. W., et al. , 1994, Astr. Ap. 292, 82.

Address of the author:

R. SCHLICKEISER, Max-Planck-Institut für Radioastronomie, Postfach 2024, D-53010 Bonn, Germany.

Diffuse γ-rays from galactic halos

Martin Pohl

1 Introduction

The γ-ray sky as we know it today is a composite of the emission from point sources and diffuse emission (see Fig.1). The most prominent feature is the galactic plane, in which interactions between cosmic rays and the thermal gas lead to γ-ray emission by π^0-decay and bremsstrahlung. On top of this diffuse emission we see point sources all over the map, part of them being pulsars, another part distant AGN, and also a significant fraction of yet unidentified sources. But we also see that the diffuse emission extends out of the galactic plane up to the poles. There appears to be an isotropic emission component which is presumably extragalactic in origin and may be understood as blend of unresolved AGN, but there is also considerable galactic emission. At higher latitudes interactions between cosmic rays and thermal gas still play a role, but inverse-Compton scattering of ambient photons by cosmic ray electrons becomes increasingly important. This emission tells us about the physical conditions in the halo as seen by the cosmic ray particles, and it may also reveal previously hidden gas, e.g. baryonic dark matter.

2 The rôle of confused point sources

Locally, any analysis of the galactic diffuse emission can be seriously hampered by unresolved galactic point sources which may have a sky distribution similar to that of gas. As is shown in Table 1, in the galactic plane the source density is significantly higher than at high latitudes, not only in total number but especially for the unidentified sources. If the sky distribution of sources were isotropic we would expect a smaller source density in the plane, since the strong background there reduces the statistical prominence of any source even for the on average higher exposure. A modelling of the sky distribution of sources under the assumption that all unidentified sources are galactic has revealed that unresolved galactic sources would account for 30-40% of the total galactic γ-ray luminosity above 100 MeV and that the sky distribution of the unresolved

EGRET $\rangle$100MeV MEM

Figure 1: The EGRET sky above 100 MeV γ-ray energy. The image is deconvolved by a Maximum-Entropy algorithm (for the method see Strong 1995). The grey scale is logarithmic with darkness indicating high intensity.

sources could resemble that of the gas since we see mainly the more distant sources (Kanbach et al. 1996). Although in reality the numbers may not be that bad, perhaps around 20%, for halo studies the point source contribution has to be taken into account.

$|b|$	[sr]	Identified	Unidentified	Identified/Total	Total/steradian
0-10	2.2	5	32	0.14$\pm$0.06	16$\pm$3.0
10-30	4.1	18	26	0.41$\pm$0.11	11$\pm$1.6
30-90	6.3	32	14	0.70$\pm$0.16	7.3$\pm$1.1

Table 1: Source statistic on the basis of the second EGRET catalog (Thompson et al. 1995). In the galactic plane the source density is higher than at high latitudes and also the fraction of identified sources is unusually small.

If there were no point sources the γ-ray intensity in the galactic plane should follow locally the gas distribution. In fact the EGRET team uses a scaling model on the basis of the gas distribution as null hypothesis in the search for point sources (Hunter et al. 1996). We can test the reliability of such models on small scales by comparing it to deconvolved data for the galactic center region where the statistic is best, i.e. small scale structures have the highest statistical significance. We will concentrate on data at higher energies where EGRET's point spread function is smaller and less photons are required to reproduce the intensity structure. The result is shown in Fig.2 and Fig.3, where despite the larger pixel size in the model (Fig.3) it is obvious that there is a wealth of structure not directly related to the gas. Part of this additional structure is listed as unidentified sources in the second EGRET catalog.

It should be noted that comparing deconvolved data to the original data is more sensitive to small scale discrepancies than the usual comparison of a convolved model to original data, especially if the global sky distribution of point sources is similar to the distribution of gas. Therefore, our finding does not contradict the statement of Hunter et al. (1996) that the gas coupling model, convolved and averaged over strips a few degrees wide, is a good representation of the observed sky distribution at γ-ray energies above 100 MeV.

3 Diffuse emission in the EGRET range

3.1 The γ-ray emissivity

In galactic halos inverse-Compton scattering (IC) of ambient photons by cosmic ray electrons is an important production process for γ-rays . Due to the large scale heights of far-infrared photons and the cosmic microwave background, the latitude distribution of this emission is much broader than in case of the bremsstrahlung and π^0-decay. Thus the IC component can tell us about the lifetime of cosmic ray electrons in the Galactic halo. Unfortunately the IC emission is generally weak and has a spectral shape similar to that of the extragalactic γ-ray background. However, the latter should be isotropic, and one may use the contrast between inner and outer Galaxy to detect the IC component.

One may also compare the spatial distribution of γ-rays not related to gas to the brightness temperature distribution in radio surveys. Depending on frequency the bulk of radio emission at high latitudes will be due to synchrotron radiation of cosmic ray electrons, and one should expect correlations between the synchrotron flux density and the IC intensity, although the typical electron energy for synchrotron emission at 408 MHz is roughly an order of magnitude less than the energy required for up-scattering infrared photons. Such work has

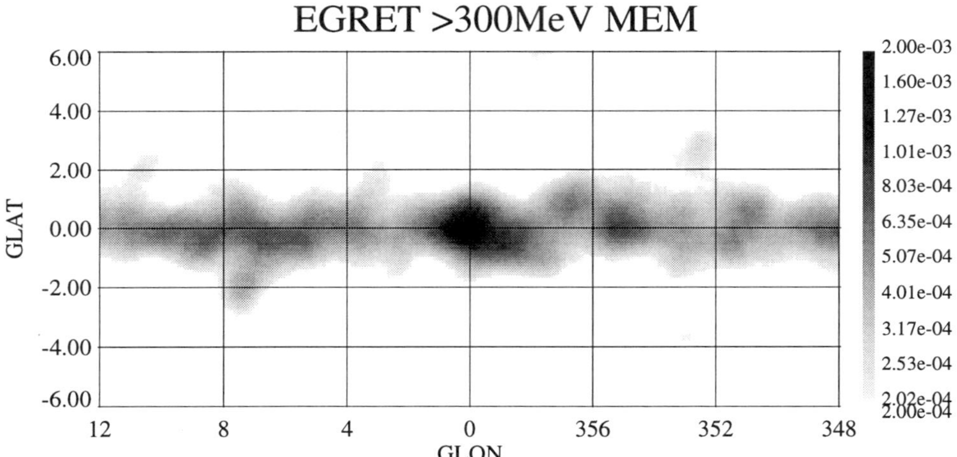

Figure 2: The galactic center region above 300 MeV γ-ray energy. The image is deconvolved by a Maximum-Entropy algorithm. The intensity scale is in units $\mathrm{cm}^{-2}\,\mathrm{sec}^{-1}\,\mathrm{sr}^{-1}$.

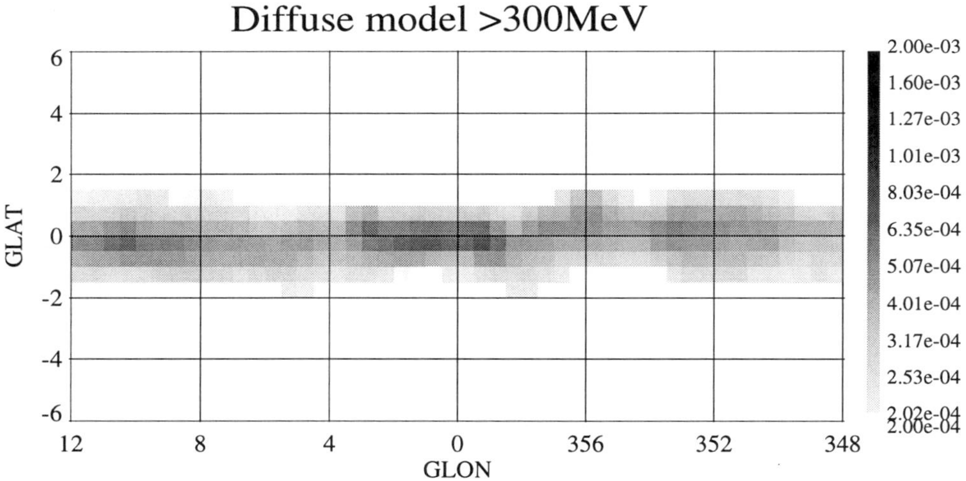

Figure 3: The model prediction based on gas distribution for the galactic center region. The scale is the same as for the deconvolved EGRET data in Fig.2. The disagreement between Fig.2 and Fig.3 is obvious, implying that point sources contribute significantly.

been done by Chen et al. (1996), who conclude that the average intensity of IC emission at high latitudes is $(5 \pm 0.8) \cdot 10^{-6}$ cm^{-2} sec^{-1} sr^{-1} for $E > 100\,$MeV (around one third of the diffuse extragalactic background) and that the photon spectral index is -1.85 ± 0.17.

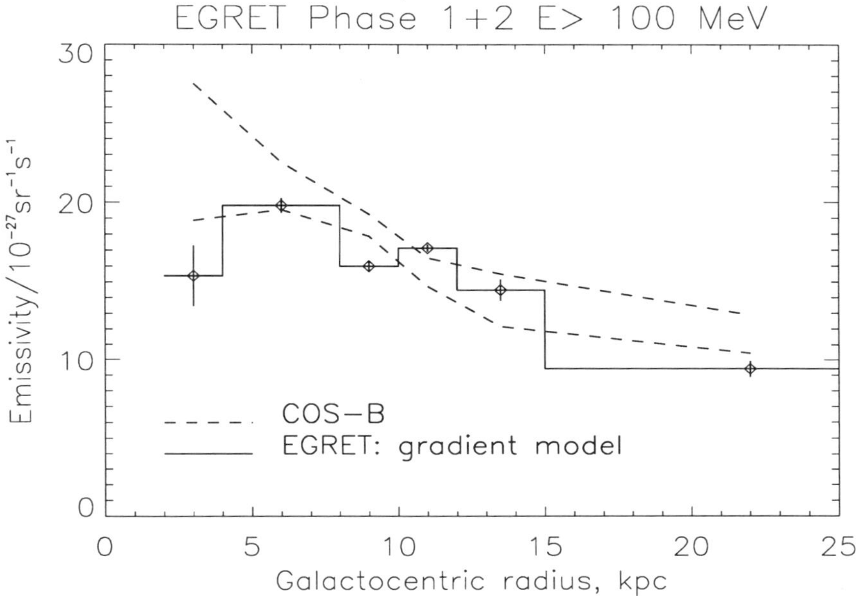

Figure 4: The γ-ray emissivity per H-atom averaged over galactocentric rings. There is weak decrease of emissivity with galactocentric radius, the so-called gradient, basically confirming the earlier COS-B result. (Strong and Mattox 1996)

To investigate the γ-ray emission originating from π^0-decay and bremsstrahlung we need some prior knowledge on the distribution of gas in the Galaxy. This includes not only HI but also H$_2$, which is indirectly traced by CO emission lines, and HII, which is traced by Hα and pulsar dispersion measurements. Even in case of the directly observable atomic hydrogen we obtain only line-of-sight integrals, albeit with some kinematical information. Any deconvolution of the velocity shifts into distance is hampered by the line broadening of individual gas clouds and by the proper motion of clouds with respect to the main rotation flow. The distance uncertainty will in general be around 1 kpc.

It may be appropriate to use only a few resolution elements and investigate the γ-ray emissivity per H-atom in galactocentric rings, i.e. to assume azimuthal symmetry for the emissivity. This kind of approach has already been successfully applied to the earlier COS-B data. Strong and Mattox (1996) have repeated the analysis on EGRET data at γ-ray energies above 100 MeV. These authors report that there is a gradient, that is a decline of the γ-ray emissivity per H-atom with galactocentric radius, confirming earlier COS-B results. This radial gradient is weak and does not exceed a factor 2 difference between inner and outer Galaxy (see also Fig.4). It should be noted that this gradient does not necessarily hold locally. A comparison of the γ-ray emissivity in the perseus arm at 3 kpc distance in the outer galaxy to the Cepheus and Polaris flare at 250 pc gave a difference of a factor 1.7 ± 0.2, by far exceeding the overall gradient (Digel et al. 1996).

3.2 Spectral information

The γ-ray production processes may not only be separable by their different sky distribution, but also by their spectral shapes. While the π^0-decay should result in a broad line centered on 68 MeV, the bremsstrahlung spectrum should follow the rather steep electron spectrum, thereby partly hiding the signature of the π^0-decay line. The emission due to IC scattering is expected to have a hard spectrum, similar to that of the diffuse extragalactic background. Strong and Mattox (1996) have performed a spectral analysis of EGRET data in the ten standard energy bands to investigate the spectrum of the isotropic extragalactic emission, the emission correlated to the gas distribution (π^0-decay and bremsstrahlung), and the emission due to IC scattering, for which the expected sky distribution was modelled. The result is shown in Fig.5 together with the overall spectrum of diffuse emission in EGRET data compared to the results of COMPTEL and the earlier mission COS-B. All sources in the second EGRET catalog have been taken into account in this analysis.

The spectrum of the isotropic extragalactic component is around E^{-2}, harmonizing with the average AGN spectrum. The IC emission has a slightly harder spectrum similar to the results of Chen et al. (1996) at higher latitudes. However, the spectrum of emission related to the gas is difficult to understand. It deviates significantly from the expected superposition of π^0-decay and bremsstrahlung. At energies above 1 GeV there is a clear excess of emission related to gas which may be somewhat relaxed by variation of the input proton spectrum to the pion production process. Part of this excess may also be explained by unresolved sources, however the small intensity at low

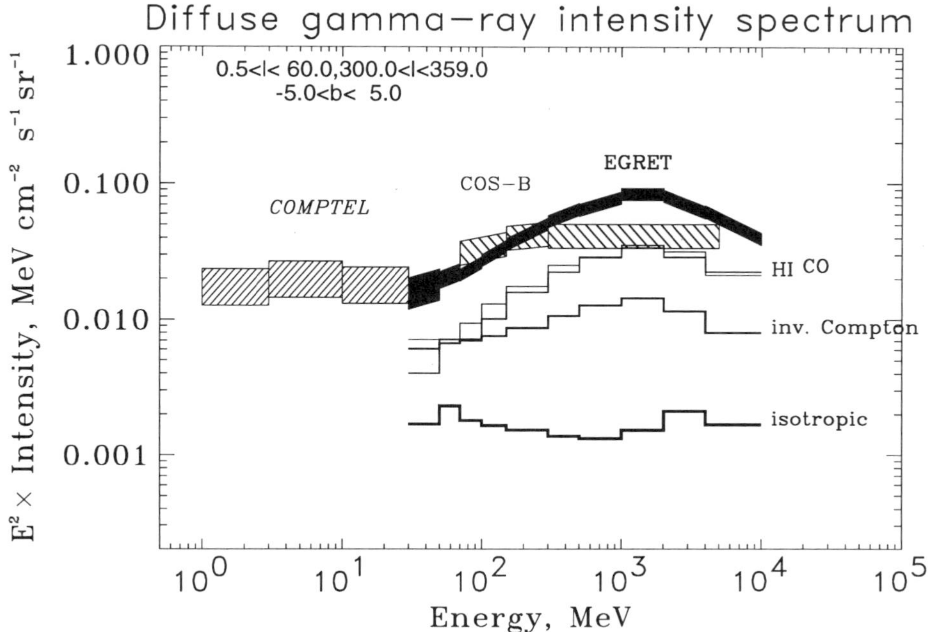

Figure 5: The γ-ray intensity in EF(E) representation. Shown is both the total spectrum (black boxes) and the spectra of the individual components (as histograms) for the inner Galaxy. For comparison the COMPTEL results at lower energies and the old COS-B data are included. The uncertainties in the spectrum of the individual components are around 10% except for the low-energy points. (Strong and Mattox 1996)

energies requires sources with very hard spectrum and possibly a cut-off at a few GeV. Only old pulsars like Geminga have a corresponding spectrum and may also have a sky distribution resembling that of gas.

The spectral hardness of the emission related to gas at low energies is also not easy to understand. The standard analysis tools of EGRET for diffuse emission implicitly assume a power-law spectrum of photon index 2.1. If the true intensity is softer than this the standard analysis will underestimate the emission at low energies, while for hard spectra the emission will be overestimated. The reason for this is that a significant part of the low energy counts are due to misinterpreted photons of higher true energy. An example of this effect is shown in Fig.6. It turns out that even in the absence of any sky emission below 100 MeV the spill-over provides roughly an E^{-1} differential photon spectrum

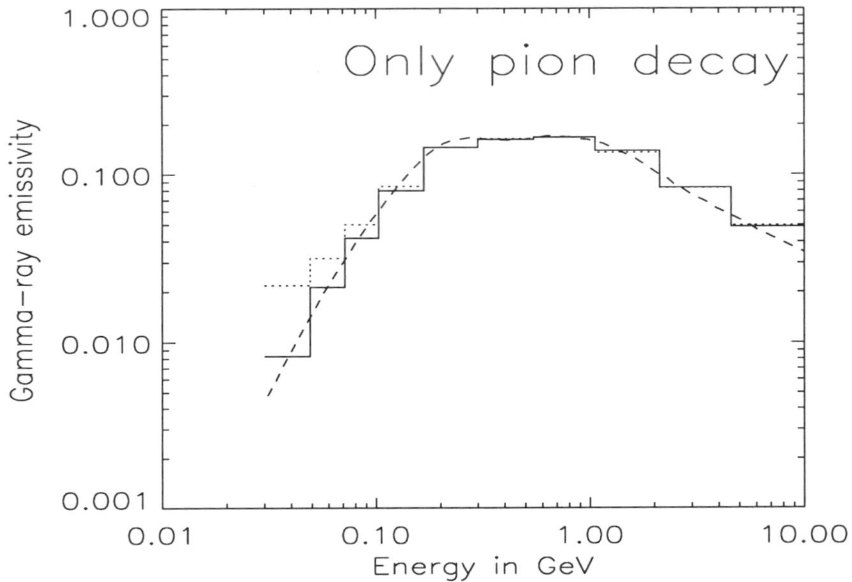

Figure 6: A demonstration of the spill-over effect at low energies. The dashed curve is the true input spectrum in EF(E) representation of arbitrary units, here a π^0-decay spectrum. The solid line histogram is the corresponding average over the standard EGRET energy bands. The dotted histogram shows what the result of the standard EGRET analysis would be. At low energies the spectrum is overestimated by a factor of three.

at these energies. The spectrum of the emission related to gas in Fig.5 is not far from this, indicating that the true sky intensity at low energies is even less. This would imply that the density of 100 MeV electrons is relatively small compared to the density of 10 GeV electrons which are responsible for the IC emission, considering also the radio synchrotron data which tell about the electron spectrum in the GeV range.

4 Diffuse emission at hard X-rays and soft γ-rays

At lower energies we have data from three instruments for the inner Galaxy: COMPTEL in the MeV range (Strong et al. 1994), OSSE in the 100 keV range (Purcell et al. 1996), and GINGA around 10 keV (Yamasaki et al. 1996). The resulting spectrum of diffuse emission from 2 keV to 30 MeV is shown in Fig.7.

The GINGA spectrum may partly be thermal in origin. In fact a thermal model plus absorbed power-law fits well to the data. Also note the prominent Fe Kα line at 6.7 keV. While the point source contribution in the COMPTEL range is probably around 20% as in case of EGRET, the confusion in the OSSE range is less easy to estimate. OSSE is a non-imaging instrument and therefore cannot identify point sources on its own. The galactic center region has been simultaneously observed by SIGMA, a coded-mask instrument designed to search for sources. It turned out that after subtracting the flux due to point sources resolved by SIGMA the residual OSSE diffuse emission was similar to the total OSSE flux at l=25°, which supports the view that the residual is extended diffuse emission (Purcell et al. 1996). We show only the residual flux in Fig.7, for which still the contamination by unresolved sources is hard to estimate, but is unlikely to account for all the emission.

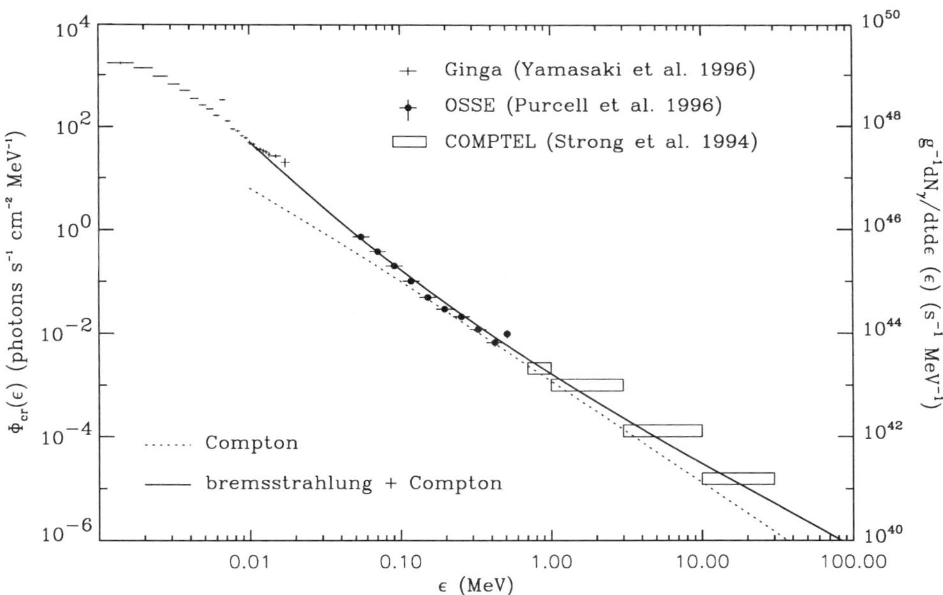

Figure 7: The γ-ray intensity in direction of the inner Galaxy from X-rays to soft γ-rays . The GINGA spectrum may partly be thermal in origin. However, the OSSE data require a very soft electron spectrum at low energies. The solid line is a model curve of bremsstrahlung and IC scattering based on such a soft electron spectrum. (Skibo et al. 1996)

The energy dependence of Coulomb and ionisation losses of electrons in the

MeV range imply that they cannot travel far from their sources and that their spectrum should be hard. Thus the OSSE result can only be explained by invoking an additional source of low energy electrons with an input power exceeding that provided by Galactic supernovae. Skibo et al. (1996) have argued that this phenomenon may be transient due to Galactic spiral density waves, thereby relaxing the energy requirement. Very promising appear models which explain the soft spectrum of low energy electrons as direct result of stochastic acceleration out of the thermal pool working against Coulomb interactions (Schlickeiser, this volume). The energy input in this class of models would come from interstellar turbulence.

The existence of a high density of low-energy electrons is an important issue for the energy and ionisation balance of the interstellar medium both in the plane and in the halo, since the bulk of the energy input is directly transferred to the thermal gas.

5 The Magellanic Clouds

The Magellanic Clouds are two irregular galaxies located at around 50 kpc distance. The Large Magellanic Cloud (LMC) has been detected by EGRET (see also Fig.8), while for the Small Magellanic Cloud (SMC) EGRET has derived only an upper limit (Sreekumar et al. 1993). Both in LMC and in SMC the γ-ray emissivity per H-atom is considerably smaller than in our Galaxy. This result implies that the bulk of GeV cosmic rays is galactic in origin. Since the gas in LMC and SMC is much less illuminated by cosmic rays than gas in the Galaxy, the cosmic ray density at some distance from the Galaxy may be very small. This has impact on dark matter studies, since this fact allows a substantial amount of baryonic dark matter to be hidden at a few 10 kpc from the Galaxy without inducing observable γ-ray emission.

There has been a debate on whether the γ-ray flux of LMC and SMC in relation to their radio emission would allow equipartition between the magnetic field and cosmic rays. It turns out that this equipartition is still possible provided one allows the cosmic ray electron-to-proton ratio to be different from that in the solar vicinity. The γ-ray data of the Magellanic Clouds are best explained when the e/p ratio is much higher so that basically only the density of cosmic ray nucleons is reduced in LMC and SMC (Pohl 1993).

All results mentioned above were derived based on the integral properties of the Magellanic Clouds. A new study including spatial and spectral information

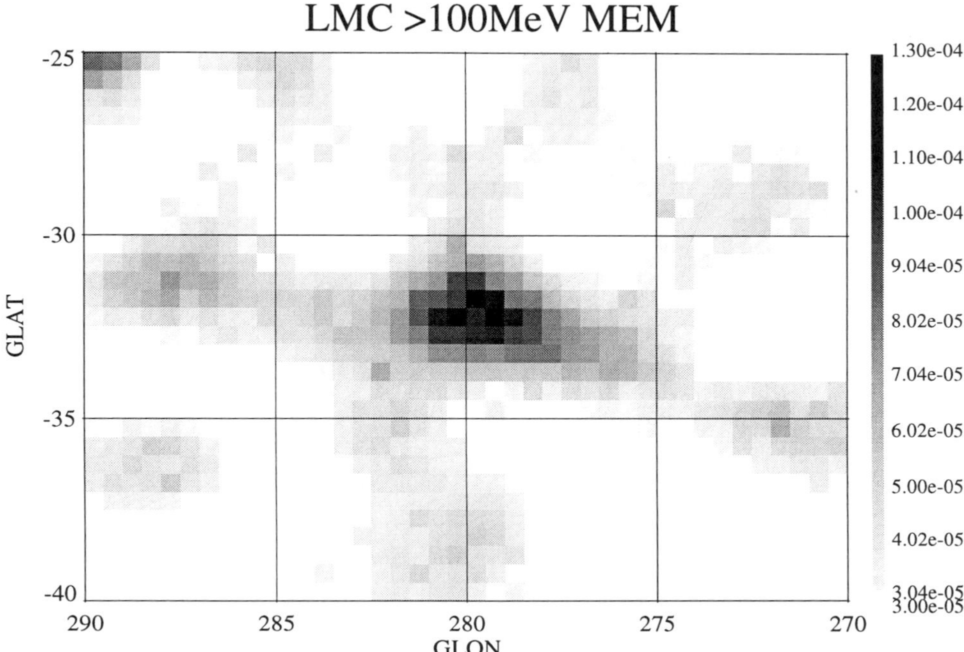

Figure 8: The γ-ray intensity above 100 MeV for the region of LMC. The data are deconvolved again and the scale is in $cm^{-2}\,sec^{-1}\,sr^{-1}$. LMC is clearly seen with its emission concentrated along the gas ridge south of 30 Doradus.

is under way.

6 What of the future?

The next step in analysing γ-ray data of diffuse emission will be based on propagation modelling of cosmic rays which better allows to link the γ-ray data to the results of radio observations and direct cosmic ray measurements. Such models are underway and they may help to answer the question of whether the sources of cosmic rays are localized entities, e.g. supernova remnants, or whether they are diffuse, and in relation to this, whether reacceleration plays a rôle. With forthcoming instruments we also may be able to set more stringent limits on the baryon fraction in a dark matter halo.

The ESA M2 mission INTEGRAL will not be ideally suited for studies of diffuse γ-ray continuum emission from the halo as it is specifically designed for high spatial and energy resolution. However, two project studies currently undertaken are highly promising. The project GLAST will use silicon-strip detectors to measure γ-rays in the energy range 50 MeV to 100 GeV a factor 100 better in sensitivity than EGRET. At low energies of 1 MeV to 50 MeV laboratory studies are undertaken to combine the Compton telescope principle with silicon-strip tracking of the scattered electron, which may also give a factor 100 in sensitivity compared to COMPTEL. So if we continue to invest in γ-ray astronomy we may go a big step forward in understanding diffuse γ-rays from Galactic halos in the next decade.

Acknowledgements: I am indebted to Andy Strong and Jeff Skibo who kindly provided figures and plots in electronic form, partly prior to publication.

References

Chen A., Dwyer J., Kaaret P.: 1996, ApJ, in press
Digel S.W., Grenier I.A., Heithausen A. et al.: 1996, ApJ, in press
Hunter S.D. et al.: 1996, A&AS, in press
Kanbach G., Bertsch D.L., Dingus B.L. et al.: 1996, A&AS, in press
Pohl M.: 1993, A&A 279, L17
Purcell W.R. et al.: 1996, A&AS, in press
Skibo J.G., Ramaty R., Purcell W.R.: 1996, A&AS, in press
Sreekumar P., Bertsch D.L., Dingus B.L. et al.: 1993, PRL 70, 127
Strong A.W.: 1995, Exper. Astron. 6, 97
Strong A.W. et al.: 1994, A&A 292, 82
Strong A.W., Mattox J.R.: 1996, A&A, in press
Thompson D.J., Bertsch D.L., Dingus B.L. et al.: 1995, ApJS 101, 259
Yamasaki N. et al.: 1996, A&AS, in press

Address of the author:

MARTIN POHL: Max-Planck-Institut für Extraterrestrische Physik, Postfach 1603, 85740 Garching, Germany; mkp@mpe-garching.mpg.de

Supernova rates derived from HI shells in Milky Way, M31 and Holmberg II

J. Palouš, S. Ehlerová

1 Introduction

The HI shells in the Milky Way (Heiles, 1979, 1984) and the holes in HI distribution in galaxies M31 (Brinks and Bajaja, 1986) and Holmberg II (Ho II) (Puche et al. 1992) are structures surrounding places of recent energy inputs into the interstellar medium. The energy originates in massive stars of OB associations or in collisions between high velocity clouds and the galactic HI plane. In the case of the Milky Way, the statistical arguments lead to the conclusion that the majority of shells is related to the star formation (Ehlerová & Palouš, 1996).

In this contribution, the total energy input needed to maintain the shells in the Milky Way, M31 and Ho II is estimated. Assuming their connection to the star formation, the corresponding supernova rates are discussed.

2 Chevalier's formula?

The energy E_e needed to form a shell of radius R_{sh} may be estimated by a formula (Chevalier, 1974):

$$\left(\frac{E_e}{erg}\right) = 5.3 \times 10^{43} \left(\frac{n_0}{cm^{-3}}\right)^{1.12} \left(\frac{R_{sh}}{pc}\right)^{3.12} \left(\frac{v_{sh}}{kms^{-1}}\right)^{1.4} \tag{1}$$

where n_0 is the density of the ambient medium and v_{sh} its expansion velocity. For the last quantity the one-dimensional rms velocity of the ambient medium can be used if there is no explicit value derived from observation.

Chevalier's formula is derived for spherically symmetric expansion in a homogeneous medium using a one dimensional hydrodynamical code with radiative cooling. The cooling is important in early stages of expansion, later its significance decreases, the evolution becomes nearly adiabatic. The spherically symmetric adiabatic expansion has been solved by Sedov (1959), who gives $E_e \sim R_{sh}^3 n_0 v_{sh}^2$.

We generalize the Sedov's expansion taking into account the influence of the galactic differential rotation. The large-scale inhomogeneities, such as the radial distribution of HI in galaxies and the stratification of HI perpendicular to the galactic symmetry plane or warping of the HI plane in the outer parts of galaxies, are also included. The small-scale inhomogeneities are not considered in this paper, their influence is discussed by Silich et al. (1996).

3   3D model

Expansion of the shocked interstellar medium from a small region, where the energy is released, is followed in 3D using the infinitesimally thin shell approximation (Ostriker & McKee, 1988; Ehlerová & Palouš, 1996). A model of the Milky Way, M31 and Ho II is adopted using rotation curves given by Wouterloot et al. (1990), Sofue & Kato (1981) and Puche et al. (1992). HI $z-$distributions perpendicular to the galactic plane are taken from Dickey & Lockman (1990), Wouterloot et al. (1990) and Puche et al. (1992).

The computer simulations with the energy E_e released at different galactocentric distances in medium with different HI densities and stratifications have been performed. The evolution of the shell gives its life-time, which is the time interval between the beginning of expansion and collapse of the most massive parts of it back to the galactic plane. It is only loosely dependent on the input energy E_e, it depends mainly on the gravitational force in the $z-$direction, on the distribution of the ambient medium and on the galactocentric distance.

4   Energy inputs and supernova rates in Milky Way, M31 and Ho II

The results are summarized in the Tab. 1: N_{tot} is the total number of the HI structures (in the case of Milky Way it is an extrapolation from those parts of the plane covered by Heiles, 1989, 1984); $\bar{\tau}$ is the mean value of the life-times in the galaxy; σ is the total shell formation rate; $\bar{E}_e$ is the mean value of the input energy needed to form an observed shell; $\dot{E}_e$ is the total energy input necessary to maintain the observed number of shells; and SNR is the corresponding supernova rate.

	MW	M31	HoII
N_{tot}	378	141	51
$\bar{\tau}(Myr)$	52	41	183
$\sigma\ (Myr^{-1})$	9.0	3.6	0.38
$\bar{E}_e\ (10^{52}\ erg)$	10.5	9.5	1.4
$\dot{E}_e(10^{45}\ erg\ yr^{-1})$	700	350	5
$SNR\ (SN\ yr^{-1})$	23×10^{-4}	12×10^{-4}	1.7×10^{-5}
$M_{tot}\ (M_\odot)$	2.5×10^{11}	2.5×10^{11}	2.0×10^{9}
$\frac{\sigma}{M_{tot}}$	3.5×10^{-11}	1.4×10^{-11}	1.9×10^{-10}
$\frac{SNR}{M_{tot}}$	9.1×10^{-15}	4.8×10^{-15}	8.5×10^{-15}

Table 1: Input energies and supernova rates for Milky Way, M31 and Ho II

The shell formation rate σ and the energy input $\dot{E}$ are derived assuming an equilibrium between their formation and extinction using a formula (Ehlerová & Palouš, 1996):

$$\sigma = \int \frac{n(R)}{\tau(R)} dR, \quad \dot{E} = \int \frac{E(R)}{\tau(R)} dR, \qquad (2)$$

where $n(R)dR$ and $E(R)dR$ are the number of shells and their total energy in a radial annulus of width dR around the median galactocentric distance R, $\tau(R)$ is the life-time at the galactocentric distance R.

Using Chevalier's formula (1) we get the energy needed to shell formation. These values are compared to estimates from the computer simulations. For the Milky Way and M31 the energies needed are $\sim 7\times$ higher than those from formula (1), in the case of Ho II both energy estimates are similar. These differences in the energy estimates are due to the flatness of the HI disc and galactic differential rotation of Milky Way and M31.

5 Conclusions

Assuming 3×10^{50} erg as the average energy inserted into the ISM in one supernova event, the supernova rate is derived from the total energy input $\dot{E}_e$. With the total mass of the galaxy M_{tot} within the volume where the HI structures reside , we get the shell formation rate per unit of mass, $\frac{\sigma}{M_{tot}}$, and supernova rate per unit of mass, $\frac{SNR}{M_{tot}}$. The average energy of observed HI structures in the Milky Way is almost one order of magnitude higher than in Ho II, which may be due to more massive stars and more numerous OB associations. However, the shell formation rate per unit of mass is higher in Ho II resulting in comparable supernova rates per unit of mass in Ho II and Milky Way, about twice as high as in M31.

References

Brinks, E., Bajaja, E., 1986, A&A, 169, 14

Chevalier, R., 1974, ApJ 188, 501

Dickey, J., Lockman, F., 1990, ARA&A 28, 215

Ehlerová, S., Palouš, J., 1996, A&A, in press

Heiles, C., 1979, ApJ, 299, 533

Heiles, C., 1984, ApJS, 55, 585

Ostriker, J.P., McKee, Ch. F., 1988, Rev. Mod. Phys. 60, 1

Puche, D., Westpfahl, D., Brinks, E., Roy, J-R, 1992, AJ, 103, 1841

Sedov, L.I., 1959, Similarity and Dimensional Methods in Mechanics, Academic
    Press, New York

Silich, S.A., Franco, J., Palouš, J., Tenorio-Tagle, G., 1996, ApJ, 468, in press

Sofue, Y, Kato, T., 1981, PASJ, 33, 449

Wouterloot, J.G.A., Brand, J., Burton, W.B., Kwee, K.K., 1990, A&A, 230,
    21

Addresses of the authors:

J. Palouš; S. Ehlerová,
Astronomical Institute, Academy of Sciences of the Czech Republic, Boční II
1401, 141 31 Praha 4, Czech Republic,
and
Max-Planck-Institut für Radioastronomie, Auf dem Hügel 69, D-5300 Bonn,
Germany.

X-ray Emission from the Halos of two Starburst Galaxies

Norbert Junkes, Gerhard Hensler

1 Introduction

We investigate luminosity and spectral distribution of the soft X-ray emission from a sample of nearby spiral galaxies with nuclear starburst activity. In the present paper we show the results for two of these galaxies (NGC2903 and NGC4569) which clearly exhibit extended X-ray emission. A counter example is NGC1808 where the bulk of soft X-ray emission is confined to the starburst nucleus, and where we could not confirm the existence of hot halo gas from soft X-ray observations (Junkes et al. 1995).

NGC2903 is a nearby Sc galaxy with a hot spot nucleus, showing six separate intensity peaks in the optical within a $20''$ nuclear region of the galaxy (Sersic & Pastoriza 1965). The central region is a powerful source in the NIR, reflecting dust which is heated by the nuclear starburst (Rieke et al. 1980). The radio emission from the central source of NGC2903 is predominantly non-thermal (caused by supernova remnants in the nuclear starburst, see Wynn-Williams & Becklin 1985). The total FIR luminosity of NGC2903 is not very high ($\approx 5 \times 10^9$ $L_\odot$); the starburst is clearly confined to the nuclear region.

NGC4569 (M90) is a bright spiral in the Virgo cluster, one of the few galaxies with negative radial velocity outside the local group. It is gas-deficient in the outer spiral arms, the neutral hydrogen content is strongly concentrated in the inner region (Cayatte et al. 1990). The bright nucleus, embedded in a normal stellar bulge, is probably the result of a recent star formation episode (Staufer et al. 1986). Based upon optical spectroscopy of its nucleus (Willner et al. 1985), the galaxy has been classified as a LINER.

2 X-ray luminosity and spectra

Fig. 1 shows the spatial distribution of the soft X-ray emission in the ROSAT band for both target galaxies after Gaussian smoothing. The bulk of the soft

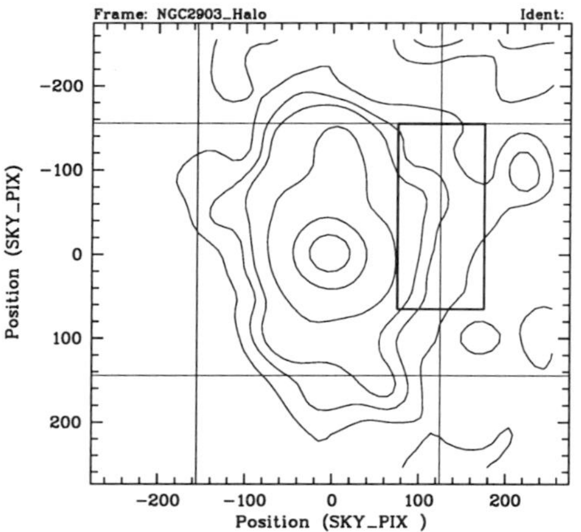
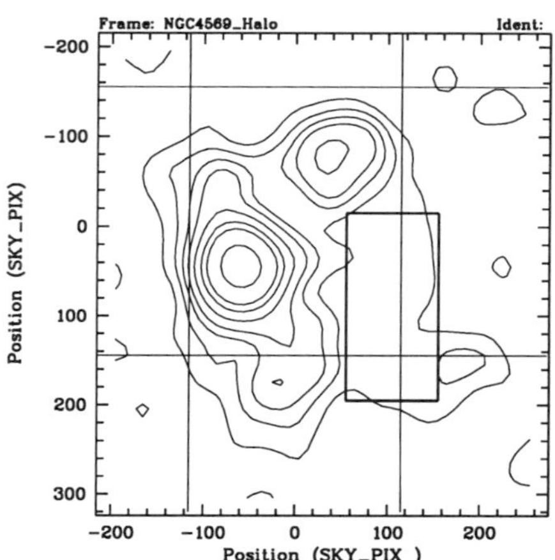

Figure 1: ROSAT PSPC images of both galaxies in the energy band 0.1–2.4 keV, smoothed to $35''$ resolution. The coordinate axes are labeled in SKY_PIX (units of $0.5''$), the reference positions are RA $= 9^{\mathrm{h}}32^{\mathrm{m}}10^{\mathrm{s}}$, DEC $= +21°30'$ for NGC2903 and RA $= 12^{\mathrm{h}}36^{\mathrm{m}}48^{\mathrm{s}}$, DEC $= +13°10'12''$ for NGC4569 (both J2000.0). Two areas are marked where hardness ratios of the soft X-ray emission outside the disk were determined.

X-ray emission arises in the nuclear starburst, and could be caused by contributions from supernova remnants as well as from high-mass X-ray binaries. The intensity peak north of the nucleus of NGC2903 is positionally coincident with a bright HII region (see figure in Jackson et al. 1991). The X-ray emission of NGC4569 outside the nucleus even traces the spiral arm patterns of this galaxy. The bright source northwest of the nucleus is probably a stellar foreground source and has been excluded for the derivation of the integrated flux of NGC4569 in the soft X-ray range.

The global X-ray spectrum of NGC2903 can be well described by a two-component fit consisting of a power-law component (predominantly from the nucleus) and a thermal plasma component (hot gas from SNRs and superbubbles). The resulting value of hydrogen absorption from the spectral fit is slightly higher than the Galactic foreground, the total luminosity in the ROSAT band is $\approx 7 \times 10^{39}$ erg s^{-1} for NGC2903 (for $d = 6.3$ Mpc).

The global X-ray spectrum of NGC4569 can be described by a similar fit.

The resulting value of hydrogen absorption is in agreement with the Galactic foreground, the total luminosity in the ROSAT band for NGC4569 is $\approx 2 \times 10^{40}$ erg s^{-1}. The value of 1.1×10^{-3} for L_X/L_{FIR} is typical for normal spirals with strong star formation predominantly in the nucleus (for $d = 17$ Mpc).

3   Hardness Ratios

Both galaxies have a similar inclination of $i \approx 65°$. They appear extended in their soft X-ray emission, and show contributions from areas outside the nucleus and, probably, outside the disk. In contrast to edge-on galaxies, where emission components from disk and halo can be clearly distinguished, we can not derive quantitative results for the suspected halo emission from NGC2903 and NGC4569. In order to account for a possible contribution from hot halo gas to the total X-ray emission, we determined the total number of counts and the hardness ratio of the emission in the ROSAT band in two areas outside the disk (see figure).

The distribution of the soft X-ray emission for both galaxies appears asymmetric; both targets show a surplus west of the central area. We marked two rectangle areas of $50'' \times 100''$ and derived hardness ratios ($0.5 - 2.1$ keV vs. $0.1 - 0.4$ keV) of -0.04 ± 0.17 and $+0.10 \pm 0.17$, respectively. The total number of counts after background subtraction is only 54 resp. 63 for both areas, hence no spectra were determined. The hardness ratios are compared with the results from model spectra. A value of HR $= 0.0$ results for a thermal plasma spectrum (e.g. Raymond & Smith 1976) with a plasma temperature of 0.2 keV.

Hence, the X-ray emission from both areas outside the disk can be explained with hot gas of a temperature of approximately 2×10^6 K.

4   Some conclusions

A comparison of the total luminosities of NGC2903 and NGC4569 in the soft X-ray and in the FIR band with a larger sample of normal and active galaxies (Green et al. 1992) shows good agreement with the normal galaxies (including LINERs and starburst galaxies) in that sample. The soft X-ray emission of both galaxies can be explained with ongoing starburst activity without the requirement of an additional (active) nuclear component.

Most of the soft X-ray emission in both galaxies comes from their nuclear regions, but additional components from the disk and, possibly, from the halo, can be found. The existence of a halo component in the X-ray emission from both, NGC2903 and NGC4569, is supported by two facts: First, extended emission is predominantly found at one side in case of both galaxies. Second, its spectral distribution (hardness ratio) can be explained with hot plasma of a temperature of a few 10^6 K, which would be expected in the halo of starburst galaxies.

Acknowledgements: The project has been supported by Deutsche Agentur für Raumfahrtangelegenheiten (DARA) GmbH with grant No. 50 OR 9302 4.

References

Cayatte V., van Gorkom J.H., Balkowski C., et al., 1990, AJ 100, 604
Green P.J., Anderson S.F. & Ward M.J., 1992, MN 254, 30
Jackson J.M., Eckart A., Cameron M., et al., 1991, ApJ 375, 105
Junkes N., Zinnecker H., Hensler G., et al., 1995, A&A 294, 8
Raymond J.C. & Smith B.W., 1977, ApJS 35, 419
Rieke G.H., Lebovski M.J., Thomson R.I., et al., 1980, ApJ 238, 24
Sersic J.L. & Pastoriza M., 1965, PASP 77, 287
Staufer J.R., Kenney J.D. & Young J.S., 1986, AJ 91, 1286
Willner S.P., Elvis M. & Fabbiano G., 1985, ApJ 299, 443
Wynn-Williams C.G. & Becklin E.E., 1985, ApJ 290, 108

Addresses of the authors:

NORBERT JUNKES[1,2]; GERHARD HENSLER[2],

[1]Astrophysikalisches Institut Potsdam, An der Sternwarte 16, D-14482 Potsdam (njunkes@aip.de)

[2]Institut für Astronomie und Astrophysik, Universität Kiel, Olshausenstraße 40, D-24098 Kiel (hensler@astrophysik.uni-kiel.d400.de)

The X-ray lobes of Virgo A

H. Rottmann, J. Kerp, K.-H. Mack

1 Introduction

Since the very beginning of X-ray astronomy, the giant elliptical galaxy M87 (Vir A, 3C274) has been observed with various instruments. Detailed studies of the nucleus, the jet, and the surrounding halo were performed with the *Einstein* X-ray telescope by Schreier et al. (1982). All investigations showed that most emission emerges from a very extended, symmetrical X-ray halo centred on M87 with only a few percent originating from an additional asymmetrical component. The orientation of this arc-like structure (hereafter referred to as X-ray lobes) is in rough coincidence with the radio lobes of Vir A (Feigelson et al. 1987). Böhringer et al. (1995) proposed, based on spectral investigations of *ROSAT* PSPC data, that the X-ray lobes are cooler than the surrounding halo plasma. However, a reanalysis of this *ROSAT* data reveals evidence for the lobes to emit hard X-ray radiation and to be considerably hotter than the ambient X-ray halo gas, in the sense of thermal plasma emission.

2 Results and discussion

We have analysed two pointed *ROSAT* PSPC (Pfeffermann et al. 1986) observations towards Vir A, with an effective integration time of about 20.500 s. The data were binned into the three standard *ROSAT* PSPC energy bands, corresponding to the mean energies of $\frac{1}{4}$, $\frac{3}{4}$, and 1.5 keV (Snowden et al. 1994). The contour diagrams of Vir A at $\frac{1}{4}$, $\frac{3}{4}$, and 1.5 keV are displayed in Fig. 1. In all three energy bands most of the observed X-ray emission originates from an extended X-ray halo centred on M87. While at $\frac{1}{4}$ keV (Fig. 1a) the halo is almost spherically symmetrical, one observes an increasing contribution of an additional asymmetric component with increasing photon energy. At 1.5 keV (Fig. 1c) this asymmetric structure forming the X-ray lobes is most prominent. Clearly, radiation originating from the X-ray lobes is harder than the emission from the ambient X-ray halo. These findings can be interpreted in two different ways. If one assumes absorption to take place in the direction of the X-ray lobes the soft X-ray emission from this region would be effectively attenuated, while at higher energies the absorbing medium would become increasingly transparent. Due to the positional correlation of the X-ray lobes with the radio lobes (comp. Fig. 2) the absorber must be located within Vir A. So far there is no observational evidence for such an absorbing feature.

Instead we believe that the observed radiation of the X-ray lobes is *excess* emission originating from an additional plasma component. In such a two component scenario we can make conclusions about the temperatures of both gas components. The X-ray halo has a similar shape and extent in all three analysed energy bands. This is an indication that we are observing emission from a plasma over a wide range of temperatures. In contrast, emission from the asymmetric X-ray lobes is missing in the soft energy bands, which implies that it originates from a gas at a significantly higher mean temperature than the ambient X-ray halo gas.

In order to investigate large-scale temperature variations in the X-ray halo we have analysed the dependance of the ratio of $\frac{1}{4}$ keV to 1.5 keV emission on the radial distance from the nucleus of M87. The images at both energies have been radially averaged, with the origin fixed to the core position of M87. At

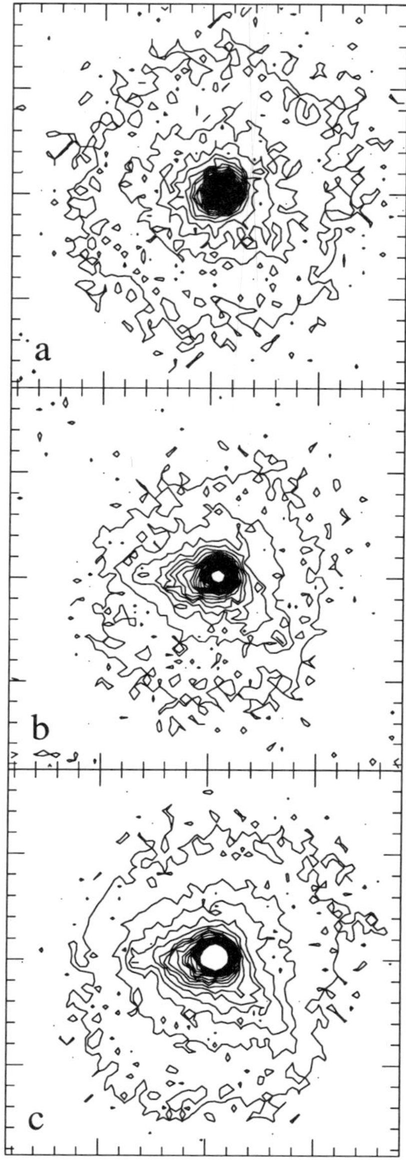

Figure 1: *ROSAT* PSPC contour diagrams of M87 at $\frac{1}{4}$ keV (a), $\frac{3}{4}$ keV (b), and 1.5 keV (c)

greater radii, the contribution of the soft X-ray component increases, sugesting a temperature gradient in the X-ray halo, with temperatures falling off with radial distance from the nucleus of M87.

The orientation of the X-ray lobes seems to be in rough coincidence with the location of the radio lobes of Vir A. For a more detailed analysis it is necessary to subtract a model for the extended X-ray halo in order to reveal the morphology of the underlying X-ray lobes. However, subtraction of a spherically symmetrical King-profile from the 1.5 keV image tends to overestimate the extended halo emission in some areas of the source. In particular, a large negative bowl to the northwest of the nucleus is created by such a method (Böhringer et al. 1995). Therefore, we have followed a different approach, by using the $\frac{1}{4}$ keV image as a model for the X-ray halo emission and subtracting it from the 1.5 keV image to uncover the hard excess emission from the X-ray lobes. This seems reasonable as the scale lengths of the X-ray halo does not differ significantly at 1.5 keV ($163''$) and at $\frac{1}{4}$ keV ($180''$). Simple scaling of the $\frac{1}{4}$ keV image intensity thus gives a good representation of a large part of the X-ray halo emission at 1.5 keV. This subtraction method removes most of the X-ray halo without creating negative artifacts. The residual image is shown in Fig. 2 superimposed onto a radio map of Vir A (Braun, priv. com.). The subtraction has disclosed the dual lobe structure of the residual X-ray emission with higher accuracy than obtained up to now. While there is a good correlation of orientation between the X-ray and the radio lobes, we find a clear anti-correlation of X-ray and synchrotron intensity. The maximum X-ray emission east to the nucleus coincides with a location on the radio jet showing a minimum of synchrotron radiation. The anti-correlation is most prominent in the region just west of the nucleus. No X-ray lobe emission is detected on the "knee" of the deflected jet where the synchrotron emission peaks. There is only one location of correlated X-ray and radio maxima viz. on the hot spot of the eastern radio lobe. The observed anti-correlation is evident in essentially all parts of the X-ray lobes. A reliable mechanism which can explain this behaviour cannot be proposed on basis of the available data at this time. We believe that such a mechanism will improve the understanding of the internal structure and the radiative processes in the outer jet of Vir A.

References

Böhringer H., Nulsen P.E.J., Braun R., Fabian A.C., 1995, MNRAS 274, L67

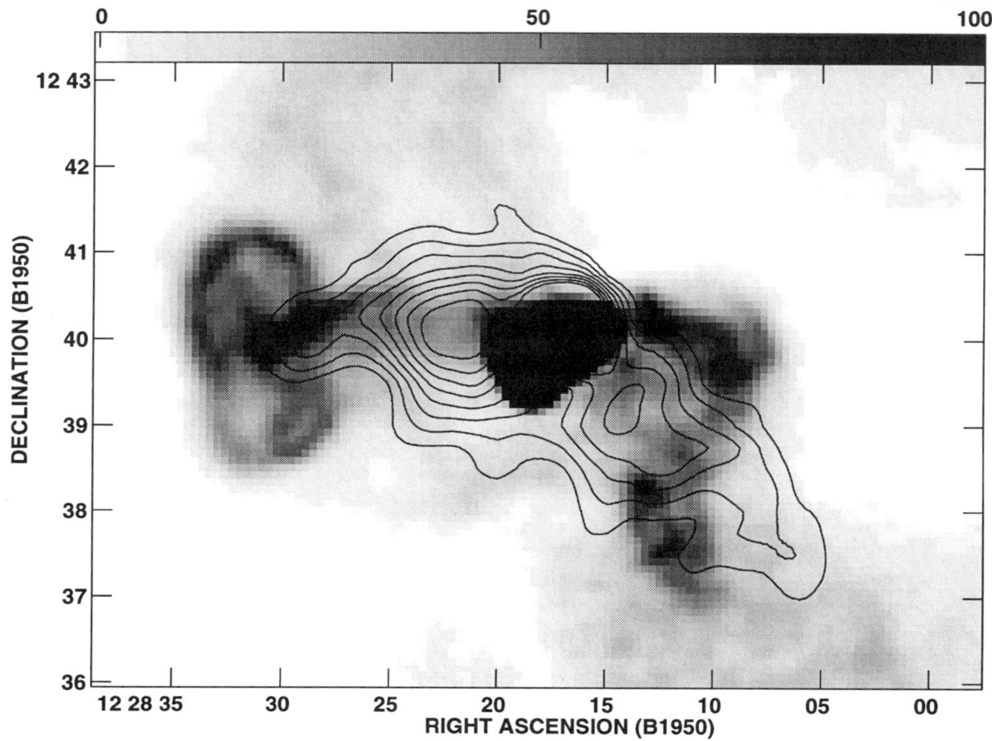

Figure 2: The grey-scale image represents the synchrotron intensity distribution of Vir A obtained with the VLA at 1667 MHz (Braun, priv. com.). The superimposed contour lines display the excess emission of the X-ray lobes of Vir A after subtraction of the contribution of the extended X-ray halo.

Feigelson E.D., Wood P.A.D., Schreier E.J., Harris D.E., Reid M.J., 1987, ApJ 312, 101

Pfeffermann E., Briel U.G., Hippmann H. et al., 1987, Proc. SPIE 733, 519

Schreier E.J., Gorenstein P., Feigelson E.D., 1982, ApJ 261, 42

Snowden S.L., McCammon D., Burrows D.N., Mendenhall J.A., 1994, ApJ 424, 714

Addresses of the authors:

H. ROTTMANN; K.-H. MACK, Radioastronomisches Institut der Universität Bonn, Auf dem Hügel 71, 53121 Bonn, Germany.
J. KERP, Max–Planck–Institut für Extraterrestrische Physik, Giessenbachstr., 85740 Garching, Germany

List of Participants

BECK, RAINER	MPI für Radioastronomie, Bonn, Germany
BENJAMIN, ROBERT A.	Dept. of Astronomy, Univ. of Minnesota, Minneapolis, U.S.A.
BERKHUIJSEN, ELLY M.	MPI für Radioastronomie, Bonn, Germany
BIERMANN, PETER L.	MPI für Radioastronomie, Bonn, Germany
BLOEMEN, HANS	SRON-Utrecht, Utrecht, The Netherlands
BREITSCHWERDT, DIETER	MPI für Kernphysik, Heidelberg, Germany
BURTON, W. BUTLER	Sterrewacht, Leiden, The Netherlands
DE BOER, KLAAS S.	Sternwarte der Univ., Bonn, Germany
DETTMAR, RALF-JÜRGEN	Astron. Institut, Ruhr-Univ., Bochum, Germany
DOMGÖRGEN, HILDE	Sternwarte der Univ., Bonn, Germany
DUHMKE, MICHAEL	MPI für Radioastronomie, Bonn, Germany
ELSTNER, DETLEV	Astrophysikalisches Insitut, Potsdam, Germany
FERRARA, ANDREA	Osservatorio Astrofisico di Arcetri, Firenze, Italy
FLEISCHER, JÜRGEN	Institut für Theor. Physik, Ruhr-Univ., Bochum, Germany
FREYBERG, MICHAEL J.	MPI für Extraterrestrische Physik, Garching, Germany
FREYER, T.	Institut für Astronomie u. Astrophysik, Kiel, Germany
GERHARD, ORTWIN	Astron. Institut, Univ. Basel, Binningen, Schweiz
GIESELER, V.	MPI für Kernphysik, Heidelberg, Germany
GOLLA, GÖTZ	Astron. Institut, Ruhr-Univ., Bochum, Germany
HENSLER, GERHARD	Institut für Astronomie u. Astrophysik, Kiel, Germany
HERBSTMEIER, UWE	MPI für Astronomie, Heidelberg, Germany
HOERNES, PHILLIP	MPI für Radioastronomie, Bonn, Germany
JUNKES, NORBERT	Astrophysikalisches Insitut, Potsdam, Germany
KAHN, FRANZ D.	Dept. of Physics & Astronomy, Univ., Manchester, U.K.
KALBERLA, PETER M.W	Radioastron. Institut der Univ., Bonn, Germany
KERP, JÜRGEN	Radioastron. Institut der Univ., Bonn, Germany
KNELLER, M.	MPI für Radioastronomie, Bonn, Germany
KRAUS, M.	Institut für Astrophysik, Univ., Bonn, Germany
KRAUSE, MARITA	MPI für Radioastronomie, Bonn, Germany
KRONBERG, PHILIPP P.	Dept. of Astronomy, Univ., Toronto, Canada

KUNDT, WOLFGANG	Institut für Astrophysik, Univ., Bonn, Germany
LESCH, HARALD	Universitäts Sternwarte, München, Germany
LOCKMANN, FELIX J.	National Radio Astronomy Observatory, Green Bank, U.S.A.
MACK, KARL-HEINZ	Radioastron. Institut, Univ., Bonn, Germany
MEBOLD, ULRICH	Radioastron. Institut, Univ., Bonn, Germany
MURPHY, ED	NRAO, Charlottesville, U.S.A.
NEININGER, NIKOLAUS	MPI für Radioastronomie, Bonn, Germany
NIKLAS, STEPHAN	MPI für Kernphysik, Heidelberg, Germany
PALOUŠ, JAN	Astronomical Institute, Academy of Sciences, Praha, Czech Republic
PIETZ, J.	Radioastron. Institut, Univ., Bonn, Germany
POHL, MARTIN	MPI für Extraterrestrische Physik, Garching, Germany
POULSON, MICHAEL	VCH Verlag, Weinheim, Germany
RASTÄTTER, LUTZ	Theor. Physik, Ruhr-Univ., Bochum, Germany
REYNOLDS, RON J.	Dept. of Astron., Univ. of Wisconsin, Madison, U.S.A.
ROTTMANN, HELGE	Radioastron. Institut der Univ., Bonn, Germany
SCHLICKEISER, REINHARD	MPI für Radioastronomie, Bonn, Germany
SHCHEKINOV, YU.	Osservatorio Astrofisico di Arcetri, Firenze, Italy
STRONG, A.	MPI für Extraterrestrische Physik, Garching, Germany
THIERBACH, M.	MPI für Radioastronomie, Bonn, Germany
VON DEN BRUCK, K.	Radioastron. Institut der Univ., Bonn, Germany
VAN DER HULST, THIJS	Kapteyn Laboratorium, Groningen, The Netherlands
VÖLK, HANS	MPI für Kernphysik, Heidelberg, Germany
WESTPHALEN, GERNOT	Radioastron. Institut der Univ., Bonn, Germany
ZIMMER, FRANK	Radioastron. Institut der Univ., Bonn, Germany

Index

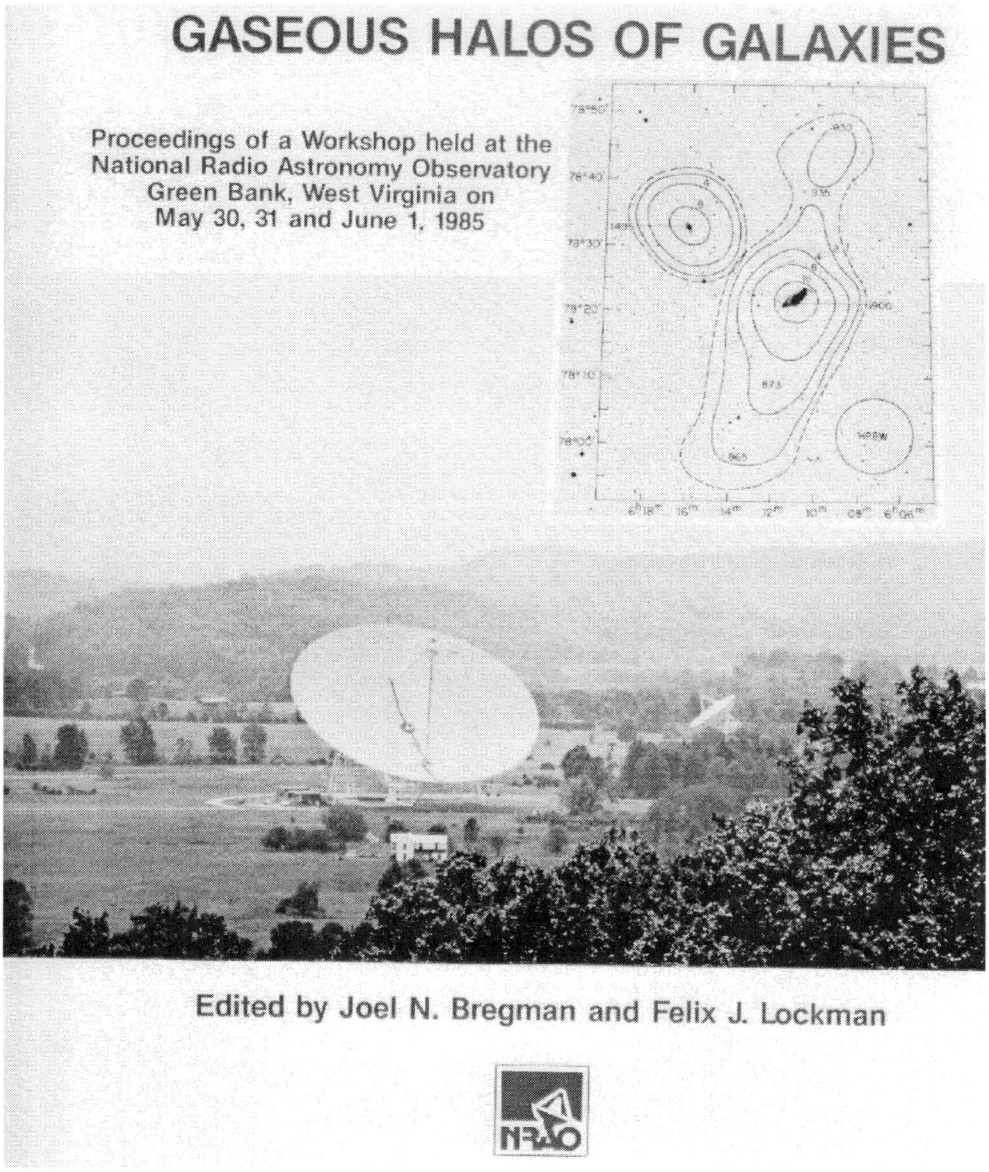

GASEOUS HALOS OF GALAXIES

Proceedings of a Workshop held at the
National Radio Astronomy Observatory
Green Bank, West Virginia on
May 30, 31 and June 1, 1985

Edited by Joel N. Bregman and Felix J. Lockman

Cover page of the Proceedings of the first workshop on Halos of Galaxies held
in 1985

WERNER VOGEL / DIRK-GUNNAR WELSCH

Lectures on Quantum Optics

1994. X, 442 pp. – 63 figs.
170 mm x 240 mm
Hardcover DM 98,– /
öS 715,– / sFr 90,–
ISBN 3-05-501387-5

Vogel and Welsch develop the theoretical concepts of modern quantum optics. Based on a general quantum-field-theoretical approach various aspects of the topic are presented in a unified manner. Passive optical systems are taken into account in the field quantization by including a space-dependent dielectric. This concept is applied to various devices such as beam splitters, spectral filters, and lossy cavities. Moreover, attention is paid to the interaction of light with atomic sources. The theory of photocounting, correlation measurements, spectral measurements, and phase-sensitive homodyne detection is developed. In particular, the book deals with fundamental aspects of quantum optics such as non-classical (antibunched, sub-Poissonian, and squeezed) light including its generation in non-linear optical processes, and possibilities of its application. The interaction of single atoms with light is studied both in the context of the quantum properties of the resonance fluorescence light and for an atom in a high-Q cavity.

From the contents:

1. Quantization of the radiation field in the presence of passive optical devices such as beam splitters, spectral filters, and lossy cavities
2. Theory of detection of light (photocounting, correlation measurements, spectral and homodyne detection)
3. Properties, generation, and application of non-classical (antibunched, sub-Poissonian, squeezed) light
4. Methods of relaxation theory
5. Fundamental interactions of atoms with light (resonance fluorescence, atom in a high-Q cavity)

Akademie Verlag

A Member of the VCH Publishing Group
Mühlenstraße 33–34 · 13187 Berlin · Germany

Please place
your order
with your bookshop.

Astronomische Nachrichten

News in Astronomy and Astrophysics

Edited by: G. Hasinger, K.-H. Rädler, E. P. J. van den Heuvel

Bibliography:
1 volume per year (6 issues /approx. 450 pp. yearly); 210 mm x 297 mm
annual list of contents
The annual subscription price 1997 (incl. postage and handling) DM 645,–
outside Germany DM 675.00 / öS 4,930.00 / sFr 643.00

Astronomische Nachrichten, founded in 1821 by H. C. Schumacher, is the oldest astronomical journal all over the world being published at present. The journal publishes original contributions to all fields of astronomy and astrophysics as well as review papers on specific topics in english language. The journal covers in particular extragalactic astronomy and cosmology, physics of stars and interstellar matter, solar and planetary physics, cosmic magnetohydrodynamics and plasma physics as well as geodetic astronomy.

Readers:
All scientists and institutions engaged in astronomy, cosmology, gravitational theory, history of astronomy, space sciences, geodesy, and geophysics

Indexed in:
Current Contents/Physical, Chemical & Earth Sciences, Science Citation Index, Astronomy and Astrophysics Abstracts, Chemical Abstracts, Zentralblatt für Mathematik

Please place your order with:
VCH Verlagsgesellschaft
Journals
Boschstr. 12
69469 Weinheim · Germany
Tel.: (0 62 01) 60 61 46
Fax: (0 62 01) 60 61 17

Akademie Verlag

A Member of the VCH Publishing Group
Mühlenstraße 33–34 · 13187 Berlin · Germany